Monographs on Statistics and Applied Probability 104

Gaussian Markov Random Fields

Theory and Applications

MONOGRAPHS ON STATISTICS AND APPLIED PROBABILITY

General Editors

V. Isham, N. Keiding, T. Louis, N. Reid, R. Tibshirani, and H. Tong

1 Stochastic Population Models in Ecology and Epidemiology *M.S. Barlett* (1960)
2 Queues *D.R. Cox and W.L. Smith* (1961)
3 Monte Carlo Methods *J.M. Hammersley and D.C. Handscomb* (1964)
4 The Statistical Analysis of Series of Events *D.R. Cox and P.A.W. Lewis* (1966)
5 Population Genetics *W.J. Ewens* (1969)
6 Probability, Statistics and Time *M.S. Barlett* (1975)
7 Statistical Inference *S.D. Silvey* (1975)
8 The Analysis of Contingency Tables *B.S. Everitt* (1977)
9 Multivariate Analysis in Behavioural Research *A.E. Maxwell* (1977)
10 Stochastic Abundance Models *S. Engen* (1978)
11 Some Basic Theory for Statistical Inference *E.J.G. Pitman* (1979)
12 Point Processes *D.R. Cox and V. Isham* (1980)
13 Identification of Outliers *D.M. Hawkins* (1980)
14 Optimal Design *S.D. Silvey* (1980)
15 Finite Mixture Distributions *B.S. Everitt and D.J. Hand* (1981)
16 Classification *A.D. Gordon* (1981)
17 Distribution-Free Statistical Methods, 2nd edition *J.S. Maritz* (1995)
18 Residuals and Influence in Regression *R.D. Cook and S. Weisberg* (1982)
19 Applications of Queueing Theory, 2nd edition *G.F. Newell* (1982)
20 Risk Theory, 3rd edition *R.E. Beard, T. Pentikäinen and E. Pesonen* (1984)
21 Analysis of Survival Data *D.R. Cox and D. Oakes* (1984)
22 An Introduction to Latent Variable Models *B.S. Everitt* (1984)
23 Bandit Problems *D.A. Berry and B. Fristedt* (1985)
24 Stochastic Modelling and Control *M.H.A. Davis and R. Vinter* (1985)
25 The Statistical Analysis of Composition Data *J. Aitchison* (1986)
26 Density Estimation for Statistics and Data Analysis *B.W. Silverman* (1986)
27 Regression Analysis with Applications *G.B. Wetherill* (1986)
28 Sequential Methods in Statistics, 3rd edition
G.B. Wetherill and K.D. Glazebrook (1986)
29 Tensor Methods in Statistics *P. McCullagh* (1987)
30 Transformation and Weighting in Regression
R.J. Carroll and D. Ruppert (1988)
31 Asymptotic Techniques for Use in Statistics
O.E. Bandorff-Nielsen and D.R. Cox (1989)
32 Analysis of Binary Data, 2nd edition *D.R. Cox and E.J. Snell* (1989)
33 Analysis of Infectious Disease Data *N.G. Becker* (1989)
34 Design and Analysis of Cross-Over Trials *B. Jones and M.G. Kenward* (1989)
35 Empirical Bayes Methods, 2nd edition *J.S. Maritz and T. Lwin* (1989)
36 Symmetric Multivariate and Related Distributions
K.T. Fang, S. Kotz and K.W. Ng (1990)
37 Generalized Linear Models, 2nd edition *P. McCullagh and J.A. Nelder* (1989)

38 Cyclic and Computer Generated Designs, 2nd edition
J.A. John and E.R. Williams (1995)

39 Analog Estimation Methods in Econometrics *C.F. Manski* (1988)

40 Subset Selection in Regression *A.J. Miller* (1990)

41 Analysis of Repeated Measures *M.J. Crowder and D.J. Hand* (1990)

42 Statistical Reasoning with Imprecise Probabilities *P. Walley* (1991)

43 Generalized Additive Models *T.J. Hastie and R.J. Tibshirani* (1990)

44 Inspection Errors for Attributes in Quality Control
N.L. Johnson, S. Kotz and X. Wu (1991)

45 The Analysis of Contingency Tables, 2nd edition *B.S. Everitt* (1992)

46 The Analysis of Quantal Response Data *B.J.T. Morgan* (1992)

47 Longitudinal Data with Serial Correlation—A State-Space Approach
R.H. Jones (1993)

48 Differential Geometry and Statistics *M.K. Murray and J.W. Rice* (1993)

49 Markov Models and Optimization *M.H.A. Davis* (1993)

50 Networks and Chaos—Statistical and Probabilistic Aspects
O.E. Barndorff-Nielsen, J.L. Jensen and W.S. Kendall (1993)

51 Number-Theoretic Methods in Statistics *K.-T. Fang and Y. Wang* (1994)

52 Inference and Asymptotics *O.E. Barndorff-Nielsen and D.R. Cox* (1994)

53 Practical Risk Theory for Actuaries
C.D. Daykin, T. Pentikäinen and M. Pesonen (1994)

54 Biplots *J.C. Gower and D.J. Hand* (1996)

55 Predictive Inference—An Introduction *S. Geisser* (1993)

56 Model-Free Curve Estimation *M.E. Tarter and M.D. Lock* (1993)

57 An Introduction to the Bootstrap *B. Efron and R.J. Tibshirani* (1993)

58 Nonparametric Regression and Generalized Linear Models
P.J. Green and B.W. Silverman (1994)

59 Multidimensional Scaling *T.F. Cox and M.A.A. Cox* (1994)

60 Kernel Smoothing *M.P. Wand and M.C. Jones* (1995)

61 Statistics for Long Memory Processes *J. Beran* (1995)

62 Nonlinear Models for Repeated Measurement Data
M. Davidian and D.M. Giltinan (1995)

63 Measurement Error in Nonlinear Models
R.J. Carroll, D. Rupert and L.A. Stefanski (1995)

64 Analyzing and Modeling Rank Data *J.J. Marden* (1995)

65 Time Series Models—In Econometrics, Finance and Other Fields
D.R. Cox, D.V. Hinkley and O.E. Barndorff-Nielsen (1996)

66 Local Polynomial Modeling and its Applications *J. Fan and I. Gijbels* (1996)

67 Multivariate Dependencies—Models, Analysis and Interpretation
D.R. Cox and N. Wermuth (1996)

68 Statistical Inference—Based on the Likelihood *A. Azzalini* (1996)

69 Bayes and Empirical Bayes Methods for Data Analysis
B.P. Carlin and T.A Louis (1996)

70 Hidden Markov and Other Models for Discrete-Valued Time Series
I.L. Macdonald and W. Zucchini (1997)

71 Statistical Evidence—A Likelihood Paradigm *R. Royall* (1997)

72 Analysis of Incomplete Multivariate Data *J.L. Schafer* (1997)

73 Multivariate Models and Dependence Concepts *H. Joe* (1997)
74 Theory of Sample Surveys *M.E. Thompson* (1997)
75 Retrial Queues *G. Falin and J.G.C. Templeton* (1997)
76 Theory of Dispersion Models *B. Jørgensen* (1997)
77 Mixed Poisson Processes *J. Grandell* (1997)
78 Variance Components Estimation—Mixed Models, Methodologies and Applications *P.S.R.S. Rao* (1997)
79 Bayesian Methods for Finite Population Sampling *G. Meeden and M. Ghosh* (1997)
80 Stochastic Geometry—Likelihood and computation *O.E. Barndorff-Nielsen, W.S. Kendall and M.N.M. van Lieshout* (1998)
81 Computer-Assisted Analysis of Mixtures and Applications—Meta-analysis, Disease Mapping and Others *D. Böhning* (1999)
82 Classification, 2nd edition *A.D. Gordon* (1999)
83 Semimartingales and their Statistical Inference *B.L.S. Prakasa Rao* (1999)
84 Statistical Aspects of BSE and vCJD—Models for Epidemics *C.A. Donnelly and N.M. Ferguson* (1999)
85 Set-Indexed Martingales *G. Ivanoff and E. Merzbach* (2000)
86 The Theory of the Design of Experiments *D.R. Cox and N. Reid* (2000)
87 Complex Stochastic Systems *O.E. Barndorff-Nielsen, D.R. Cox and C. Klüppelberg* (2001)
88 Multidimensional Scaling, 2nd edition *T.F. Cox and M.A.A. Cox* (2001)
89 Algebraic Statistics—Computational Commutative Algebra in Statistics *G. Pistone, E. Riccomagno and H.P. Wynn* (2001)
90 Analysis of Time Series Structure—SSA and Related Techniques *N. Golyandina, V. Nekrutkin and A.A. Zhigljavsky* (2001)
91 Subjective Probability Models for Lifetimes *Fabio Spizzichino* (2001)
92 Empirical Likelihood *Art B. Owen* (2001)
93 Statistics in the 21st Century *Adrian E. Raftery, Martin A. Tanner, and Martin T. Wells* (2001)
94 Accelerated Life Models: Modeling and Statistical Analysis *Vilijandas Bagdonavicius and Mikhail Nikulin* (2001)
95 Subset Selection in Regression, Second Edition *Alan Miller* (2002)
96 Topics in Modelling of Clustered Data *Marc Aerts, Helena Geys, Geert Molenberghs, and Louise M. Ryan* (2002)
97 Components of Variance *D.R. Cox and P.J. Solomon* (2002)
98 Design and Analysis of Cross-Over Trials, 2nd Edition *Byron Jones and Michael G. Kenward* (2003)
99 Extreme Values in Finance, Telecommunications, and the Environment *Bärbel Finkenstädt and Holger Rootzén* (2003)
100 Statistical Inference and Simulation for Spatial Point Processes *Jesper Møller and Rasmus Plenge Waagepetersen* (2004)
101 Hierarchical Modeling and Analysis for Spatial Data *Sudipto Banerjee, Bradley P. Carlin, and Alan E. Gelfand* (2004)
102 Diagnostic Checks in Time Series *Wai Keung Li* (2004)
103 Stereology for Statisticians *Adrian Baddeley and Eva B. Vedel Jensen* (2004)
104 Gaussian Markov Random Fields: Theory and Applications *Havard Rue and Leonard Held* (2005)

Monographs on Statistics and Applied Probability 104

Gaussian Markov Random Fields

Theory and Applications

Håvard Rue
Leonhard Held

CRC Press
Taylor & Francis Group
Boca Raton London New York

CRC Press is an imprint of the
Taylor & Francis Group, an **informa** business
A CHAPMAN & HALL BOOK

Published in 2005 by
Chapman & Hall/CRC
Taylor & Francis Group
6000 Broken Sound Parkway NW
Boca Raton, FL 33487-2742

First issued in paperback 2022

ISBN 13: 978-1-03-247790-9 (pbk)
ISBN 13: 978-1-58488-432-3 (hbk)

DOI: 10.1201/9780203492024

Library of Congress Card Number 2004061870

Publisher's Note
The publisher has gone to great lengths to ensure the quality of this reprint but points out that some imperfections in the original copies may be apparent.

Library of Congress Cataloging-in-Publication Data

Rue, Havard.
Gaussian Markov random fields : theory and applications / Havard Rue & Leonhard Held.
p. cm. -- (Monographs on statistics and applied probability ; 104)
Includes bibliographical references and index.
ISBN 1-58488-432-0 (alk. paper)
1. Gaussian Markov random fields. I. Held, Leonhard. II. Title. III. Series.

QA274.R84 2005
519.2'33--dc22 2004061870

To Mona and Ulrike

Contents

Preface

1 Introduction **1**
1.1 Background 1
1.1.1 An introductory example 1
1.1.2 Conditional autoregressions 3
1.2 The scope of this monograph 5
1.2.1 Numerical methods for sparse matrices 6
1.2.2 Statistical inference in hierarchical models 8
1.3 Applications of GMRFs 10

2 Theory of Gaussian Markov random fields **15**
2.1 Preliminaries 15
2.1.1 Matrices and vectors 15
2.1.2 Lattice and torus 16
2.1.3 General notation and abbreviations 17
2.1.4 Conditional independence 17
2.1.5 Undirected graphs 18
2.1.6 Symmetric positive-definite matrices 19
2.1.7 The normal distribution 20
2.2 Definition and basic properties of GMRFs 21
2.2.1 Definition 21
2.2.2 Markov properties of GMRFs 24
2.2.3 Conditional properties of GMRFs 26
2.2.4 Specification through full conditionals 28
2.2.5 Multivariate GMRFs★ 30
2.3 Simulation from a GMRF 31
2.3.1 Some basic numerical linear algebra 32
2.3.2 Unconditional simulation of a GMRF 33
2.3.3 Conditional simulation of a GMRF 36
2.4 Numerical methods for sparse matrices 40
2.4.1 Factorizing a sparse matrix 41
2.4.2 Bandwidth reduction 45
2.4.3 Nested dissection 48

2.5 A numerical case study of typical GMRFs 52
2.5.1 GMRF models in time 53
2.5.2 Spatial GMRF models 54
2.5.3 Spatiotemporal GMRF models 56
2.6 Stationary GMRFs★ 57
2.6.1 Circulant matrices 57
2.6.2 Block-circulant matrices 62
2.6.3 GMRFs with circulant precision matrices 65
2.6.4 Toeplitz matrices and their approximations 67
2.6.5 Stationary GMRFs on infinite lattices 72
2.7 Parameterization of GMRFs★ 75
2.7.1 The valid parameter space 76
2.7.2 Diagonal dominance 81
2.8 Bibliographic notes 82

3 Intrinsic Gaussian Markov random fields 85
3.1 Preliminaries 85
3.1.1 Some additional definitions 85
3.1.2 Forward differences 87
3.1.3 Polynomials 87
3.2 GMRFs under linear constraints 89
3.3 IGMRFs of first order 93
3.3.1 IGMRFs of first order on the line 94
3.3.2 IGMRFs of first order on lattices 101
3.4 IGMRFs of higher order 108
3.4.1 IGMRFs of higher order on the line 109
3.4.2 IGMRFs of higher order on regular lattices★ 114
3.4.3 Nonpolynomial IGMRFs of higher order 120
3.5 Continuous-time random walks★ 123
3.6 Bibliographic notes 130

4 Case studies in hierarchical modeling 133
4.1 MCMC for hierarchical GMRF models 135
4.1.1 A brief introduction to MCMC 135
4.1.2 Blocking strategies 137
4.2 Normal response models 144
4.2.1 Example: Drivers data 144
4.2.2 Example: Munich rental guide 150
4.3 Auxiliary variable models 153
4.3.1 Scale mixtures of normals 154
4.3.2 Hierarchical-t formulations 156
4.3.3 Binary regression models 157
4.3.4 Example: Tokyo rainfall data 160

4.3.5 Example: Mapping cancer incidence 164
4.4 Nonnormal response models 167
4.4.1 The GMRF approximation 167
4.4.2 Example: Joint disease mapping 172
4.5 Bibliographic notes 180

5 Approximation techniques 183
5.1 GMRFs as approximations to Gaussian fields 183
5.1.1 Gaussian fields 184
5.1.2 Fitting GMRFs to Gaussian fields 186
5.1.3 Results 190
5.1.4 Regular lattices and boundary conditions 195
5.1.5 Example: Swiss rainfall data 199
5.2 Approximating hidden GMRFs 203
5.2.1 Constructing non-Gaussian approximations 204
5.2.2 Example: A stochastic volatility model 209
5.2.3 Example: Reanalyzing Tokyo rainfall data 211
5.3 Bibliographic notes 215

Appendices 217

A Common distributions 219

B The library GMRFLib 221
B.1 The graph object and the function Qfunc 221
B.2 Sampling from a GMRF 225
B.3 Implementing block-updating algorithms for hierarchical GMRF models 231

References 237

Author index 255

Subject index 259

Preface

This monograph describes Gaussian Markov random fields (GMRFs) and some of its applications in statistics. At first sight, this seems to be a rather specialized topic, as the wider class of Markov random fields is probably known only to researchers in spatial statistics and image analysis. However, GMRFs have applications far beyond these two areas, for example in structural time-series analysis, analysis of longitudinal and survival data, spatiotemporal statistics, graphical models, and semi-parametric statistics.

Despite the wide range of application, there is a unified framework for both representing, understanding and computing with GMRFs using the graph formulation. Our main motivation to write this monograph is to provide the first comprehensive account of the main properties of GMRFs, to emphasize the strong connection between GMRFs and numerical methods for sparse matrices, and to outline various applications of GMRFs for statistical inference.

Complex hierarchical models are at the core of modern statistics, and GMRFs play a central role in this framework to describe the spatial and temporal dynamics of nature and real systems. Statistical inference in hierarchical models, however, can typically only be done using simulation, in particular through Markov chain Monte Carlo (MCMC) methods. Thus we emphasize computational issues, which allow us to construct fast and reliable algorithms for (Bayesian) inference in hierarchical models with GMRF components. We emphasize the concept of *blocking*, i.e., updating all or nearly all of the parameters jointly, which we believe to be perhaps the only way to overcome problems with convergence and mixing of ordinary MCMC algorithms. We hope that the reader will share our enthusiasm and that the examples provided in this book will stimulate further research in this area.

The book can be loosely categorized as follows. We begin in Chapter 1 by introducing GMRFs through two simple examples, an autoregressive model in time and a conditional autoregressive model in space. We then briefly discuss numerical methods for sparse matrices, and why they are important for simulation-based inference in GMRF models. We illustrate this through a simple hierarchical model. We finally describe various areas where GMRFs are used in statistics.

Chapter 2 is the main theoretical chapter, describing the most important results for GMRFs. It starts by introducing the necessary notation and describing the central concept of conditional independence. GMRFs are then defined and studied in detail. Efficient direct simulation from a GMRF is described using numerical techniques for sparse matrices. A numerical case study illustrates the performance of the algorithms in different scenarios. Finally, two optional sections follow: The first describes the theory of *stationary* GMRFs, where circulant and block circulant matrices become important. Lastly we discuss the problem on how to parameterize the precision matrix, the inverse covariance matrix, of a GMRF without destroying positive definiteness.

In Chapter 3 we give a detailed discussion of intrinsic GMRFs (IGMRFs). IGMRFs do have precision matrices which are no longer of full rank. They are of central importance in Bayesian hierarchical models, where they are often used as a nonstationary prior distribution for dependent parameters in space or in time. A key concept to understanding IGMRFs is the conditional distribution of a proper GMRF under linear constraints. We then describe IGMRFs of various kinds, on the line, the lattice, the torus, and on irregular graphs. A final optional section is devoted to the representation of integrated Wiener process priors as IGMRFs.

In Chapter 4 we discuss various applications of GMRFs for hierarchical modeling. We outline how to use MCMC algorithms in hierarchical models with GMRF components. We start with some general comments regarding MCMC via blocking. We then discuss models with normal observations, auxiliary variable models for probit and logistic regression and nonnormal regression models, all with latent GMRF components. The GMRFs may have a temporal or a spatial component, or they relate to particular covariate effects in a semiparametric regression framework.

Finally, in Chapter 5 we first describe how GMRFs can be used to approximate so-called *Gaussian fields*, i.e., normal distributed random vectors where the covariance matrix rather than its inverse, the precision matrix, is specified. The final section in Chapter 5 is devoted to the problem of how to construct improved and non-GMRF approximations to hidden GMRFs.

Appendices A and B describe the distributions we use and the implementation of the algorithms in the public-domain library GMRFLib.

Chapters 2 and 3 are fairly self-contained and do not require much prior knowledge from the reader, except for some familiarity with probability theory and linear algebra. Chapters 4 and 5 assume that the reader is experienced in the area of Bayesian hierarchical models and their statistical analysis via MCMC, perhaps at the level of standard textbooks such as Carlin and Louis (1996), Gilks et al. (1996), Robert and Casella (1999), or Gelman et al. (2004).

This monograph can be read chronologically. Sections marked with a '$\star$' indicate more advanced material which can be skipped at first reading. We might ask too much of some readers patience in Chapter 2 and 3, which are motivated from the various applications of GMRFs for hierarchical modeling in Chapter 4. It might therefore be useful to skim through Chapter 4 before reading Chapter 2 and 3 in detail.

This book was conceived in the spring of 2003 but the main body of work was done in the first half of 2004. We are indebted to Julian Besag, who read his seminal paper on Markov random fields (Besag, 1974) 30 years ago to the Royal Statistical Society, his seminal contributions to this field since then, and for introducing LH to MRFs in 1995/1996 during a visit to the University of Washington. We also appreciate his comments on the initial draft and sending us a copy of Mondal and Besag (2004).

We thank Hans R. Künsch for sharing his wisdom with HR during a visit to the ETH Zürich in February 2004, Ludwig Fahrmeir, Stefan Lang, and Håkon Tjelmeland for many good discussions, and Dag Myrhaug for providing a quiet working environment for HR. The interaction and collaboration with (past) Ph.D. students about this theme have been valuable, thanks to Sveinung Erland, Turid Follestad, Oddvar K. Husby, Günter Raßer, Volker Schmid, and Ingelin Steinsland. The support of the German Research Foundation (DFG, Sonderforschungsbereich 386) and the department of mathematical sciences at NTNU is also appreciated. HR also thanks Anne Kajander for all administrative help.

Håkon Tjelmeland and Geir Storvik read carefully through the initial drafts and provided numerous comments and critical questions. Thank you! Also the comments from Arnoldo Frigessi, Martin Sköld and Hanne T. Wist were much appreciated. The collaboration with Chapman & Hall/CRC was always smooth and constructive.

We look forward to returning to everyday life and enjoying our families, Kristine and Mona, Valentina and Ulrike. Thank you for your patience!

HÅVARD RUE	Trondheim
LEONHARD HELD	Munich
	Summer 2004

CHAPTER 1

Introduction

1.1 Background

This monograph considers *Gaussian Markov random fields* (GMRFs) covering both theory and applications. A GMRF is really a simple construct: It is just a (finite-dimensional) random vector following a multivariate normal (or Gaussian) distribution. However, we will be concerned with more restrictive versions where the GMRF satisfies additional *conditional independence* assumptions, hence the term *Markov.*

Conditional independence is a powerful concept. Let $\boldsymbol{x} = (x_1, x_2, x_3)^T$ be a random vector, then x_1 and x_2 are conditionally independent given x_3 if, for known value of x_3, discovering x_2 tells you nothing new about the distribution of x_1. Under this condition the joint density $\pi(\boldsymbol{x})$ must have the representation

$$\pi(\boldsymbol{x}) = \pi(x_1 \mid x_3)\, \pi(x_2 \mid x_3)\, \pi(x_3),$$

which is a simplification of a general representation

$$\pi(\boldsymbol{x}) = \pi(x_1 \mid x_2, x_3)\, \pi(x_2 \mid x_3)\, \pi(x_3).$$

The conditional independence property implies that $\pi(x_1|x_2, x_3)$ is simplified to $\pi(x_1|x_3)$, which is easier to understand, to represent, and to interpret.

1.1.1 An introductory example

As a simple example of a GMRF, consider an autoregressive process of order 1 with standard normal errors, which if often expressed as

$$x_t = \phi x_{t-1} + \epsilon_t, \quad \epsilon_t \overset{\text{iid}}{\sim} \mathcal{N}(0, 1), \quad |\phi| < 1 \tag{1.1}$$

where the index t represents time. Assumptions about conditional independence are not stated explicitly here, but show up more clearly if we express (1.1) in the conditional form

$$x_t \mid x_1, \ldots, x_{t-1} \;\sim\; \mathcal{N}(\phi x_{t-1},\, 1) \tag{1.2}$$

for $t = 2, \ldots, n$. In this model x_s and x_t with $1 \leq s < t \leq n$ are conditionally independent given $\{x_{s+1}, \ldots, x_{t-1}\}$ if $t - s > 1$.

In addition to (1.2), let us now assume that the marginal distribution of x_1 is normal with mean zero and variance $1/(1-\phi^2)$, which is simply the stationary distribution of this process. Then the joint density of $\boldsymbol{x}$ is

$$\begin{aligned} \pi(\boldsymbol{x}) &= \pi(x_1)\,\pi(x_2 \mid x_1) \cdots \pi(x_n \mid x_{n-1}) \\ &= \frac{1}{(2\pi)^{n/2}} |\boldsymbol{Q}|^{1/2} \exp\left(-\frac{1}{2}\boldsymbol{x}^T \boldsymbol{Q} \boldsymbol{x}\right), \end{aligned} \tag{1.3}$$

where the *precision matrix* $\boldsymbol{Q}$ is the tridiagonal matrix

$$\boldsymbol{Q} = \begin{pmatrix} 1 & -\phi & & & & \\ -\phi & 1+\phi^2 & -\phi & & & \\ & & \ddots & \ddots & \ddots & \\ & & & -\phi & 1+\phi^2 & -\phi \\ & & & & -\phi & 1 \end{pmatrix}$$

with zero entries outside the diagonal and first off-diagonals. The conditional independence assumptions impose certain restrictions on the precision matrix. The tridiagonal form is due to the fact that x_i and x_j are conditionally independent for $|i-j| > 1$, given the rest. This also holds in general for any GMRF: If $Q_{ij} = 0$ for $i \neq j$, then x_i and x_j are conditionally independent given the other variables $\{x_k : k \neq i \text{ and } k \neq j\}$ and vice versa. The sparse structure of $\boldsymbol{Q}$ prepares the ground for fast computations of GMRFs to which we return in Section 1.2.1.

The simple relationship between conditional independence and the zero structure of the precision matrix is not evident in the covariance matrix $\boldsymbol{\Sigma} = \boldsymbol{Q}^{-1}$, which is a (completely) dense matrix with entries

$$\sigma_{ij} = \frac{1}{1-\phi^2}\phi^{|i-j|}.$$

For example, for $n = 7$,

$$\boldsymbol{\Sigma} = \frac{1}{1-\phi^2}\begin{pmatrix} 1 & \phi & \phi^2 & \phi^3 & \phi^4 & \phi^5 & \phi^6 \\ \phi & 1 & \phi & \phi^2 & \phi^3 & \phi^4 & \phi^5 \\ \phi^2 & \phi & 1 & \phi & \phi^2 & \phi^3 & \phi^4 \\ \phi^3 & \phi^2 & \phi & 1 & \phi & \phi^2 & \phi^3 \\ \phi^4 & \phi^3 & \phi^2 & \phi & 1 & \phi & \phi^2 \\ \phi^5 & \phi^4 & \phi^3 & \phi^2 & \phi & 1 & \phi \\ \phi^6 & \phi^5 & \phi^4 & \phi^3 & \phi^2 & \phi & 1 \end{pmatrix}.$$

It is therefore difficult to derive conditional independence properties from the structure of $\boldsymbol{\Sigma}$. Clearly, the entries in $\boldsymbol{\Sigma}$ only give (direct) information about the *marginal* dependence structure, not the conditional one. For example, in the autoregressive model, x_s and x_t are *marginally dependent* for any finite s and t as long as $\phi \neq 0$.

Simplifications due to conditional independence do not only appear for the *directed* conditional distributions as in (1.2), but also for the *undirected* conditional distributions, often called *full conditionals* $\{\pi(x_t|\boldsymbol{x}_{-t})\}$, where $\boldsymbol{x}_{-t}$ denotes all elements in $\boldsymbol{x}$ but x_t. In the autoregressive example,

$$x_t \mid \boldsymbol{x}_{-t} \sim \begin{cases} \mathcal{N}\left(\phi x_{t+1},\ 1\right) & t = 1, \\ \mathcal{N}\left(\frac{\phi}{1+\phi^2}(x_{t-1} + x_{t+1}),\ \frac{1}{1+\phi^2}\right) & 1 < t < n, \\ \mathcal{N}\left(\phi x_{n-1},\ 1\right) & t = n, \end{cases} \tag{1.4}$$

so x_t depends in general both on x_{t-1} and x_{t+1}. Equation (1.4) is important as it allows for an alternative specification of the first-order autoregressive models through the full conditionals $\pi(x_t|\boldsymbol{x}_{-t})$ for $t = 1, \ldots, n$. In fact, by starting with these full conditionals, we obtain an alternative and completely equivalent representation of this model with the same joint density for $\boldsymbol{x}$. This is not so obvious as for the directed conditional distributions (1.2), where the joint density is simply the product of the densities corresponding to (1.2) for $t = 2, \ldots, n$ times the (marginal) density of x_1.

1.1.2 Conditional autoregressions

We now make the discussion more general, leaving autoregressive models. Let $\boldsymbol{x}$ be associated with observations or some property of points or regions in the spatial domain. For example, x_i could be the value of pixel i in an image, the height of tile i in a tessellation or the relative risk for some disease in the ith district. Now there is no natural ordering of the indices and (1.3) is no longer useful to specify the joint density of $\boldsymbol{x}$. A common approach is then to specify the joint density of a zero mean GMRF implicitly by specifying each of the n full conditionals

$$x_i \mid \boldsymbol{x}_{-i} \sim \mathcal{N}\left(\sum_{j:j\neq i} \beta_{ij} x_j, \kappa_i^{-1}\right), \tag{1.5}$$

which was pioneered by Besag (1974, 1975). These models are also known by the name *conditional autoregressions*, abbreviated as *CAR* models. There is also an alternative and more restrictive approach to CAR models, the so-called *simultaneous autoregressions* (SAR), which we will not discuss specifically. This approach dates back to Whittle (1954), see for example, Cressie (1993) for further details.

The n full conditionals (1.5) must satisfy some consistency conditions to ensure that a joint normal density exists with these full conditionals.

These reduces to require that $\boldsymbol{Q} = (Q_{ij})$ with elements

$$Q_{ij} = \begin{cases} \kappa_i & i = j \\ -\kappa_i \beta_{ij} & i \neq j, \end{cases}$$

is symmetric and positive definite. Symmetry is ensured by $\kappa_i \beta_{ij} = \kappa_j \beta_{ji}$ for all $i \neq j$, while positive definiteness requires $\kappa_i > 0$ for all $i = 1, \ldots, n$, but imposes further (and often quite complicated) constraints on the β_{ij}'s. A common (perhaps too common!) approach to ensure positive definiteness is to require that $\boldsymbol{Q}$ is *diagonal dominant*, which means that, in each row (or column) of $\boldsymbol{Q}$, the diagonal entry is larger than the sum of the absolute off-diagonal entries. This is a sufficient but not necessary condition for positive definiteness.

The conditional independence properties of this GMRF can now be found by simply checking if Q_{ij} is zero or not. If $Q_{ij} = 0$ then x_i and x_j are conditionally independent given the rest, and if $Q_{ij} \neq 0$ then they are conditionally dependent. It is useful to represent these findings using an *undirected graph* with nodes $\{1, \ldots, n\}$ and an edge between node i and $j \neq i$ if and only if $Q_{ij} \neq 0$. We then say that $\boldsymbol{x}$ is a GMRF with respect to this graph. The neighbors to node i are all nodes $j \neq i$ with $\beta_{ij} \neq 0$, hence all nodes on which the full conditional (1.5) depends. Going back to the autoregressive model (1.4), the neighbors of i are $\{i-1, i+1\}$ for $i = 2, \ldots, n-1$, and $\{2\}$ and $\{n-1\}$ of node 1 and n, respectively.

In general the neighbors of i are often those that are, in one way or the other, in the 'proximity' of node i. The common approach is first to specify the graph by choosing a suitable set of neighbors to each node, and then to choose β_{ij} for each pair $i \sim j$ of neighboring nodes i and j.

Figure 1.1 displays two such graphs, (a) a linear graph corresponding to (1.2) with $n = 50$ and (b) the graph corresponding to the 16 states of Germany where two states are neighbors if they share a common border. The graph in (b) is not drawn to mimic the map of Germany but only to visualize the graph itself. The number of neighbors in (b) varies between 2 and 9.

Figure 1.2 displays a graph constructed similarly to Figure 1.1(b), but which now corresponds to the 366 regions in Sardinia. The neighborhood structure is now slightly more complex and the number of neighbors varies between 1 and 13 with a median of 5. This is a typical (but simple) graph for applications of GMRF models.

The case where $\boldsymbol{Q}$ is symmetric and positive semidefinite is of particular interest. This class is known under the name *intrinsic conditional autoregressions* or *intrinsic GMRF*s (IGMRFs). The density of $\boldsymbol{x}$ is then improper but, by construction, $\boldsymbol{x}$ defines a proper distribution on a specific lower-dimensional space. For example, if each row (or column) of $\boldsymbol{Q}$ sums up to zero, then $\boldsymbol{Q}$ has rank $n-1$ and the (improper) density

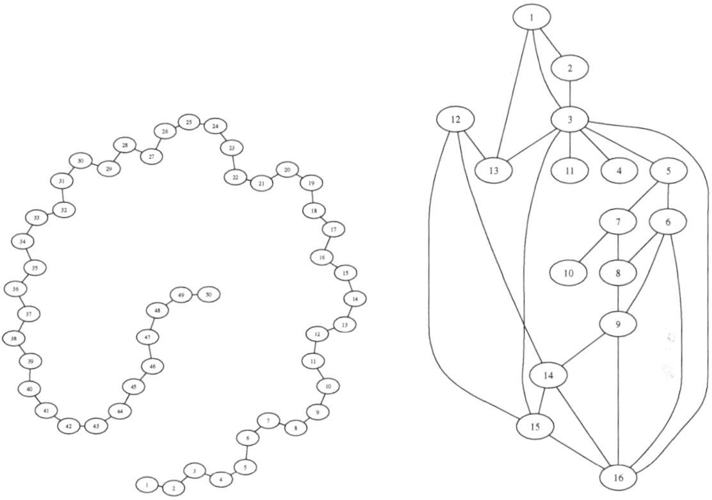

(a) (b)

Figure 1.1 *(a) The linear graph corresponding to an autoregressive process of order* 1*, (b) the graph of the* 16 *states of Germany where two states sharing a common border are considered to be neighbors.*

of $\boldsymbol{x}$ is invariant to the addition of a constant to all components in $\boldsymbol{x}$. This is of benefit if the level of $\boldsymbol{x}$ is unknown or perhaps not constant but varies smoothly over the region of interest. More generally, one can for example construct IGMRFs that are invariant to the addition of polynomials. IGMRFs play a central role in *hierarchical models*, which we discuss later.

1.2 The scope of this monograph

The main scope of this monograph is as follows:

- To provide a systematic presentation of the main theoretical results for GMRFs and intrinsic GMRFs. We will focus mainly on finite GMRFs, but also discuss GMRFs on infinite lattices.
- To present and discuss numerical methods for sparse matrices and how these can be used to simulate from a GMRF and how to evaluate the log density of a GMRF. Both tasks also can be done under various forms of conditioning and linear constraints.
- To discuss hierarchical GMRF models, which illustrates the use of GMRFs in various areas using different choices for the distribution of the observed data.

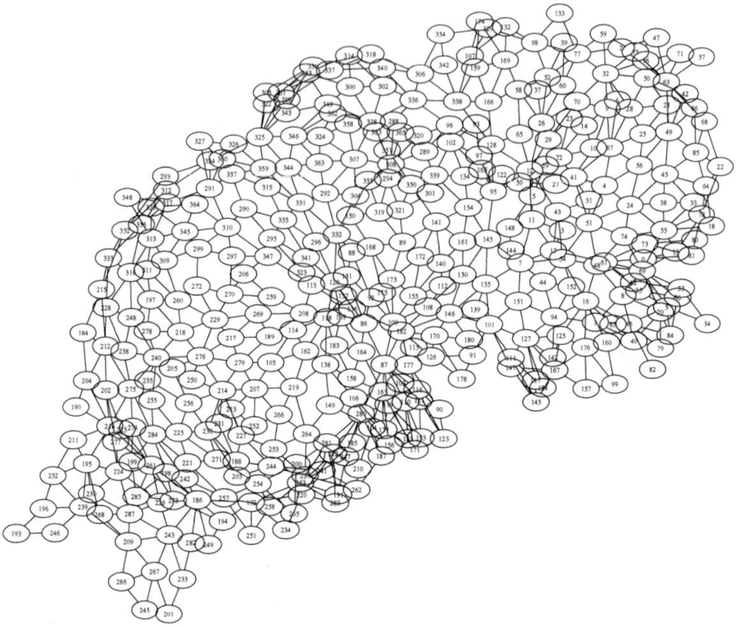

Figure 1.2 *The graph of the* 366 *administrative regions in Sardinia where two regions sharing a common border are neighbors. Neighbors in the graph are linked by edges or indicated by overlapping nodes.*

- To provide a unified framework for fast and reliable Bayesian inference in hierarchical GMRF models based on Markov chain Monte Carlo (MCMC) simulation. Typically, all or nearly all unknown parameters are updated simultaneously or at least in large blocks. An important part of the algorithms is to use fast numerical methods for sparse matrices.

Perhaps the two most innovative methodological contributions in this monograph are the connection between GMRF and numerical methods for sparse matrices, and the fast and reliable MCMC block algorithms for Bayesian inference in hierarchical models with GMRF components. We briefly describe the main ideas in the following.

1.2.1 Numerical methods for sparse matrices

Sparse matrices appear naturally for GMRFs as $Q_{ij} \neq 0$ only if i and j are neighbors. *By construction* (most) precision matrices for GMRFs are sparse where only $\mathcal{O}(n)$ of the terms in $\boldsymbol{Q}$ are nonzero. We can take advantage of this for computing the Cholesky factorization of $\boldsymbol{Q}$,

$$\boldsymbol{Q} = \boldsymbol{L}\boldsymbol{L}^T,$$

where $\boldsymbol{L}$ is a lower-triangular matrix. It turns out that $\boldsymbol{L}$ can inherit the nonzero pattern of $\boldsymbol{Q}$ so it can be sparse as well. However, *how sparse* $\boldsymbol{L}$ is depends heavily on the *ordering* of the indices of the GMRF $\boldsymbol{x}$. Therefore the indices are permuted in advance to obtain a matrix $\boldsymbol{L}$ with as few as possible nonzero entries. The computational savings stem from the simple fact that we do not need to compute terms that are known to be zero. Hence, only the nonzero terms in $\boldsymbol{L}$ are computed and stored. For larger GMRFs, for example with $10,000$ - $100,000$ nodes, this results in a *huge* speedup and low memory usage. The classical approach to obtain such a matrix $\boldsymbol{L}$ is to construct a permutation of the indices of $\boldsymbol{x}$ such that the permuted $\boldsymbol{Q}$ becomes a band matrix. Then a band-Cholesky factorization can be used to compute $\boldsymbol{L}$. In this case $\boldsymbol{L}$ will be a (lower) band matrix with the same bandwidth as $\boldsymbol{Q}$. For an introduction to numerical methods for sparse matrices, see Dongarra et al. (1998), Duff et al. (1989), George and Liu (1981), and Gupta (2002) for a comparison.

As an illustration, suppose we want to simulate from a GMRF. Simulation-based inference via MCMC is typically the only way for (Bayesian) inference in complex hierarchical models, and efficient simulation of GMRFs is therefore one of the central themes of the book. To simulate from a zero mean GMRF with precision matrix $\boldsymbol{Q}$, we compute the Cholesky triangle $\boldsymbol{L}$ and then *solve*

$$\boldsymbol{L}^T\boldsymbol{x} = \boldsymbol{z}, \tag{1.6}$$

where $\boldsymbol{z}$ is a vector of independent standard normal variables. The sparse structure of $\boldsymbol{L}$ will also make this step more efficient. It is easy to show that the solution of (1.6) has precision matrix $\boldsymbol{Q}$ as required. The generalization to arbitrary mean $\boldsymbol{\mu}$ is trivial. Algorithms for conditional simulation of GMRFs can also be constructed such that all sparse matrices involved are taken advantage of. The same is true if one is interested in the evaluation of the log density of the GMRF. Roughly, the cost is $\mathcal{O}(n)$, $\mathcal{O}(n^{3/2})$, $\mathcal{O}(n^2)$ for GMRFs in time, space, space $\times$ time, respectively.

The symbiotic connection between GMRFs and numerical methods sparse matrices has been known implicitly and for special cases for a long time. For autoregressive models and state-space models in general, fast $\mathcal{O}(n)$ algorithms exist, derived from the Kalman filter and its variants. The *forward-filtering backward-sampling* algorithm (Carter and Kohn, 1994, Frühwirth-Schnatter, 1994) uses intermediate results from the Kalman filter to sample from a (hidden) GMRF, but reduces to factorizing a positive definite (block-)tridiagonal matrix (Knorr-Held and Rue, 2002, Appendix). Lavine (1999) uses this algorithm for a two- and three-dimensional GMRF on a regular lattice and derives algorithms for the evaluation of the log density and for sampling a GMRF. This algorithm is similar to the one derived by Moura and Balram (1992). However, the extension from the regular lattice to a general graph is difficult using this approach. Pace and Barry (1997) propose using general numerical methods for sparse matrices to evaluate the log density. Rue (2001) derives algorithms for conditional sampling, evaluation of the corresponding log density and demonstrate how to use these numerical methods to construct block-updating MCMC algorithms further developed by Knorr-Held and Rue (2002). Rue and Follestad (2003) provide additional details of Rue (2001, Appendix) and a statistical interpretation of numerical methods for sparse matrices and various permutation approaches.

A nice feature about modern techniques for sparse matrices is that the permutation adapts to the graph of the GMRF under study, hence such algorithms provide close-to-optimal algorithms for most cases of interest. This is of great advantage as it allows us to *merge* the different GMRFs usually involved in a hierarchical GMRF model into a larger one, which makes it possible to construct a unified approach to MCMC-based inference for hierarchical GMRF models. This will be sketched in the following section.

1.2.2 Statistical inference in hierarchical models

GMRFs are frequently used in hierarchical models in order to allow for stochastic dependence between a set of unknown parameters. A typical setup uses three stages where unknown *hyperparameters* $\boldsymbol{\theta}$ specify a GMRF $\boldsymbol{x}$. The field $\boldsymbol{x}$ is now connected to data $\boldsymbol{y}$, which are commonly assumed to be conditionally independent given $\boldsymbol{x}$. In the simplest case, each observation y_i depends only on a corresponding ith element x_i in $\boldsymbol{x}$, so $\boldsymbol{y}$ and $\boldsymbol{x}$ have the same dimension. Hence the three stages are specified as

$$\begin{aligned}
\boldsymbol{\theta} &\sim \pi(\boldsymbol{\theta}) \\
\boldsymbol{x} &\sim \pi(\boldsymbol{x} \mid \boldsymbol{\theta}) \\
y_i &\overset{\text{iid}}{\sim} \pi(y_i \mid x_i), \quad i = 1, \ldots, n.
\end{aligned}$$

The posterior distribution is

$$\pi(\boldsymbol{x}, \boldsymbol{\theta} \mid \boldsymbol{y}) \propto \pi(\boldsymbol{\theta})\, \pi(\boldsymbol{x} \mid \boldsymbol{\theta}) \prod_{i=1}^{n} \pi(y_i \mid x_i).$$

For example, y_i could be a normal variable with mean x_i, Bernoulli with mean $1/(1+\exp(x_i))$ or Poisson with mean $\exp(x_i)$. Consider for example the Poisson case. In many applications there will be so-called *extra-Poisson variation*, and a common approach to deal with this is to add independent zero mean normal random effects v_i to the model so that y_i is now Poisson with mean $\exp(x_i + v_i)$.

Since $\boldsymbol{v}$ is also a GMRF with a diagonal precision matrix and $\boldsymbol{x}$ and $\boldsymbol{v}$ are assumed to be independent, they form a joint GMRF of size $2n$. However, conditional on $\boldsymbol{y}$, both x_i and v_i will depend on y_i. An alternative approach is to parameterize the model from $\boldsymbol{v}$ to $\boldsymbol{u} = \boldsymbol{x} + \boldsymbol{v}$, which defines a GMRF $\boldsymbol{w}$ of size 2n,

$$\boldsymbol{w} = \begin{pmatrix} \boldsymbol{x} \\ \boldsymbol{u} \end{pmatrix}. \tag{1.7}$$

The graph of $\boldsymbol{w}$ is displayed in Figure 1.3 where the graph $\boldsymbol{x}$ corresponds either to Figure 1.1(a) or (b) and the gray nodes in the graph correspond to $\boldsymbol{u}$. Using the new GMRF $\boldsymbol{w}$, each observations y_i is now only connected to w_{i+n} and the posterior distribution has the form

$$\pi(\boldsymbol{w}, \boldsymbol{\theta} \mid \boldsymbol{y}) \propto \pi(\boldsymbol{\theta})\; \pi(\boldsymbol{w} \mid \boldsymbol{\theta}) \prod_{i=1}^{n} \pi(y_i \mid w_{n+i}).$$

This is a typical example where MCMC is the only way for statistical inference, but where the choice of the particular MCMC algorithm is crucial. In Section 4, we will describe an MCMC algorithm that jointly updates the GMRF $\boldsymbol{w}$ and the hyperparameters $\boldsymbol{\theta}$ (here the unknown precisions of $\boldsymbol{x}$ and $\boldsymbol{v}$) in one block, thus ensuring good mixing and convergence properties of the algorithm.

Suppose now there is also *covariate information* $\boldsymbol{z}_i$ available for each observation y_i, here $\boldsymbol{z}_i$ is of dimension p, say. A common approach is to assume now that y_i is Poisson with mean $\exp(x_i + v_i + \boldsymbol{z}_i^T \boldsymbol{\beta})$, where $\boldsymbol{\beta}$ is a vector of unknown regression parameters, with a multivariate normal prior with some mean and some precision matrix, which can be zero. We do not give the exact details here, but $\boldsymbol{\beta}$ can also be merged with the GMRF $\boldsymbol{w}$ to a larger GMRF of dimension $2n + p$, which still inherits the sparse structure. Furthermore, block updates of the enlarged field, jointly with unknown hyperparameters, is still possible.

Merging two or more GMRFs into a larger one typically preserves the local features of the GMRF and simplifies the *structure* of the model. This is important mainly for computational reasons, as we can then construct efficient MCMC algorithms. Note that if $\boldsymbol{\theta}$ is fixed and y_i is normal, this will correspond to independent simulation from the posterior, no matter how large the dimension of the GMRF. For nonnormal observations, as in the above Poisson case, we will use the Metropolis-Hastings algorithm combined with Taylor expansions to construct appropriate GMRF block proposals for the posterior distribution. However, for binary responses we will introduce so-called *auxiliary variables* in the model, which avoid the use of Taylor expansions.

1.3 Applications of GMRFs

GMRFs have an *enormous* list of applications, dating back to 1880, at least, with Thiele's first-order random walk model for time-series analysis, see Lauritzen (1981). We will now briefly describe some main areas of application of GMRFs, not mutually disjoint, where GMRFs are being used, pointing the interested reader to some key references.

Structural time-series analysis Autoregressive models are GMRFs on a linear graph that is part of the standard literature in time series. Extensions to state-space models add normal observations that makes the conditional distribution of the hidden state $\boldsymbol{x}$ also a GMRF. Some of the theoretical results derived in this area depend particularly on the linear graph and its sequential representation. Computational algorithms used are based on the Kalman filter and its variants. Approximate inference for state-space models with nonnormal observations is discussed in Fahrmeir (1992). Simulation-based inference for normal state-space models is described in Carter and Kohn (1994), Frühwirth-Schnatter (1994), and Shephard (1994), while simulation-based inference for state-space models with nonnormal observations is proposed in Shephard and Pitt (1997) and Knorr-Held (1999). The connection of these algorithms to our more general graph-oriented approach will be discussed in Chapter 4, see also Knorr-Held and Rue (2002, Appendix A). Good references to time-series analysis and state-space models are Brockwell and Davis (1987), Harvey (1989) and West and Harrison (1997).

Analysis of longitudinal and survival data GMRF priors, in particular their temporal versions, are used extensively to analyze longitudinal and survival data. Some key references for state-space approaches are Fahrmeir (1994), Gamerman and West (1987), Jones (1993), see also Fahrmeir and Knorr-Held (2000, Sec. 18.3.3). The analysis of longitudinal or survival data with additional GMRFs on spatial components is described in Banerjee et al. (2003), Carlin and Banerjee (2003), Crook et al. (2003), Knorr-Held (2000a), Knorr-Held and Besag (1998), Knorr-Held and Richardson (2003), and Banerjee et al. (2004) among others. Analysis of rates with several time scales is described in Berzuini and Clayton (1994), Besag et al. (1995), Knorr-Held and Rainer (2001), and Bray (2002). Finally, applications of GMRF priors to longitudinal data in sports are described in Glickman

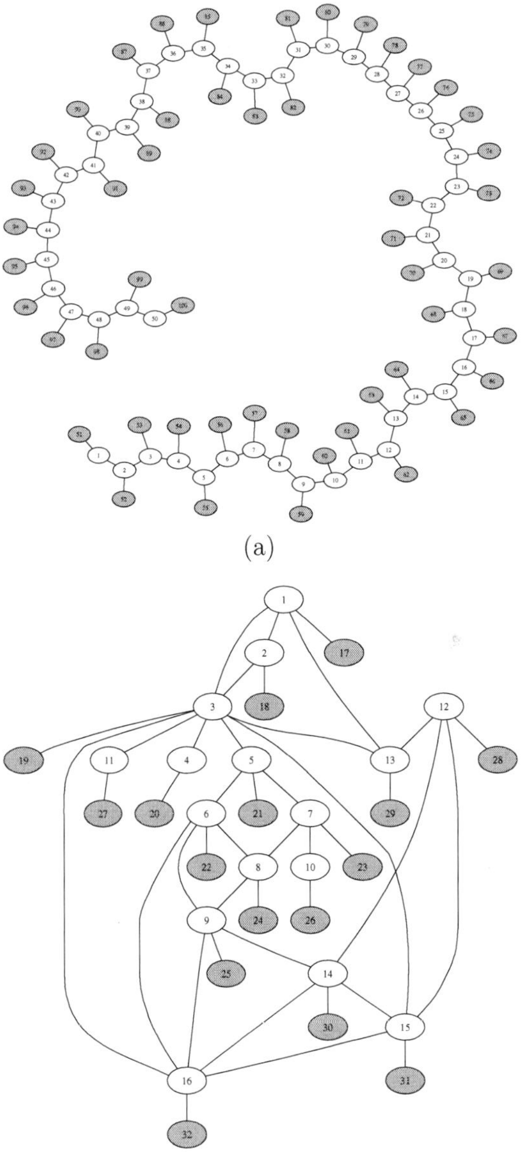

(a)

(b)

Figure 1.3 *The graph of* $\boldsymbol{w}$ *(1.7) where the graph of* $\boldsymbol{x}$ *is in Figure 1.1(a) and (b), respectively. The nodes corresponding to* $\boldsymbol{u}$ *are displayed in gray.*

and Stern (1998), Knorr-Held (2000b), Rue and Salvesen (2000), and Held and Vollnhals (2005).

Graphical models GMRFs are central in the area of graphical models. One problem is to estimate $\boldsymbol{Q}$ and its (associated) graph from data, see Dempster (1972), Giudici and Green (1999), Whittaker (1990), and Dobra et al. (2003) for an application in statistical genetics. More generally, GMRFs are used in a larger setting involving not only undirected but also directed or chain graphs, and perhaps nonnormal or discrete random variables. Some theoretical results regarding GMRFs that we do not cover in this monograph can be found in this area, see for example, Speed and Kiiveri (1986) and the books by Whittaker (1990) and Lauritzen (1996). Exact propagation algorithms for graphical models also include algorithms for GMRFs, see Lauritzen and Jensen (2001). Wilkinson and Yeung (2002, 2004) discuss propagation algorithms and the connection to the sparse matrix approach taken in this monograph.

Semiparametric regression and splines A similar task appearing in both semiparametric statistics and spline models is to describe a smooth curve in time or a surface in space, see for example, Fahrmeir and Lang (2001a,c), Heikkinen and Arjas (1998). A semiparametric approach is often based on intrinsic GMRF models using either a first- or second-order random walk model in time or on the line. Spline models are formulated differently, but Wahba (1978) derived the connection between the posterior expectation of a diffuse integrated Wiener process and polynomial splines. Second-order random walk models, as they are commonly defined, can be seen as an approximation to a discretely observed integrated Wiener process. However, this connection can be made rigorous as we will discuss later using results of Wecker and Ansley (1983) and Jones (1981). A more recent approach taken by Lang and Brezger (2004) is to use IGMRF models for the coefficients of B-splines. The presentation of statistical modeling approaches using generalized linear models by Fahrmeir and Tutz (2001) illustrates the use of GMRFs and splines for semi-parametric regression in various settings.

Image analysis Image analysis is perhaps the first main area of application of spatial GMRFs, see for example, techniques for image restoration using the Wiener filter (Hunt, 1973), texture modeling, and texture discrimination (Chellappa and Chatterjee, 1985, Chellappa et al., 1985, Cross and Jain, 1983, Descombes et al., 1999, Rellier et al., 2002). Further applications of GMRFs in image analysis include modeling stationary fields (Chellappa and Jain, 1993, Chellappa and Kashyap, 1982, Dubes and Jain, 1989, Kashyap and Chellappa,

1983), modeling inhomogeneous fields (Aykroyd, 1998, Dreesman and Tutz, 2001), segmentation (Dryden et al., 2003, Manjunath and Chellappa, 1991), low-level vision (Marroquin et al., 2001), blind restoration (Jeffs et al., 1998), deformable templates (Amit et al., 1991, Grenander, 1993, Grenander and Miller, 1994, Hobolth et al., 2002, Hurn et al., 2001, Kent et al., 2000, 1996, Ripley and Sutherland, 1990, Rue and Husby, 1998), object identification (Rue and Hurn, 1999), 3D reconstruction (Lindgren, 1997, Lindgren et al., 1997), restoring ultrasound images (Husby et al., 2001, Husby and Rue, 2004) and adjusted maximum likelihood and pseudolikelihood estimation (Besag, 1975, 1977a,b, Dryden et al., 2002). GMRFs are also used in edge-preserving restoration using auxiliary variables, see Geman and Yang (1995). This field is simply to large to be treated fairly, see also Hurn et al. (2003) for a statistically oriented (but still incomplete) overview.

Spatial statistics The use of GMRF in this field is large, see for example, Banerjee et al. (2004), Cressie (1993), and the references therein. Some more recent applications include the analysis of spatial binary data (Pettitt et al., 2002, Weir and Pettitt, 1999, 2000), non-stationary models (Dreesman and Tutz, 2001), geostatistical applications using GMRF approximations for Gaussian fields (Allcroft and Glasbey, 2003, Follestad and Rue, 2003, Hrafnkelsson and Cressie, 2003, Husby and Rue, 2004, Rue and Follestad, 2003, Rue et al., 2004, Rue and Tjelmeland, 2002, Steinsland and Rue, 2003, Werner, 2004), analysis of data in social science, see Fotheringham et al. (2002), Haining (1990) and the references therein, spatial econometrics, see Anselin and Florax (1995) and the references therein, multivariate GMRFs (Gamerman et al., 2003, Gelfand and Vounatsou, 2003, Mardia, 1988), space-varying regression models (Assunção et al., 1998, Gamerman et al., 2003), analysis of agricultural field experiments (Bartlett, 1978, Besag et al., 1995, Besag and Higdon, 1999), applications in spatial and space-time epidemiology (Besag et al., 1991, Cressie and Chan, 1989, Knorr-Held, 2000a, Knorr-Held and Besag, 1998, Knorr-Held et al., 2002, Knorr-Held and Rue, 2002, Mollié, 1996, Natario and Knorr-Held, 2003, Schmid and Held, 2004), in environmental statistics (Huerta et al., 2004, Lindgren and Rue, 2004, Wikle et al., 1998), to inverse problems (Higdon et al., 2003) and so on. The list seems endless.

CHAPTER 2

Theory of Gaussian Markov random fields

In this chapter, we will present the basic properties of a GMRF. As a GMRF is normal, all results valid for a normal distribution are also valid for a GMRF. However, in order to apply GMRFs in Bayesian hierarchical models, we need to sample from GMRFs and to compute certain properties of GMRFs under various conditions. What makes GMRFs extremely useful in practice, is that the things we often need to compute are particularly fast to compute for a GMRF. The key is naturally the sparseness of the precision matrix and the structure of its nonzero terms. It will be useful to represent GMRFs on a graph representing the nonzero pattern of the precision matrix. This representation serves two purposes. First, it will provide a unified way of interpreting and understanding a GMRF through conditional independence, either for a GMRF in time, on a lattice, or on some more general structure. Secondly, this representation will also provide a unified way to actually compute various properties for a GMRF and to generate samples from it, by using numerical methods for sparse matrices.

2.1 Preliminaries

2.1.1 Matrices and vectors

Vectors and matrices are typeset in bold, like $\boldsymbol{x}$ and $\boldsymbol{A}$. The transpose of $\boldsymbol{A}$ is denoted by $\boldsymbol{A}^T$. The notation $\boldsymbol{A} = (A_{ij})$ means that the element in the ith row and jth column of $\boldsymbol{A}$ is A_{ij}. For a vector we use the same notation, $\boldsymbol{x} = (x_i)$. We denote by $\boldsymbol{x}_{i:j}$ the vector $(x_i, x_{i+1}, \ldots, x_j)^T$. For an $n \times m$ matrix $\boldsymbol{A}$ with columns $\boldsymbol{A}_1, \boldsymbol{A}_2, \ldots, \boldsymbol{A}_m$, $\text{vec}(\boldsymbol{A})$ denotes the vector obtained by stacking the columns one above the other, $\text{vec}(\boldsymbol{A}) = (\boldsymbol{A}_1^T, \boldsymbol{A}_2^T, \ldots, \boldsymbol{A}_m^T)^T$. A *submatrix* of $\boldsymbol{A}$ is obtained by deleting some rows and/or columns of $\boldsymbol{A}$. A submatrix of an $n \times n$ matrix $\boldsymbol{A}$ is called a *principal submatrix*, if it can be obtained by deleting rows and columns of the same index, so for example, $\boldsymbol{B} = \begin{pmatrix} A_{11} & A_{13} \\ A_{31} & A_{33} \end{pmatrix}$ is a principal submatrix of $\boldsymbol{A}$. An $r \times r$ submatrix is called a *leading principal submatrix* of $\boldsymbol{A}$, if it can be obtained by deleting the last $n - r$ rows and columns.

We use the notation $\text{diag}(\boldsymbol{A})$ and $\text{diag}(\boldsymbol{a})$, where $\boldsymbol{A}$ is an $n \times n$ matrix and $\boldsymbol{a}$ a vector of length n, for the $n \times n$ diagonal matrices

$$\begin{pmatrix} A_{11} & & \\ & \ddots & \\ & & A_{nn} \end{pmatrix} \quad \text{and} \quad \begin{pmatrix} a_1 & & \\ & \ddots & \\ & & a_n \end{pmatrix},$$

respectively. We denote by $\boldsymbol{I}$ the identity matrix.

The matrix $\boldsymbol{A}$ is called *upper triangular* if $A_{ij} = 0$ whenever $i > j$ and *lower triangular* if $A_{ij} = 0$ whenever $i < j$. The *bandwidth* of a matrix $\boldsymbol{A}$ is $\max\{|i-j| : A_{ij} \neq 0\}$. The *lower bandwidth* is $\max\{|i-j| : A_{ij} \neq 0 \text{ and } i > j\}$. The *determinant* of an $n \times n$ matrix $\boldsymbol{A}$ is denoted by $|\boldsymbol{A}|$ and equals the product of the eigenvalues of $\boldsymbol{A}$. The *rank* of $\boldsymbol{A}$, denoted by $\text{rank}(\boldsymbol{A})$, is the number of linearly independent rows or columns of the matrix. The *trace* of $\boldsymbol{A}$ is the sum of the diagonal elements, $\text{trace}(\boldsymbol{A}) = \sum_i A_{ii}$. For elementwise multiplication of two matrices of size $n \times m$, we use the symbol '$\odot$', i.e.,

$$\boldsymbol{A} \odot \boldsymbol{B} = \begin{pmatrix} A_{11}B_{11} & \dots & A_{1m}B_{1m} \\ \vdots & \ddots & \vdots \\ A_{n1}B_{n1} & \dots & A_{nm}B_{nm} \end{pmatrix}.$$

Similarly, '$\oslash$' denotes elementwise division. We will use the symbol '$\circledwedge$' for raising each element of a matrix $\boldsymbol{A}$ to a scalar power a, i.e., element ij of $\boldsymbol{A} \circledwedge a$ is A_{ij}^a.

2.1.2 Lattice and torus

We denote by $\mathcal{I}_{\boldsymbol{n}}$ a (regular) *lattice* (or grid) of size $\boldsymbol{n} = (n_1, n_2)$ (for a two-dimensional lattice). The location of *pixel* or *site* ij is denoted by (i, j). Let $\boldsymbol{x}$ take values on $\mathcal{I}_{\boldsymbol{n}}$ and denote by x_{ij} the value of $\boldsymbol{x}$ at site ij, for $i = 1, \dots, n_1$ and $j = 1, \dots, n_2$. We add where needed a ',' in the indices, like $x_{11,1}$, to avoid confusion. On an *infinite lattice* $\mathcal{I}_\infty$ the sites ij are numbered as $i = 0, \pm 1, \pm 2, \dots$, and $j = 0, \pm 1, \pm 2, \dots$.

A *torus* is a lattice with cyclic (or toroidal) boundary conditions and denoted by $\mathcal{T}_{\boldsymbol{n}}$. By notational convenience, the dimension is $\boldsymbol{n} = (n_1, n_2)$ (for a two-dimensional torus) and all indices are modulus $\boldsymbol{n}$ and run from 0 to $n_1 - 1$ and $n_2 - 1$, respectively. If a GMRF $\boldsymbol{x}$ is defined on $\mathcal{T}_{\boldsymbol{n}}$ the toroidal boundary conditions imply that x_{-2,n_2} equals $x_{n_1-2,0}$ as $-2 \bmod n_1$ equals $n_1 - 2$ and $n_2 \bmod n_2$ equals 0. Figure 2.1 (a) illustrates the form of a torus.

With an *irregular lattice*, a slightly imprecise term, we mean a spatial configuration of regions $i = 1, \dots, n$, where (most often) the regions share common borders. A typical examples is displayed in Figure 2.1(b)

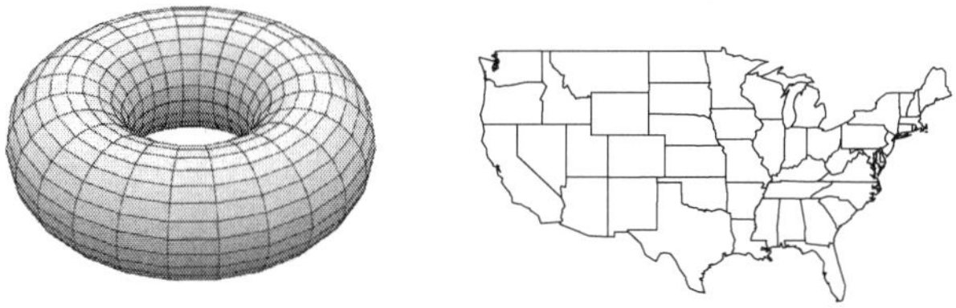

(a) (b)

Figure 2.1 *(a) Illustration of a torus obtained on a two-dimensional lattice with cyclic boundary conditions, (b) the states of the United States is an illustration of an irregular lattice.*

showing (most of) the states of the USA. Each state represents a region and they share common borders.

2.1.3 General notation and abbreviations

For $C \in \mathcal{I} = \{1, \ldots, n\}$, define $\boldsymbol{x}_C = \{x_i : i \in C\}$. With $-C$ we denote the set $\mathcal{I} - C$, so that $\boldsymbol{x}_{-C} = \{x_i : i \in -C\}$. For two sets A and B, then $A \setminus B = \{i : i \in A \text{ and } i \notin B\}$.

We will make no notational difference between a random variable and a specific realization of a random variable. The notation $\pi(\cdot)$ is a generic notation for the density of its arguments, like $\pi(\boldsymbol{x})$ for the density of $\boldsymbol{x}$ and $\pi(\boldsymbol{x}_A|\boldsymbol{x}_{-A})$ for the conditional density or $\boldsymbol{x}_A$, given a realization of $\boldsymbol{x}_{-A}$. By '$\sim$' we mean 'distributed as', so if $x \sim \mathcal{L}$ then x is distributed according to the law $\mathcal{L}$. We denote generically the expected value by $\mathrm{E}(\cdot)$, the variance by $\mathrm{Var}(\cdot)$, the covariance by $\mathrm{Cov}(\cdot)$, the precision by $\mathrm{Prec}(\cdot) = \mathrm{Cov}(\cdot)^{-1}$, and the correlation by $\mathrm{Corr}(\cdot, \cdot)$.

We use the shortcut *iff* for 'if and only if', *wrt* for 'with respect to', and *lhs* (*rhs*) for the left- or right-hand side of an equation.

One *flop* is defined as one floating-point operation. For example, evaluating `x + a*b` requires two flops: one multiplication and one addition.

2.1.4 Conditional independence

To compute the conditional density of $\boldsymbol{x}_A$, given $\boldsymbol{x}_{-A}$, we will repeatedly use that

$$\pi(\boldsymbol{x}_A \mid \boldsymbol{x}_{-A}) = \frac{\pi(\boldsymbol{x}_A, \boldsymbol{x}_{-A})}{\pi(\boldsymbol{x}_{-A})} \propto \pi(\boldsymbol{x}). \tag{2.1}$$

This is true since the denominator does not depend on $\boldsymbol{x}_A$.

A key concept for understanding GMRFs is *conditional independence*. Clearly, two random variables x and y are independent iff $\pi(x, y) =$

$\pi(x)\pi(y)$. We write this as $x \perp y$. Two variables x and y are called *conditionally independent* given z, iff $\pi(x, y|z) = \pi(x|z)\pi(y|z)$. We write this as

$$x \perp y \mid z.$$

Note that x and y might be (marginally) dependent, although they are conditionally independent given z. Using the following *factorization criterion for conditional independence*, it is easy to verify conditional independence.

Theorem 2.1

$$x \perp y \mid z \iff \pi(x, y, z) = f(x, z)g(y, z) \tag{2.2}$$

for some functions f and g, and for all z with $\pi(z) > 0$.

Example 2.1 *For $\pi(x, y, z) \propto \exp(x+xz+yz)$, on some bounded region, we see that $x \perp y|z$. However, this is not the case for $\pi(x, y, z) \propto \exp(xyz)$.*

The concept of conditional independence easily extends to the multivariate case, where $\boldsymbol{x}$ and $\boldsymbol{y}$ are called conditionally independent given $\boldsymbol{z}$, iff $\pi(\boldsymbol{x}, \boldsymbol{y}|\boldsymbol{z}) = \pi(\boldsymbol{x}|\boldsymbol{z})\pi(\boldsymbol{y}|\boldsymbol{z})$, which we write as $\boldsymbol{x} \perp \boldsymbol{y}|\boldsymbol{z}$. The factorization theorem still holds in this case.

2.1.5 Undirected graphs

We will use undirected graphs for representing the conditional independence structure in a GMRF. An undirected graph $\mathcal{G}$ is a tuple $\mathcal{G} = (\mathcal{V}, \mathcal{E})$, where $\mathcal{V}$ is the set of nodes in the graph, and $\mathcal{E}$ is the set of edges $\{i, j\}$, where $i, j \in \mathcal{V}$ and $i \neq j$. If $\{i, j\} \in \mathcal{E}$, there is an undirected edge from node i to node j, otherwise, there is no edge between node i to node j. A graph is *fully connected* if $\{i, j\} \in \mathcal{E}$ for all $i, j \in \mathcal{V}$ with $i \neq j$. In most cases we will assume that $\mathcal{V} = \{1, 2, \ldots, n\}$, in which case the graph is called *labelled*. A simple example of an undirected graph is shown in Figure 2.2.

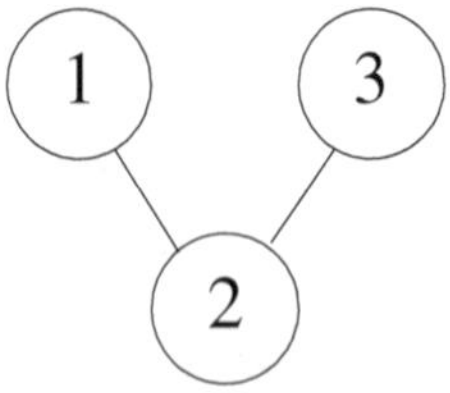

Figure 2.2 *An example of an undirected labelled graph with $n = 3$ nodes, here $\mathcal{V} = \{1, 2, 3\}$ and $\mathcal{E} = \{\{1, 2\}, \{2, 3\}\}$. We also see that $ne(1) = 2$, $ne(2) = \{1, 3\}$, $ne(\{1, 2\}) = 3$, and 2 separates 1 and 3.*

The *neighbors* of node i are all nodes in $\mathcal{G}$ having an edge to node i,

$$\text{ne}(i) = \{j \in \mathcal{V} \, : \, \{i, j\} \in \mathcal{E}\} \, .$$

We can extend this definition to a set $A \subset \mathcal{V}$, where we define the neighbors of A as

$$\text{ne}(A) = \bigcup_{i \in A} \text{ne}(i) \setminus A.$$

The neighbors of A are all nodes not in A, but adjacent to a node in A. Figure 2.2 illustrates this definition.

A *path* from i_1 to i_m is a sequence of distinct nodes in $\mathcal{V}$, $i_1, i_2, \ldots, i_m$, for which $(i_j, i_{j+1}) \in \mathcal{E}$ for $j = 1, \ldots, m-1$. A subset $C \subset \mathcal{V}$ *separates* two nodes $i \notin C$ and $j \notin C$, if every path from i to j contains at least one node from C. Two disjoint sets $A \subset \mathcal{V} \setminus C$ and $B \subset \mathcal{V} \setminus C$ are separated by C, if all $i \in A$ and $j \in B$ are separated by C, i.e., we cannot walk on the graph starting somewhere in A ending somewhere in B without passing through C.

We write $i \overset{\mathcal{G}}{\sim} j$ if node i and j are neighbors in graph $\mathcal{G}$, or just $i \sim j$ where the graph is implicit. A direct consequence of the definition is that $i \sim j \Leftrightarrow j \sim i$.

We need the notion of a *subgraph*. Let A be a subset of $\mathcal{V}$. Then $\mathcal{G}^A$ denotes the graph restricted to A, i.e., the graph we obtain after removing all nodes not belonging to A and all edges where at least one node does not belong to A. Precisely, $\mathcal{G}^A = \{\mathcal{V}^A, \mathcal{E}^A\}$, where $\mathcal{V}^A = A$ and

$$\mathcal{E}^A = \{\{i, j\} \in \mathcal{E} \text{ and } \{i, j\} \in A \times A\} \, .$$

For example, if we let $\mathcal{G}$ be the graph in Figure 2.2 and $\mathcal{A} = \{1, 2\}$, then $\mathcal{V}^A = \{1, 2\}$ and $\mathcal{E}^A = \{\{1, 2\}\}$.

2.1.6 *Symmetric positive-definite matrices*

An $n \times n$ matrix $\boldsymbol{A}$ is *positive definite* iff

$$\boldsymbol{x}^T \boldsymbol{A} \boldsymbol{x} > 0, \quad \forall \boldsymbol{x} \neq \boldsymbol{0}.$$

If $\boldsymbol{A}$ is also symmetric, then it is called a symmetric positive-definite (SPD) matrix. We only consider SPD matrices and sometimes use the notation '$\boldsymbol{A} > 0$' for an SPD matrix $\boldsymbol{A}$.

Some of the properties of a SPD matrix $\boldsymbol{A}$ are the following.

1. $\text{rank}(\boldsymbol{A}) = n$.
2. $|\boldsymbol{A}| > 0$.
3. $A_{ii} > 0$.
4. $A_{ii}A_{jj} - A_{ij}^2 > 0$, for $i \neq j$.

5. $A_{ii} + A_{jj} - 2|A_{ij}| > 0$, for $i \neq j$.
6. $\max A_{ii} > \max_{i \neq j} |A_{ij}|$.
7. $\boldsymbol{A}^{-1}$ is SPD.
8. All principal submatrices of $\boldsymbol{A}$ are SPD.

If $\boldsymbol{A}$ and $\boldsymbol{B}$ are SPD, then so is $\boldsymbol{A}+\boldsymbol{B}$, but the converse is not true in general. If $\boldsymbol{A}$ and $\boldsymbol{B}$ are SPD and $\boldsymbol{AB} = \boldsymbol{BA}$, then $\boldsymbol{AB}$ is SPD.

The following conditions are all sufficient and necessary for a symmetric matrix $\boldsymbol{A}$ to be SPD:

1. All the eigenvalues $\lambda_1, \ldots, \lambda_n$ of $\boldsymbol{A}$ are strictly positive.
2. There exists a matrix $\boldsymbol{C}$ such that $\boldsymbol{A} = \boldsymbol{C}\boldsymbol{C}^T$. If $\boldsymbol{C}$ is lower triangular it is called the *Cholesky triangle* of $\boldsymbol{A}$.
3. All leading principal submatrices have strictly positive determinants.

A sufficient but not necessary condition for a (symmetric) matrix to be SPD is the *diagonal dominance* criterion:

$$A_{ii} - \sum_{j:j\neq i} |A_{ij}| > 0, \ \forall i.$$

An $n \times n$ matrix $\boldsymbol{A}$ is called *positive semidefinite* iff

$$\boldsymbol{x}^T \boldsymbol{A} \boldsymbol{x} \geq 0, \quad \forall \boldsymbol{x} \neq \boldsymbol{0}.$$

If $\boldsymbol{A}$ is also symmetric, then it is called a *symmetric positive semidefinite* (SPSD) matrix. A SPSD matrix $\boldsymbol{A}$ is sometimes denoted by '$\boldsymbol{A} \geq 0$'.

2.1.7 The normal distribution

We now recall the multivariate normal distribution and give some of its basic properties. This makes the difference to a GMRF more clear. Other distributions are defined in Appendix A.

The density of a normal random variable $\boldsymbol{x} = (x_1, \ldots, x_n)^T$, $n < \infty$, with mean $\boldsymbol{\mu}$ ($n \times 1$ vector) and SPD covariance matrix $\boldsymbol{\Sigma}$ ($n \times n$ matrix), is

$$\pi(\boldsymbol{x}) = (2\pi)^{-n/2} |\boldsymbol{\Sigma}|^{-1/2} \exp\left(-\frac{1}{2}(\boldsymbol{x}-\boldsymbol{\mu})^T \boldsymbol{\Sigma}^{-1} (\boldsymbol{x}-\boldsymbol{\mu}) \right), \quad \boldsymbol{x} \in \mathbb{R}^n. \tag{2.3}$$

Here, $\mu_i = \text{E}(x_i)$, $\Sigma_{ij} = \text{Cov}(x_i, x_j)$, $\Sigma_{ii} = \text{Var}(x_i) > 0$ and $\text{Corr}(x_i, x_j) = \Sigma_{ij}/(\Sigma_{ii}\Sigma_{jj})^{1/2}$. We write this as $\boldsymbol{x} \sim \mathcal{N}(\boldsymbol{\mu}, \boldsymbol{\Sigma})$. A standard normal distribution is obtained if $n = 1$, $\mu = 0$ and $\Sigma_{11} = 1$.

We now divide $\boldsymbol{x}$ into two parts, $\boldsymbol{x} = (\boldsymbol{x}_A^T, \boldsymbol{x}_B^T)^T$, and split $\boldsymbol{\mu}$ and $\boldsymbol{\Sigma}$ accordingly:

$$\boldsymbol{\mu} = \begin{pmatrix} \boldsymbol{\mu}_A \\ \boldsymbol{\mu}_B \end{pmatrix} \quad \text{and} \quad \boldsymbol{\Sigma} = \begin{pmatrix} \boldsymbol{\Sigma}_{AA} & \boldsymbol{\Sigma}_{AB} \\ \boldsymbol{\Sigma}_{BA} & \boldsymbol{\Sigma}_{BB} \end{pmatrix}.$$

Here are some basic properties of the normal distribution.

1. $\boldsymbol{x}_A \sim \mathcal{N}(\boldsymbol{\mu}_A, \boldsymbol{\Sigma}_{AA})$
2. $\boldsymbol{\Sigma}_{AB} = \boldsymbol{0}$ iff $\boldsymbol{x}_A$ and $\boldsymbol{x}_B$ are independent.
3. The conditional distribution $\pi(\boldsymbol{x}_A|\boldsymbol{x}_B)$ is $\mathcal{N}(\boldsymbol{\mu}_{A|B}, \boldsymbol{\Sigma}_{A|B})$, where

$$\begin{aligned}\boldsymbol{\mu}_{A|B} &= \boldsymbol{\mu}_A + \boldsymbol{\Sigma}_{AB}\boldsymbol{\Sigma}_{BB}^{-1}(\boldsymbol{x}_B - \boldsymbol{\mu}_B) \quad \text{and} \\ \boldsymbol{\Sigma}_{A|B} &= \boldsymbol{\Sigma}_{AA} - \boldsymbol{\Sigma}_{AB}\boldsymbol{\Sigma}_{BB}^{-1}\boldsymbol{\Sigma}_{BA}.\end{aligned}$$

4. If $\boldsymbol{x} \sim \mathcal{N}(\boldsymbol{\mu}, \boldsymbol{\Sigma})$ and $\boldsymbol{x}' \sim \mathcal{N}(\boldsymbol{\mu}', \boldsymbol{\Sigma}')$ are independent, then $\boldsymbol{x} + \boldsymbol{x}' \sim \mathcal{N}(\boldsymbol{\mu} + \boldsymbol{\mu}', \boldsymbol{\Sigma} + \boldsymbol{\Sigma}')$.

2.2 Definition and basic properties of GMRFs

2.2.1 Definition

Let $\boldsymbol{x} = (x_1, \ldots, x_n)^T$ have a normal distribution with mean $\boldsymbol{\mu}$ and covariance matrix $\boldsymbol{\Sigma}$. Define the labelled graph $\mathcal{G} = (\mathcal{V}, \mathcal{E})$, where $\mathcal{V} = \{1, \ldots, n\}$ and $\mathcal{E}$ be such that there is no edge between node i and j iff $x_i \perp x_j|\boldsymbol{x}_{-ij}$, where $\boldsymbol{x}_{-ij}$ is short for $\boldsymbol{x}_{-\{i,j\}}$. Then we say that $\boldsymbol{x}$ is a GMRF wrt $\mathcal{G}$.

Before we define a GMRF formally, let us investigate the connection between the graph $\mathcal{G}$ and the parameters of the normal distribution. Since the mean $\boldsymbol{\mu}$ does not have any influence on the pairwise conditional independence properties of $\boldsymbol{x}$, we can deduce that this information must be 'hidden' solely in the covariance matrix $\boldsymbol{\Sigma}$. It turns out that the inverse covariance matrix, the *precision matrix* $\boldsymbol{Q} = \boldsymbol{\Sigma}^{-1}$ plays the key role.

Theorem 2.2 *Let $\boldsymbol{x}$ be normal distributed with mean $\boldsymbol{\mu}$ and precision matrix $\boldsymbol{Q} > 0$. Then for $i \neq j$,*

$$x_i \perp x_j \mid \boldsymbol{x}_{-ij} \quad \Longleftrightarrow \quad Q_{ij} = 0.$$

This is a nice and useful result. It simply says that the nonzero pattern of $\boldsymbol{Q}$ determines $\mathcal{G}$, so we can read off from $\boldsymbol{Q}$ whether x_i and x_j are conditionally independent. We will return to this in a moment. On the other hand, for a given graph $\mathcal{G}$, we know the nonzero terms in $\boldsymbol{Q}$. This can be used to provide a parameterization of $\boldsymbol{Q}$, being aware that we also require $\boldsymbol{Q} > 0$.

Before providing the proof of Theorem 2.2, we state the formal definition of a GMRF.

Definition 2.1 (GMRF) *A random vector $\boldsymbol{x} = (x_1, \ldots, x_n)^T \in \mathbb{R}^n$ is called a GMRF wrt a labelled graph $\mathcal{G} = (\mathcal{V}, \mathcal{E})$ with mean $\boldsymbol{\mu}$ and* *precision matrix* $\boldsymbol{Q} > 0$, *iff its density has the form*

$$\pi(\boldsymbol{x}) = (2\pi)^{-n/2}|\boldsymbol{Q}|^{1/2} \exp\left(-\frac{1}{2}(\boldsymbol{x}-\boldsymbol{\mu})^T\boldsymbol{Q}(\boldsymbol{x}-\boldsymbol{\mu})\right) \tag{2.4}$$

and

$$Q_{ij} \neq 0 \quad \Longleftrightarrow \quad \{i,j\} \in \mathcal{E} \quad \textit{for all} \quad i \neq j.$$

If $\boldsymbol{Q}$ is a completely dense matrix then $\mathcal{G}$ is fully connected. This implies that any normal distribution with SPD covariance matrix is also a GMRF and vice versa. We will focus on the case when $\boldsymbol{Q}$ is sparse, as it is here that the nice properties of GMRFs are really useful.

Proof. [Theorem 2.2] We partition $\boldsymbol{x}$ as $(x_i, x_j, \boldsymbol{x}_{-ij})$ and then use the multivariate version of the factorization criterion (Theorem 2.1) on $\pi(x_i, x_j, \boldsymbol{x}_{-ij})$. Fix $i \neq j$ and assume $\boldsymbol{\mu} = \boldsymbol{0}$ without loss of generality. From (2.4) we get

$$\begin{aligned}\pi(x_i, x_j, \boldsymbol{x}_{-ij}) &\propto \exp\Big(-\frac{1}{2}\sum_{k,l} x_k Q_{kl} x_l\Big)\\ &\propto \exp\Big(-\frac{1}{2}\underbrace{x_i x_j (Q_{ij}+Q_{ji})}_{\text{term 1}} \underbrace{-\frac{1}{2}\sum_{\{k,l\}\neq\{i,j\}} x_k Q_{kl} x_l}_{\text{term 2}}\Big).\end{aligned}$$

Term 2 does not involve $x_i x_j$ while term 1 involves $x_i x_j$ iff $Q_{ij} \neq 0$. Comparing with (2.2) in Theorem 2.1, we see that

$$\pi(x_i, x_j, \boldsymbol{x}_{-ij}) = f(x_i, \boldsymbol{x}_{-ij})g(x_j, \boldsymbol{x}_{-ij})$$

for some functions f and g, iff $Q_{ij} = 0$. The claim then follows. $\square$

We have argued that the natural way to describe a GMRF is by its precision matrix $\boldsymbol{Q}$. The elements of $\boldsymbol{Q}$ have nice conditional interpretations.

Theorem 2.3 *Let* $\boldsymbol{x}$ *be a GMRF wrt* $\mathcal{G} = (\mathcal{V}, \mathcal{E})$ *with mean* $\boldsymbol{\mu}$ *and precision matrix* $\boldsymbol{Q} > 0$, *then*

$$E(x_i \mid \boldsymbol{x}_{-i}) = \mu_i - \frac{1}{Q_{ii}}\sum_{j:j\sim i} Q_{ij}(x_j - \mu_j), \tag{2.5}$$

$$\mathit{Prec}(x_i \mid \boldsymbol{x}_{-i}) = Q_{ii} \quad \textit{and} \tag{2.6}$$

$$\mathit{Corr}(x_i, x_j \mid \boldsymbol{x}_{-ij}) = -\frac{Q_{ij}}{\sqrt{Q_{ii}Q_{jj}}}, \quad i \neq j. \tag{2.7}$$

The diagonal elements of $\boldsymbol{Q}$ are the conditional precisions of x_i given $\boldsymbol{x}_{-i}$, while the off-diagonal elements, with a proper scaling, provide information about the conditional correlation between x_i and x_j, given $\boldsymbol{x}_{-ij}$. These results should be compared to the interpretation of the elements of the covariance matrix $\boldsymbol{\Sigma} = (\Sigma_{ij})$; As $\mathrm{Var}(x_i) = \Sigma_{ii}$ and $\mathrm{Corr}(x_i, x_j) = \Sigma_{ij}/\sqrt{\Sigma_{ii}\Sigma_{jj}}$, the covariance matrix gives information about the marginal variance of x_i and the marginal correlation between x_i and x_j. The marginal interpretation given by $\boldsymbol{\Sigma}$ is intuitive and directly informative, as it reduces the interpretation from an n-dimensional distribution to a one- or two-dimensional distribution. The interpretation provided by $\boldsymbol{Q}$ is hard (or nearly impossible) to interpret marginally, as we have to integrate out $\boldsymbol{x}_{-i}$ or $\boldsymbol{x}_{-ij}$ from the joint distribution parameterized in terms of $\boldsymbol{Q}$. In matrix terms this is immediate; by definition $\boldsymbol{Q}^{-1} = \boldsymbol{\Sigma}$, and Σ_{ii} depends generally on all elements in $\boldsymbol{Q}$, and visa versa.

Proof. [Theorem 2.3] First recall that a univariate normal random variable x_i with mean γ and precision κ has density proportional to

$$\exp\left(-\frac{1}{2}\kappa x_i^2 + \kappa x_i \gamma\right). \tag{2.8}$$

Assume for the moment that $\boldsymbol{\mu} = \mathbf{0}$ and apply (2.1) to (2.4):

$$\begin{aligned} \pi(x_i \mid \boldsymbol{x}_{-i}) &\propto \exp\left(-\frac{1}{2}\boldsymbol{x}^T \boldsymbol{Q} \boldsymbol{x}\right) \\ &\propto \exp\left(-\frac{1}{2}x_i^2 Q_{ii} - x_i \sum_{j:j\sim i} Q_{ij} x_j\right). \end{aligned} \tag{2.9}$$

Comparing (2.8) and (2.9) we see that $\pi(x_i|\boldsymbol{x}_{-i})$ is normal. Comparing the coefficients for the quadratic term, we obtain (2.6). Comparing the coefficients for the linear term, we obtain

$$\mathrm{E}(x_i \mid \boldsymbol{x}_{-i}) = -\frac{1}{Q_{ii}} \sum_{j:j\sim i} Q_{ij} x_j.$$

If $\boldsymbol{x}$ has mean $\boldsymbol{\mu}$, then $\boldsymbol{x} - \boldsymbol{\mu}$ has mean zero, hence replacing x_i and x_j by $x_i - \mu_i$ and $x_j - \mu_j$, respectively, gives (2.5). To show (2.7), we proceed similarly and consider

$$\pi(x_i, x_j \mid \boldsymbol{x}_{-ij}) \propto \exp\left(-\frac{1}{2}(x_i, x_j)\begin{pmatrix} Q_{ii} & Q_{ij} \\ Q_{ji} & Q_{jj} \end{pmatrix}\begin{pmatrix} x_i \\ x_j \end{pmatrix} + \text{linear terms}\right). \tag{2.10}$$

We compare this density with the density of the bivariate normal random variable $(x_i, x_j)^T$ with covariance matrix $\boldsymbol{\Sigma} = (\Sigma_{ij})$, which has density proportional to

$$\exp\left(-\frac{1}{2}(x_i, x_j)\begin{pmatrix} \Sigma_{ii} & \Sigma_{ij} \\ \Sigma_{ji} & \Sigma_{jj} \end{pmatrix}^{-1}\begin{pmatrix} x_i \\ x_j \end{pmatrix} + \text{linear terms}\right). \tag{2.11}$$

Comparing (2.10) with (2.11), we obtain

$$\begin{pmatrix} Q_{ii} & Q_{ij} \\ Q_{ji} & Q_{jj} \end{pmatrix}^{-1} = \begin{pmatrix} \Sigma_{ii} & \Sigma_{ij} \\ \Sigma_{ji} & \Sigma_{jj} \end{pmatrix},$$

which implies that $\Sigma_{ii} = Q_{jj}/\Delta$, $\Sigma_{jj} = Q_{ii}/\Delta$, and $\Sigma_{ij} = -Q_{ij}/\Delta$ where $\Delta = Q_{ii}Q_{jj} - Q_{ij}^2$. Using these expressions and the definition of conditional correlation we obtain

$$\begin{aligned} \operatorname{Corr}(x_i, x_j \mid \boldsymbol{x}_{-ij}) &= -\frac{Q_{ij}/\Delta}{\sqrt{(Q_{jj}/\Delta)(Q_{ii}/\Delta)}} \\ &= -\frac{Q_{ij}}{\sqrt{Q_{ii}Q_{jj}}}. \end{aligned}$$

□

2.2.2 Markov properties of GMRFs

We have defined the graph $\mathcal{G}$ from checking if $x_i \perp x_j | \boldsymbol{x}_{-ij}$ or not. Theorem 2.2 says this is the same as checking if the corresponding off-diagonal entry of the precision matrix, Q_{ij}, is zero or not. Hence $\mathcal{G}$ is constructed from the nonzero pattern of $\boldsymbol{Q}$. An interesting and useful property of a GMRF is that more information regarding conditional independence can be extracted from $\mathcal{G}$. We consider now the *local Markov property* and the *global Markov property*, additional to the *pairwise Markov property* used to define $\mathcal{G}$. It turns out that all these properties are equivalent for a GMRF.

Theorem 2.4 *Let $\boldsymbol{x}$ be a GMRF wrt $\mathcal{G} = (\mathcal{V}, \mathcal{E})$. Then the following are equivalent.*

The pairwise Markov property:

$$x_i \perp x_j \mid \boldsymbol{x}_{-ij} \quad \textit{if } \{i,j\} \notin \mathcal{E} \textit{ and } i \neq j.$$

The local Markov property:

$$x_i \perp \boldsymbol{x}_{-\{i,ne(i)\}} \mid \boldsymbol{x}_{ne(i)} \quad \textit{for every } i \in \mathcal{V}.$$

The global Markov property:

$$\boldsymbol{x}_A \perp \boldsymbol{x}_B \mid \boldsymbol{x}_C \tag{2.12}$$

for all disjoint sets A, B and C where C separates A and B, and A and B are non-empty.

Figure 2.3 illustrates Theorem 2.4. The proof is a consequence of a more general result, stating the equivalence of the various Markov properties under some conditions satisfied for GMRFs. A simpler proof can be constructed in the Gaussian case (Speed and Kiiveri, 1986), but we omit it here.

The global Markov property immediately implies the local and pairwise Markov property, but the converse is a bit surprising. Note that the union of A, B, and C does not need to be $\mathcal{V}$, so properties of the marginal distribution can also be derived from $\mathcal{G}$.

If C in (2.12) is empty, then $\boldsymbol{x}_A$ and $\boldsymbol{x}_B$ are independent.

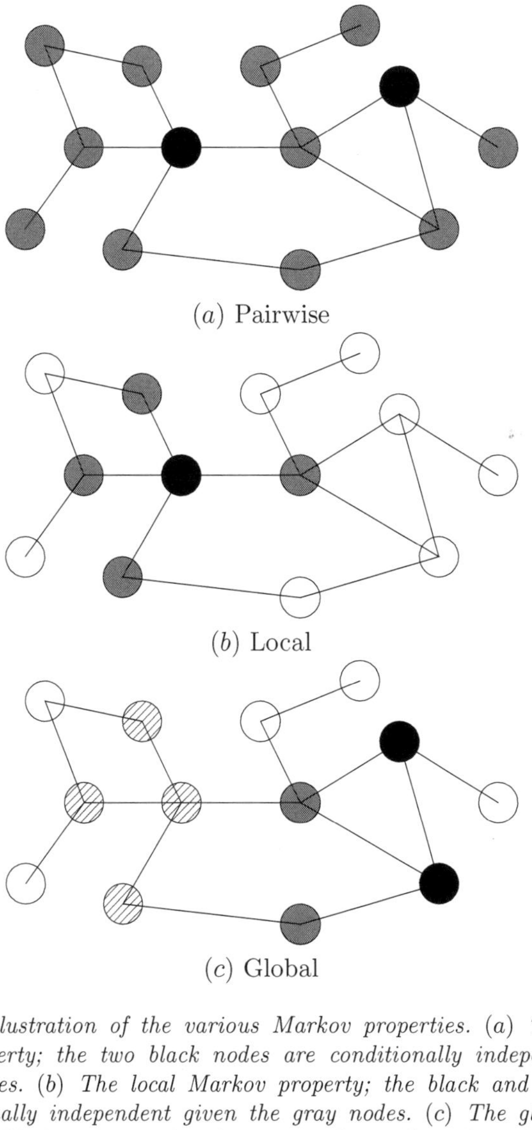

Figure 2.3 *Illustration of the various Markov properties. (a) The pairwise Markov property; the two black nodes are conditionally independent given the gray nodes. (b) The local Markov property; the black and white nodes are conditionally independent given the gray nodes. (c) The global Markov property; the black and striped nodes are conditionally independent given the gray nodes.*

2.2.3 *Conditional properties of GMRFs*

We will now discuss an important result of GMRFs; the conditional distribution for a subset $\boldsymbol{x}_A$ of $\boldsymbol{x}$ given the rest $\boldsymbol{x}_{-A}$. In this context the canonical parameterization will be useful, a parameterization that is easily updated under successive conditioning. Although all computations can be expressed with matrices, we will also consider a more graph-oriented view in Appendix B, which allows for efficient computation of the conditional densities.

Conditional distribution

We split the indices into the nonempty sets A and denote by B the set $-A$, so that

$$\boldsymbol{x} = \begin{pmatrix} \boldsymbol{x}_A \\ \boldsymbol{x}_B \end{pmatrix}. \tag{2.13}$$

Partition the mean and precision accordingly,

$$\boldsymbol{\mu} = \begin{pmatrix} \boldsymbol{\mu}_A \\ \boldsymbol{\mu}_B \end{pmatrix}, \quad \text{and} \quad \boldsymbol{Q} = \begin{pmatrix} \boldsymbol{Q}_{AA} & \boldsymbol{Q}_{AB} \\ \boldsymbol{Q}_{BA} & \boldsymbol{Q}_{BB} \end{pmatrix}. \tag{2.14}$$

Our next result, is a generalization of Theorem 2.3.

Theorem 2.5 *Let $\boldsymbol{x}$ be a GMRF wrt $\mathcal{G} = (\mathcal{V}, \mathcal{E})$ with mean $\boldsymbol{\mu}$ and precision matrix $\boldsymbol{Q} > 0$. Let $A \subset \mathcal{V}$ and $B = \mathcal{V} \setminus A$ where $A, B \neq \emptyset$. The conditional distribution of $\boldsymbol{x}_A | \boldsymbol{x}_B$ is then a GMRF wrt the subgraph $\mathcal{G}^A$ with mean $\boldsymbol{\mu}_{A|B}$ and precision matrix $\boldsymbol{Q}_{A|B} > 0$, where*

$$\boldsymbol{\mu}_{A|B} = \boldsymbol{\mu}_A - \boldsymbol{Q}_{AA}^{-1} \boldsymbol{Q}_{AB} (\boldsymbol{x}_B - \boldsymbol{\mu}_B) \tag{2.15}$$

and

$$\boldsymbol{Q}_{A|B} = \boldsymbol{Q}_{AA}.$$

This is a powerful result for two reasons. First, we have explicit knowledge of $\boldsymbol{Q}_{A|B}$ through the principal matrix $\boldsymbol{Q}_{AA}$, so no computation is needed to obtain the conditional precision matrix. Constructing the subgraph $\mathcal{G}^A$ does not change the structure; it just removes all nodes not in A and the corresponding edges. This is important for the computational issues that will be discussed in Section 2.3. Secondly, since Q_{ij} is zero unless $j \in \text{ne}(i)$, the conditional mean only depends on values of $\boldsymbol{\mu}$ and $\boldsymbol{Q}$ in $A \cup \text{ne}(A)$. This is a great advantage if A is a small subset of $\mathcal{V}$ and in striking contrast to the corresponding general result for the normal distribution, see Section 2.1.7.

Example 2.2 *To illustrate Theorem 2.5, we compute the mean and precision of x_i given $\boldsymbol{x}_{-i}$, which are found using $A = \{i\}$ as (2.5) and (2.6). This result is frequently used for single-site Gibbs sampling in GMRF models, to which we return in Section 4.1.*

Proof. [Theorem 2.5] The proof is similar to Theorem 2.3, but uses matrices. Assume $\boldsymbol{\mu} = \mathbf{0}$ and compute the conditional density,

$$\begin{aligned}\pi(\boldsymbol{x}_A \mid \boldsymbol{x}_B) &\propto \exp\Big(-\frac{1}{2}(\boldsymbol{x}_A, \boldsymbol{x}_B)\begin{pmatrix}\boldsymbol{Q}_{AA} & \boldsymbol{Q}_{AB}\\ \boldsymbol{Q}_{BA} & \boldsymbol{Q}_{BB}\end{pmatrix}\begin{pmatrix}\boldsymbol{x}_A\\ \boldsymbol{x}_B\end{pmatrix}\Big)\\ &\propto \exp\Big(-\frac{1}{2}\boldsymbol{x}_A^T\boldsymbol{Q}_{AA}\boldsymbol{x}_A - (\boldsymbol{Q}_{AB}\boldsymbol{x}_B)^T\boldsymbol{x}_A\Big).\end{aligned}$$

Comparing this with the density of a normal with precision $\boldsymbol{P}$ and mean $\boldsymbol{\gamma}$,

$$\pi(\boldsymbol{z}) \propto \exp\Big(-\frac{1}{2}\boldsymbol{z}^T\boldsymbol{P}\boldsymbol{z} + (\boldsymbol{P}\boldsymbol{\gamma})^T\boldsymbol{z}\Big),$$

we see that $\boldsymbol{Q}_{AA}$ is the conditional precision matrix and the conditional mean is given by the solution of

$$\boldsymbol{Q}_{AA}\boldsymbol{\mu}_{A|B} = -\boldsymbol{Q}_{AB}\boldsymbol{x}_B.$$

Note that $\boldsymbol{Q}_{AA} > 0$ since $\boldsymbol{Q} > 0$. If $\boldsymbol{x}$ has mean $\boldsymbol{\mu}$ then $\boldsymbol{x} - \boldsymbol{\mu}$ has mean zero, hence (2.15) follows. The subgraph $\mathcal{G}^A$ follows from the nonzero elements of $\boldsymbol{Q}_{AA}$. □

To compute the conditional mean $\boldsymbol{\mu}_{A|B}$, we need to *solve* the linear system

$$\boldsymbol{Q}_{AA}(\boldsymbol{\mu}_{A|B} - \boldsymbol{\mu}_A) = -\boldsymbol{Q}_{AB}(\boldsymbol{x}_B - \boldsymbol{\mu}_B)$$

but *not* necessarily invert $\boldsymbol{Q}_{AA}$. We postpone the discussion of this numerical issue until Section 2.3.

The canonical parameterization

The canonical parameterization for a GMRF will be useful for successive conditioning.

Definition 2.2 (Canonical parameterization) *A GMRF $\boldsymbol{x}$ wrt $\mathcal{G}$ with canonical parameters $\boldsymbol{b}$ and $\boldsymbol{Q} > 0$ has density*

$$\pi(\boldsymbol{x}) \propto \exp\Big(-\frac{1}{2}\boldsymbol{x}^T\boldsymbol{Q}\boldsymbol{x} + \boldsymbol{b}^T\boldsymbol{x}\Big),$$

i.e., the precision matrix is $\boldsymbol{Q}$ and the mean is $\boldsymbol{\mu} = \boldsymbol{Q}^{-1}\boldsymbol{b}$. We write the canonical parameterization as

$$\boldsymbol{x} \sim \mathcal{N}_C(\boldsymbol{b}, \boldsymbol{Q}).$$

The relation to the normal distribution, is that $\mathcal{N}(\boldsymbol{\mu}, \boldsymbol{Q}^{-1}) = \mathcal{N}_C(\boldsymbol{Q}\boldsymbol{\mu}, \boldsymbol{Q})$.

Partition the indices into two nonempty sets A and B, and partition $\boldsymbol{x}$, $\boldsymbol{b}$ and $\boldsymbol{Q}$ accordingly as in (2.13) and (2.14). Two lemmas follow easily.

Lemma 2.1 *Let $\boldsymbol{x} \sim \mathcal{N}_C(\boldsymbol{b}, \boldsymbol{Q})$, then*

$$\boldsymbol{x}_A \mid \boldsymbol{x}_B \sim \mathcal{N}_C(\boldsymbol{b}_A - \boldsymbol{Q}_{AB}\boldsymbol{x}_B, \boldsymbol{Q}_{AA}).$$

Lemma 2.2 *Let* $\boldsymbol{x} \sim \mathcal{N}_C(\boldsymbol{b}, \boldsymbol{Q})$ *and* $\boldsymbol{y}|\boldsymbol{x} \sim \mathcal{N}(\boldsymbol{x}, \boldsymbol{P}^{-1})$*, then*

$$\boldsymbol{x} \mid \boldsymbol{y} \sim \mathcal{N}_C(\boldsymbol{b} + \boldsymbol{P}\boldsymbol{y}, \boldsymbol{Q} + \boldsymbol{P}). \tag{2.16}$$

These results are useful for computing conditional densities with several sources of conditioning, for example, conditioning on observed data and a subset of variables. We can successively update the canonical parameterization, without explicitly computing the mean, until we actually need it. Computing the mean requires the solution of $\boldsymbol{Q}\boldsymbol{\mu} = \boldsymbol{b}$, but only matrix-vector products are required to update the canonical parameterization.

2.2.4 Specification through full conditionals

An alternative to specifying a GMRF by its mean and precision matrix, is to specify it implicitly through the full conditionals $\{\pi(x_i|\boldsymbol{x}_{-i})\}$. This approach was pioneered by Besag (1974, 1975) and the models are also known by the name *conditional autoregressions*, abbreviated as *CAR* models. We will now discuss this possibility and the specific conditions we must impose on the full conditionals to correspond to a valid GMRF.

Suppose we specify the full conditionals as normals with

$$\mathrm{E}(x_i \mid \boldsymbol{x}_{-i}) = \mu_i - \sum_{j:j\sim i} \beta_{ij}(x_j - \mu_j) \quad \text{and} \tag{2.17}$$

$$\mathrm{Prec}(x_i \mid \boldsymbol{x}_{-i}) = \kappa_i > 0 \tag{2.18}$$

for $i = 1, \ldots, n$, for some $\{\beta_{ij}, i \neq j\}$, and vectors $\boldsymbol{\mu}$ and $\boldsymbol{\kappa}$. Clearly, $\sim$ is defined implicitly by the nonzero terms of $\{\beta_{ij}\}$. These full conditionals must be consistent so that there exists a joint density $\pi(\boldsymbol{x})$ that will give rise to these full conditional distributions. Since $\sim$ is symmetric, this immediate gives the requirement that if $\beta_{ij} \neq 0$ then $\beta_{ji} \neq 0$. Comparing term by term with (2.5) and (2.6), we see that if we choose the entries of the precision matrix $\boldsymbol{Q}$ as

$$Q_{ii} = \kappa_i, \quad \text{and} \quad Q_{ij} = \kappa_i\beta_{ij}$$

and also require that $\boldsymbol{Q}$ is symmetric, i.e.,

$$\kappa_i\beta_{ij} = \kappa_j\beta_{ji},$$

then we have a candidate for a joint density giving the specified full conditionals provided $\boldsymbol{Q} > 0$. The next result says that this candidate is unique.

Theorem 2.6 *Given the* n *normal full conditionals with conditional mean and precision as in (2.17) and (2.18), then* $\boldsymbol{x}$ *is a GMRF wrt a labelled graph* $\mathcal{G} = (\mathcal{V}, \mathcal{E})$ *with mean* $\boldsymbol{\mu}$ *and precision matrix* $\boldsymbol{Q} = (Q_{ij})$*,* *where*

$$Q_{ij} = \begin{cases} \kappa_i \beta_{ij} & i \neq j \\ \kappa_i & i = j \end{cases}$$

provided $\kappa_i \beta_{ij} = \kappa_j \beta_{ji}, i \neq j$, *and* $\boldsymbol{Q} > 0$.

To prove this result we need *Brook's lemma.*

Lemma 2.3 (Brook's lemma) *Let* $\pi(\boldsymbol{x})$ *be the density for* $\boldsymbol{x} \in \mathbb{R}^n$ *and define* $\Omega = \{\boldsymbol{x} \in \mathbb{R}^n : \pi(\boldsymbol{x}) > 0\}$. *Let* $\boldsymbol{x}, \boldsymbol{x}' \in \Omega$, *then*

$$\frac{\pi(\boldsymbol{x})}{\pi(\boldsymbol{x}')} = \prod_{i=1}^{n} \frac{\pi(x_i | x_1, \ldots, x_{i-1}, x'_{i+1}, \ldots, x'_n)}{\pi(x'_i | x_1, \ldots, x_{i-1}, x'_{i+1}, \ldots, x'_n)} \tag{2.19}$$

$$= \prod_{i=1}^{n} \frac{\pi(x_i | x'_1, \ldots, x'_{i-1}, x_{i+1}, \ldots, x_n)}{\pi(x'_i | x'_1, \ldots, x'_{i-1}, x_{i+1}, \ldots, x_n)}. \tag{2.20}$$

If we fix $\boldsymbol{x}'$ then (2.19) (and (2.20)) represents $\pi(\boldsymbol{x})$, up to a constant of proportionality, using the set of full conditionals $\{\pi(x_i|\boldsymbol{x}_{-i})\}$. The constant of proportionality is found using that $\pi(\boldsymbol{x})$ integrates to unity.

Proof. [Brook's lemma] Start with the identity

$$\frac{\pi(x_n | x_1, \ldots, x_{n-1})\, \pi(x_1, \ldots, x_{n-1})}{\pi(x'_n | x_1, \ldots, x_{n-1})\, \pi(x_1, \ldots, x_{n-1})} = \frac{\pi(x_1, \ldots, x_{n-1}, x_n)}{\pi(x_1, \ldots, x_{n-1}, x'_n)}$$

from which it follows that

$$\pi(x_1, \ldots, x_n) = \frac{\pi(x_n | x_1, \ldots, x_{n-1})}{\pi(x'_n | x_1, \ldots, x_{n-1})} \pi(x_1, \ldots, x_{n-1}, x'_n).$$

Express the last term on the rhs similarly to obtain

$$\begin{aligned} \pi(x_1, \ldots, x_n) = & \frac{\pi(x_n | x_1, \ldots, x_{n-1})}{\pi(x'_n | x_1, \ldots, x_{n-1})} \\ & \times \frac{\pi(x_{n-1} | x_1, \ldots, x_{n-2}, x'_n)}{\pi(x'_{n-1} | x_1, \ldots, x_{n-2}, x'_n)} \\ & \times \pi(x_1, \ldots, x_{n-2}, x'_{n-1}, x'_n). \end{aligned}$$

By repeating this process (2.19) follows. The alternative (2.20) is proved similarly starting with

$$\pi(x_1, \ldots, x_n) = \frac{\pi(x_1 | x_2, \ldots, x_n)}{\pi(x'_1 | x_2, \ldots, x_n)} \pi(x'_1, x_2, \ldots, x_n)$$

and proceeding forward. □

Proof. [Theorem 2.6] Assume $\boldsymbol{\mu} = \boldsymbol{0}$ and fix $\boldsymbol{x}' = \boldsymbol{0}$. Then (2.19) simplifies to

$$\log \frac{\pi(\boldsymbol{x})}{\pi(\boldsymbol{0})} = -\frac{1}{2} \sum_{i=1}^{n} \kappa_i x_i^2 - \sum_{i=2}^{n} \sum_{j=1}^{i-1} \kappa_i \beta_{ij} x_i x_j. \tag{2.21}$$

Using (2.20) we obtain

$$\log \frac{\pi(\boldsymbol{x})}{\pi(\boldsymbol{0})} = -\frac{1}{2}\sum_{i=1}^{n} \kappa_i x_i^2 - \sum_{i=1}^{n-1}\sum_{j=i+1}^{n} \kappa_i \beta_{ij} x_i x_j. \tag{2.22}$$

Since (2.21) and (2.22) must be identical it follows that $\kappa_i\beta_{ij} = \kappa_j\beta_{ji}$ for $i \neq j$. The density of $\boldsymbol{x}$ can then be expressed as

$$\log \pi(\boldsymbol{x}) = \text{const} - \frac{1}{2}\sum_{i=1}^{n} \kappa_i x_i^2 - \frac{1}{2}\sum_{i\neq j} \kappa_i \beta_{ij} x_i x_j;$$

hence $\boldsymbol{x}$ is zero mean multivariate normal provided $\boldsymbol{Q} > 0$. The precision matrix has elements $Q_{ij} = \kappa_i\beta_{ij}$ for $i \neq j$ and $Q_{ii} = \kappa_i$. □

In matrix terms (defining $\beta_{ii} = 0$), the precision matrix is

$$\boldsymbol{Q} = \text{diag}(\boldsymbol{\kappa})\left(\boldsymbol{I} + (\beta_{ij})\right);$$

hence $\boldsymbol{Q} > 0 \iff (\boldsymbol{I} + (\beta_{ij})) > 0$.

2.2.5 Multivariate GMRFs★

A multivariate GMRF (MGMRF) is a multivariate extension of a GMRF that has been shown to be useful in applications. To motivate its construction, let $\boldsymbol{x}$ be a GMRF wrt to $\mathcal{G}$. The Markov property implies that

$$\pi(x_i \mid \boldsymbol{x}_{-i}) = \pi(x_i \mid \{x_j : j \sim i\}).$$

We associate x_i as the value related to node i. The nodes have often a physical interpretation like a pixel in a lattice or an administrative region of a country, and this may also be used to define the neighbors to node i. For an illustration, see Figure 2.1(b). The extension is now to associate a vector with dimension p, $\boldsymbol{x}_i$, with each of the n nodes, leading to a GMRF of size np. We denote such an MGMRF by $\boldsymbol{x} = (\boldsymbol{x}_1^T, \ldots, \boldsymbol{x}_n^T)^T$. The Markov property in terms of the nodes is then preserved, meaning that

$$\pi(\boldsymbol{x}_i \mid \boldsymbol{x}_{-i}) = \pi(\boldsymbol{x}_i \mid \{\boldsymbol{x}_j : j \sim i\}).$$

where $\sim$ is wrt *the same graph* $\mathcal{G}$. Let $\boldsymbol{\mu} = (\boldsymbol{\mu}_1^T, \ldots, \boldsymbol{\mu}_n^T)^T$ be the mean of $\boldsymbol{x}$ where $\text{E}(\boldsymbol{x}_i) = \boldsymbol{\mu}_i$, and $\widetilde{\boldsymbol{Q}} = (\widetilde{\boldsymbol{Q}}_{ij})$ its precision matrix. Note that each element $\widetilde{\boldsymbol{Q}}_{ij}$ is a $p \times p$ matrix.

It follows directly from Theorem 2.2, that

$$\boldsymbol{x}_i \perp \boldsymbol{x}_j \mid \boldsymbol{x}_{-ij} \quad \iff \quad \widetilde{\boldsymbol{Q}}_{ij} = \boldsymbol{0}.$$

The definition of an MGMRF with dimension p is an extension of the definition of a GMRF (Definition 2.1).

Definition 2.3 (MGMRF$_p$) *A random vector* $\boldsymbol{x} = (\boldsymbol{x}_1^T, \ldots, \boldsymbol{x}_n^T)^T$ *where* $dim(\boldsymbol{x}_i) = p$, *is called a* $MGMRF_p$ *wrt* $\mathcal{G} = (\mathcal{V} = \{1, \ldots, n\}, \mathcal{E})$ *with mean* $\boldsymbol{\mu}$ *and precision matrix* $\widetilde{\boldsymbol{Q}} > 0$, *iff its density has the form*

$$\begin{aligned}\pi(\boldsymbol{x}) &= (\frac{1}{2\pi})^{np/2}|\widetilde{\boldsymbol{Q}}|^{1/2}\exp\left(-\frac{1}{2}(\boldsymbol{x}-\boldsymbol{\mu})^T\widetilde{\boldsymbol{Q}}(\boldsymbol{x}-\boldsymbol{\mu})\right)\\ &= (\frac{1}{2\pi})^{np/2}|\widetilde{\boldsymbol{Q}}|^{1/2}\exp\left(-\frac{1}{2}\sum_{ij}(\boldsymbol{x}_i-\boldsymbol{\mu}_i)^T\widetilde{\boldsymbol{Q}}_{ij}(\boldsymbol{x}_j-\boldsymbol{\mu}_j)\right)\end{aligned}$$

and

$$\widetilde{\boldsymbol{Q}}_{ij} \neq \mathbf{0} \Longleftrightarrow \{i,j\} \in \mathcal{E} \quad \textit{for all} \quad i \neq j.$$

An MGMRF$_p$ is also a GMRF with dimension np with identical mean vector and precision matrix. All results valid for a GMRF are then also valid for an MGMRF$_p$, with obvious changes as the graph for an MGMRF$_p$ is of size n and defined wrt $\{\boldsymbol{x}_i\}$, while for a GMRF it is of size np and defined wrt $\{x_i\}$.

Interpretation of $\widetilde{\boldsymbol{Q}}_{ii}$ and $\widetilde{\boldsymbol{Q}}_{ij}$ can be derived from the full conditional $\pi(\boldsymbol{x}_i|\boldsymbol{x}_{-i})$. The extensions of (2.5) and (2.6) are

$$\begin{aligned}\mathrm{E}(\boldsymbol{x}_i \mid \boldsymbol{x}_{-i}) &= \boldsymbol{\mu}_i - \widetilde{\boldsymbol{Q}}_{ii}^{-1}\sum_{j:j\sim i}\widetilde{\boldsymbol{Q}}_{ij}(\boldsymbol{x}_j-\boldsymbol{\mu}_j)\\ \mathrm{Prec}(\boldsymbol{x}_i \mid \boldsymbol{x}_{-i}) &= \widetilde{\boldsymbol{Q}}_{ii}.\end{aligned}$$

In some applications, the full conditionals

$$\begin{aligned}\mathrm{E}(\boldsymbol{x}_i \mid \boldsymbol{x}_{-i}) &= \boldsymbol{\mu}_i - \sum_{j:j\sim i}\boldsymbol{\beta}_{ij}(\boldsymbol{x}_j-\boldsymbol{\mu}_j)\\ \mathrm{Prec}(\boldsymbol{x}_i \mid \boldsymbol{x}_{-i}) &= \boldsymbol{\kappa}_i > 0,\end{aligned}$$

are used to define the MGMRF$_p$, for some $p \times p$-matrices $\{\boldsymbol{\beta}_{ij}, i \neq j\}$, $\{\boldsymbol{\kappa}_i\}$, and vectors $\boldsymbol{\mu}_i$. Again, $\sim$ is defined implicitly by the nonzero matrices $\{\boldsymbol{\beta}_{ij}\}$. The requirements for the joint density to exist are similar to those for $p = 1$ (see Theorem 2.6): $\boldsymbol{\kappa}_i\boldsymbol{\beta}_{ij} = \boldsymbol{\beta}_{ji}^T\boldsymbol{\kappa}_j$ for $i \neq j$ and $\widetilde{\boldsymbol{Q}} > 0$. The $p \times p$-elements of $\widetilde{\boldsymbol{Q}}$ are

$$\widetilde{\boldsymbol{Q}}_{ij} = \begin{cases}\boldsymbol{\kappa}_i\boldsymbol{\beta}_{ij} & i \neq j\\ \boldsymbol{\kappa}_i & i = j\end{cases};$$

hence $\widetilde{\boldsymbol{Q}} > 0 \Longleftrightarrow (\boldsymbol{I} + (\boldsymbol{\beta}_{ij})) > 0$.

2.3 Simulation from a GMRF

This chapter will be more computationally oriented, presenting algorithms for

- Simulation of a GMRF
- Evaluation of the log density
- Calculating conditional densities
- Simulation conditional on a subset of a GMRF, a hard constraint, or a soft constraint, and the corresponding evaluation of the log conditional densities.

We will formulate all these tasks as simple matrix operations on the precision matrix $\boldsymbol{Q}$, which we know is sparse, hence easier to store and faster to compute. One example is the *Cholesky factorization* $\boldsymbol{Q} = \boldsymbol{L}\boldsymbol{L}^T$, where $\boldsymbol{L}$ is a lower triangular matrix referred to as the *Cholesky triangle*. It turns out that $\boldsymbol{L}$ can be sparse as well and thus inherits the (somewhat modified) nonzero pattern from $\boldsymbol{Q}$. In general, computing this factorization requires $\mathcal{O}(n^3)$ *flops*, while a sparse $\boldsymbol{Q}$ will typically reduce this to $\mathcal{O}(n)$ for temporal, $\mathcal{O}(n^{3/2})$ for spatial and $\mathcal{O}(n^2)$ for spatiotemporal GMRFs. Similarly, solving for example $\boldsymbol{L}\boldsymbol{x} = \boldsymbol{b}$, will also be faster as $\boldsymbol{L}$ is sparse.

We postpone the discussion of numerical methods for sparse matrices to Section 2.4, and will now discuss how simulation and evaluation of the log density can be done based on $\boldsymbol{Q}$.

2.3.1 Some basic numerical linear algebra

We start with some basic facts on numerical linear algebra.

Let $\boldsymbol{A}$ be an $n \times n$ SPD matrix, then there exists a unique *Cholesky triangle* $\boldsymbol{L}$ such that $\boldsymbol{L}$ is a lower triangular matrix where $L_{ii} > 0 \; \forall i$ and $\boldsymbol{A} = \boldsymbol{L}\boldsymbol{L}^T$. Computing $\boldsymbol{L}$ costs $n^3/3$ flops. This factorization is the basis for *solving* systems like $\boldsymbol{A}\boldsymbol{x} = \boldsymbol{b}$ or $\boldsymbol{A}\boldsymbol{X} = \boldsymbol{B}$ for k right-hand sides, or equivalently, computing $\boldsymbol{x} = \boldsymbol{A}^{-1}\boldsymbol{b}$ or $\boldsymbol{X} = \boldsymbol{A}^{-1}\boldsymbol{B}$. For example, we solve $\boldsymbol{A}\boldsymbol{x} = \boldsymbol{b}$ using Algorithm 2.1. Clearly, $\boldsymbol{x}$ is the solution of $\boldsymbol{A}\boldsymbol{x} = \boldsymbol{b}$

Algorithm 2.1 Solving $\boldsymbol{A}\boldsymbol{x} = \boldsymbol{b}$ where $\boldsymbol{A} > 0$

1: Compute the Cholesky factorization, $\boldsymbol{A} = \boldsymbol{L}\boldsymbol{L}^T$
2: Solve $\boldsymbol{L}\boldsymbol{v} = \boldsymbol{b}$
3: Solve $\boldsymbol{L}^T\boldsymbol{x} = \boldsymbol{v}$
4: **Return** $\boldsymbol{x}$

because $\boldsymbol{x} = (\boldsymbol{L}^{-1})^T\boldsymbol{v} = \boldsymbol{L}^{-T}(\boldsymbol{L}^{-1}\boldsymbol{b}) = (\boldsymbol{L}\boldsymbol{L}^T)^{-1}\boldsymbol{b} = \boldsymbol{A}^{-1}\boldsymbol{b}$ as required.

Step 2 is called *forward substitution*, as the solution $\boldsymbol{v}$ is computed in a forward loop (recall that $\boldsymbol{L}$ is lower triangular),

$$v_i = \frac{1}{L_{ii}}(b_i - \sum_{j=1}^{i-1} L_{ij}v_j), \quad i = 1, \ldots, n.$$

The cost is in general n^2 flops. Step 3 is called *back substitution*, as the solution $\boldsymbol{x}$ is computed in a backward loop (recall that $\boldsymbol{L}^T$ is upper triangular),

$$x_i = \frac{1}{L_{ii}}(v_i - \sum_{j=i+1}^{n} L_{ji}x_j), \quad i = n, \ldots, 1. \tag{2.23}$$

If we need to compute $\boldsymbol{A}^{-1}\boldsymbol{B}$, where $\boldsymbol{B}$ is a $n \times k$ matrix, we do this by computing the solution $\boldsymbol{X}$ of $\boldsymbol{AX} = \boldsymbol{B}$ for each column of $\boldsymbol{X}$. More specifically, we solve $\boldsymbol{AX}_j = \boldsymbol{B}_j$, where $\boldsymbol{X}_j$ is the jth column of $\boldsymbol{X}$ and $\boldsymbol{B}_j$ is the jth column of $\boldsymbol{B}$, see Algorithm 2.2.

Algorithm 2.2 Solving $\boldsymbol{AX} = \boldsymbol{B}$ where $\boldsymbol{A} > 0$

1: Compute the Cholesky factorization, $\boldsymbol{A} = \boldsymbol{LL}^T$
2: **for** $j = 1$ to k **do**
3: Solve $\boldsymbol{Lv} = \boldsymbol{B}_j$
4: Solve $\boldsymbol{L}^T\boldsymbol{X}_j = \boldsymbol{v}$
5: **end for**
6: **Return** $\boldsymbol{X}$

Note that choosing $k = n$ and $\boldsymbol{B} = \boldsymbol{I}$, we obtain $\boldsymbol{X} = \boldsymbol{A}^{-1}$. Hence, solving $\boldsymbol{Ax} = \boldsymbol{b}$ in comparison to computing $\boldsymbol{x} = \boldsymbol{A}^{-1}\boldsymbol{b}$, gives a speedup of 4. There is no need to compute explicitly the inverse $\boldsymbol{A}^{-1}$.

If $\boldsymbol{A}$ is a general invertible $n \times n$ matrix, but no longer SPD, then similar algorithms apply with only minor modifications: We compute the LU decomposition, as $\boldsymbol{A} = \boldsymbol{LU}$, where $\boldsymbol{L}$ is lower triangular and $\boldsymbol{U}$ is upper triangular, and replace $\boldsymbol{L}^T$ by $\boldsymbol{U}$ in Algorithm 2.1 and Algorithm 2.2.

2.3.2 Unconditional simulation of a GMRF

In this section we discuss simulation from GMFRs for the different parameterizations.

Sample $\boldsymbol{x} \sim \mathcal{N}(\boldsymbol{\mu}, \boldsymbol{\Sigma})$

We start with Algorithm 2.3, the most commonly used algorithm for sampling from a multivariate normal random variable $\boldsymbol{x} \sim \mathcal{N}(\boldsymbol{\mu}, \boldsymbol{\Sigma})$. Then $\boldsymbol{x}$ has the required distribution, as

$$\text{Cov}(\boldsymbol{x}) = \text{Cov}(\widetilde{\boldsymbol{L}}\boldsymbol{z}) = \widetilde{\boldsymbol{L}}\widetilde{\boldsymbol{L}}^T = \boldsymbol{\Sigma} \tag{2.24}$$

and $\text{E}(\boldsymbol{x}) = \boldsymbol{\mu}$. To obtain repeated samples, we do step 1 only once.

Algorithm 2.3 Sampling $\boldsymbol{x} \sim \mathcal{N}(\boldsymbol{\mu}, \boldsymbol{\Sigma})$

1: Compute the Cholesky factorization, $\boldsymbol{\Sigma} = \widetilde{\boldsymbol{L}}\widetilde{\boldsymbol{L}}^T$
2: Sample $\boldsymbol{z} \sim \mathcal{N}(\boldsymbol{0}, \boldsymbol{I})$
3: Compute $\boldsymbol{v} = \widetilde{\boldsymbol{L}}\boldsymbol{z}$
4: Compute $\boldsymbol{x} = \boldsymbol{\mu} + \boldsymbol{v}$
5: **Return** $\boldsymbol{x}$

The log density is computed using (2.3), where

$$\frac{1}{2}\log|\boldsymbol{\Sigma}| = \sum_{i=1}^{n} \log \widetilde{L}_{ii}$$

because $|\boldsymbol{\Sigma}| = |\widetilde{\boldsymbol{L}}\widetilde{\boldsymbol{L}}^T| = |\widetilde{\boldsymbol{L}}||\widetilde{\boldsymbol{L}}^T| = |\widetilde{\boldsymbol{L}}|^2$. Hence we obtain

$$\log \pi(\boldsymbol{x}) = -\frac{n}{2}\log(2\pi) - \sum_{i=1}^{n} \log \widetilde{L}_{ii} - \frac{1}{2}\boldsymbol{u}^T\boldsymbol{u}, \tag{2.25}$$

where $\boldsymbol{u}$ is the solution of $\widetilde{\boldsymbol{L}}\boldsymbol{u} = \boldsymbol{x} - \boldsymbol{\mu}$. If $\boldsymbol{x}$ is sampled using Algorithm 2.3 then $\boldsymbol{u} = \boldsymbol{z}$.

For a GMRF, we assume $\boldsymbol{Q}$ is known and $\boldsymbol{\Sigma}$ known only implicitly, hence we aim at deriving an algorithm similar to (2.24) but using a factorization of the precision matrix $\boldsymbol{Q} = \boldsymbol{L}\boldsymbol{L}^T$. In Section 2.4 we will discuss how to compute this factorization rapidly taking the sparsity of $\boldsymbol{Q}$ into account, and discover that the sparsity of $\boldsymbol{Q}$ may also be inherited by $\boldsymbol{L}$.

Sample $\boldsymbol{x} \sim \mathcal{N}(\boldsymbol{\mu}, \boldsymbol{Q}^{-1})$

To sample $\boldsymbol{x} \sim \mathcal{N}(\boldsymbol{\mu}, \boldsymbol{Q}^{-1})$, where $\boldsymbol{Q} = \boldsymbol{L}\boldsymbol{L}^T$, we use the following result: If $\boldsymbol{z} \sim \mathcal{N}(\boldsymbol{0}, \boldsymbol{I})$, then the solution of $\boldsymbol{L}^T\boldsymbol{x} = \boldsymbol{z}$ has covariance matrix

$$\mathrm{Cov}(\boldsymbol{x}) = \mathrm{Cov}(\boldsymbol{L}^{-T}\boldsymbol{z}) = (\boldsymbol{L}\boldsymbol{L}^T)^{-1} = \boldsymbol{Q}^{-1}.$$

Hence we obtain Algorithm 2.4. For repeated samples, we do step 1 only

Algorithm 2.4 Sampling $\boldsymbol{x} \sim \mathcal{N}(\boldsymbol{\mu}, \boldsymbol{Q}^{-1})$

1: Compute the Cholesky factorization, $\boldsymbol{Q} = \boldsymbol{L}\boldsymbol{L}^T$
2: Sample $\boldsymbol{z} \sim \mathcal{N}(\boldsymbol{0}, \boldsymbol{I})$
3: Solve $\boldsymbol{L}^T\boldsymbol{v} = \boldsymbol{z}$
4: Compute $\boldsymbol{x} = \boldsymbol{\mu} + \boldsymbol{v}$
5: **Return** $\boldsymbol{x}$

once. Step 3 *solves* the linear system $\boldsymbol{L}^T\boldsymbol{v} = \boldsymbol{z}$ using back substitution

(2.23), from which we obtain the following result as a by-product, giving some interpretation to the elements of $\boldsymbol{L}$.

Theorem 2.7 *Let $\boldsymbol{x}$ be a GMRF wrt to the labelled graph $\mathcal{G}$, with mean $\boldsymbol{\mu}$ and precision matrix $\boldsymbol{Q} > 0$. Let $\boldsymbol{L}$ be the Cholesky triangle of $\boldsymbol{Q}$. Then for $i \in \mathcal{V}$,*

$$\begin{aligned} E(x_i \mid \boldsymbol{x}_{(i+1):n}) &= \mu_i - \frac{1}{L_{ii}} \sum_{j=i+1}^{n} L_{ji}(x_j - \mu_j) \qquad \textit{and} \\ \mathit{Prec}(x_i \mid \boldsymbol{x}_{(i+1):n}) &= L_{ii}^2. \end{aligned}$$

Theorem 2.7 provides an alternative representation of a GMRF as a non-homogeneous autoregressive process defined backward in the indices (or a virtual time). It will be shown later in Section 5.2 that this is a useful representation. The following corollary is immediate when we compare $L_{ii}^2 = \text{Prec}(x_i \mid \boldsymbol{x}_{(i+1):n})$ with $Q_{ii} = \text{Prec}(x_i \mid \boldsymbol{x}_{-i})$.

Corollary 2.1 *$Q_{ii} \geq L_{ii}^2$ for all i.*

In matrix terms, this is a direct consequence of $\boldsymbol{Q} = \boldsymbol{L}\boldsymbol{L}^T$, which gives $Q_{ii} = L_{ii}^2 + \sum_{j=1}^{i-1} L_{ij}^2$.

Sample $\boldsymbol{x} \sim \mathcal{N}_C(\boldsymbol{b}, \boldsymbol{Q})$

To sample from a GMRF defined from its canonical representation (see Definition 2.2) we use Algorithm 2.5. This algorithm sample from $\mathcal{N}(\boldsymbol{Q}^{-1}\boldsymbol{b}, \boldsymbol{Q}^{-1})$, see Definition 2.2. The mean $\boldsymbol{Q}^{-1}\boldsymbol{b}$ is computed using Algorithm 2.1.

Algorithm 2.5 Sampling $\boldsymbol{x} \sim \mathcal{N}_C(\boldsymbol{b}, \boldsymbol{Q})$

1: Compute the Cholesky factorization, $\boldsymbol{Q} = \boldsymbol{L}\boldsymbol{L}^T$
2: Solve $\boldsymbol{L}\boldsymbol{w} = \boldsymbol{b}$
3: Solve $\boldsymbol{L}^T\boldsymbol{\mu} = \boldsymbol{w}$
4: Sample $\boldsymbol{z} \sim \mathcal{N}(\boldsymbol{0}, \boldsymbol{I})$
5: Solve $\boldsymbol{L}^T\boldsymbol{v} = \boldsymbol{z}$
6: Compute $\boldsymbol{x} = \boldsymbol{\mu} + \boldsymbol{v}$
7: **Return** $\boldsymbol{x}$

For repeated samples, we do steps 1–3 only once. This algorithm requires three back or forward substitutions compared to only one when the mean is known.

The log density of a sample

The log density of a sample $\boldsymbol{x} \sim \mathcal{N}(\boldsymbol{\mu}, \boldsymbol{Q}^{-1})$ or $\boldsymbol{x} \sim \mathcal{N}_C(\boldsymbol{b}, \boldsymbol{Q})$ is easily calculated using (2.4), where

$$\frac{1}{2}\log|\boldsymbol{Q}| = \sum_{i=1}^{n} \log L_{ii}$$

because $|\boldsymbol{Q}| = |\boldsymbol{L}\boldsymbol{L}^T| = |\boldsymbol{L}||\boldsymbol{L}^T| = |\boldsymbol{L}|^2$. Hence we obtain

$$\log \pi(\boldsymbol{x}) = -\frac{n}{2}\log(2\pi) + \sum_{i=1}^{n} \log L_{ii} - \frac{1}{2}q, \tag{2.26}$$

where

$$q = (\boldsymbol{x} - \boldsymbol{\mu})^T \boldsymbol{Q}(\boldsymbol{x} - \boldsymbol{\mu}). \tag{2.27}$$

If $\boldsymbol{x}$ is generated via Algorithm 2.4 or (2.5), q simplifies to $q = \boldsymbol{z}^T\boldsymbol{z}$. Otherwise we use (2.27) and first compute $\boldsymbol{\mu}$, if necessary, and then $\boldsymbol{v} = \boldsymbol{x} - \boldsymbol{\mu}$, $\boldsymbol{w} = \boldsymbol{Q}\boldsymbol{v}$, and $q = \boldsymbol{v}^T\boldsymbol{w}$.

2.3.3 Conditional simulation of a GMRF

Sampling from $\pi(\boldsymbol{x}_A|\boldsymbol{x}_{-A})$ where $\boldsymbol{x} \sim \mathcal{N}(\boldsymbol{\mu}, \boldsymbol{Q}^{-1})$

From Theorem 2.5 we know that the conditional distribution $\pi(\boldsymbol{x}_A|\boldsymbol{x}_B)$, where $\boldsymbol{x}_B = \boldsymbol{x}_{-A}$, is

$$\boldsymbol{x}_A \mid \boldsymbol{x}_B \sim \mathcal{N}(\boldsymbol{\mu}_A - \boldsymbol{Q}_{AA}^{-1}\boldsymbol{Q}_{AB}(\boldsymbol{x}_B - \boldsymbol{\mu}_B), \boldsymbol{Q}_{AA}^{-1}).$$

To sample from $\pi(\boldsymbol{x}_A|\boldsymbol{x}_B)$, it is convenient to first subtract the marginal mean $\boldsymbol{\mu}_A$ and to write $\boldsymbol{x}_A - \boldsymbol{\mu}_A|\boldsymbol{x}_B$ in the canonical parameterization:

$$\boldsymbol{x}_A - \boldsymbol{\mu}_A \mid \boldsymbol{x}_B \sim \mathcal{N}_C(-\boldsymbol{Q}_{AB}(\boldsymbol{x}_B - \boldsymbol{\mu}_B), \boldsymbol{Q}_{AA}).$$

Hence we can use Algorithm 2.5 to sample from $\pi(\boldsymbol{x}_A - \boldsymbol{\mu}_A|\boldsymbol{x}_B)$, and then we simply add $\boldsymbol{\mu}_A$. Some more insight will be given to the term $\boldsymbol{Q}_{AB}(\boldsymbol{x}_B - \boldsymbol{\mu}_B)$ in Appendix B.

Sampling from $\pi(\boldsymbol{x}|\boldsymbol{A}\boldsymbol{x} = \boldsymbol{e})$ where $\boldsymbol{x} \sim \mathcal{N}(\boldsymbol{\mu}, \boldsymbol{Q}^{-1})$

We now consider the important case, where we want to sample from a GMRF under an additional linear constraint

$$\boldsymbol{A}\boldsymbol{x} = \boldsymbol{e},$$

where $\boldsymbol{A}$ is a $k \times n$ matrix, $0 < k < n$, with rank k, and $\boldsymbol{e}$ is a vector of length k. We will denote this problem sampling under a *hard constraint*. This problem occurs quite frequently in practice, for example we might require that the sum of the x_i's is zero, which corresponds to $k = 1$, $\boldsymbol{A} = \boldsymbol{1}^T$ and $\boldsymbol{e} = 0$.

The linear constraint ensures that the conditional distribution is normal, but singular as the rank of the constrained covariance matrix is $n-k$. For this reason, more care must be taken when sampling from this distribution. One approach is to compute the mean and the covariance from the joint distribution of $\boldsymbol{x}$ and $\boldsymbol{Ax}$, which is normal with moments

$$\mathrm{E}\begin{pmatrix}\boldsymbol{x}\\ \boldsymbol{Ax}\end{pmatrix} = \begin{pmatrix}\boldsymbol{\mu}\\ \boldsymbol{A\mu}\end{pmatrix} \quad \text{and} \quad \mathrm{Cov}\begin{pmatrix}\boldsymbol{x}\\ \boldsymbol{Ax}\end{pmatrix} = \begin{pmatrix}\boldsymbol{Q}^{-1} & \boldsymbol{Q}^{-1}\boldsymbol{A}^T\\ \boldsymbol{AQ}^{-1} & \boldsymbol{AQ}^{-1}\boldsymbol{A}^T\end{pmatrix}.$$

We condition on $\boldsymbol{Ax} = \boldsymbol{e}$, which leads to the conditional moments $\boldsymbol{\mu}^* = \mathrm{E}(\boldsymbol{x}|\boldsymbol{Ax})$ and $\boldsymbol{\Sigma}^* = \mathrm{Cov}(\boldsymbol{x}|\boldsymbol{Ax})$, where

$$\boldsymbol{\mu}^* = \boldsymbol{\mu} - \boldsymbol{Q}^{-1}\boldsymbol{A}^T(\boldsymbol{AQ}^{-1}\boldsymbol{A}^T)^{-1}(\boldsymbol{A\mu} - \boldsymbol{e}) \tag{2.28}$$

and

$$\boldsymbol{\Sigma}^* = \boldsymbol{Q}^{-1} - \boldsymbol{Q}^{-1}\boldsymbol{A}^T(\boldsymbol{AQ}^{-1}\boldsymbol{A}^T)^{-1}\boldsymbol{AQ}^{-1}. \tag{2.29}$$

We can sample from this distribution as follows. As the conditional covariance matrix $\boldsymbol{\Sigma}^*$ is singular, we first compute the eigenvalues and eigenvectors, and factorize $\boldsymbol{\Sigma}^*$ as $\boldsymbol{V\Lambda V}^T$ where $\boldsymbol{V}$ has the eigenvectors on each column and $\boldsymbol{\Lambda}$ is a diagonal matrix with the corresponding eigenvalues on the diagonal. This is a different factorization than the one used in Algorithm 2.3, but any matrix $\boldsymbol{C} = \boldsymbol{V\Lambda}^{1/2}$, which satisfies $\boldsymbol{CC}^T = \boldsymbol{\Sigma}^*$ will do. Note that k of the eigenvalues are zero. We can now generate a sample by computing $\boldsymbol{v} = \boldsymbol{Cz}$, where $\boldsymbol{z} \sim \mathcal{N}(\boldsymbol{0}, \boldsymbol{I})$, and then add the conditional mean. We can compute the log density as

$$\begin{aligned}\log \pi(\boldsymbol{x} \mid \boldsymbol{Ax} = \boldsymbol{e}) &= -\frac{n-k}{2}\log 2\pi - \frac{1}{2}\sum_{i:\Lambda_{ii}>0} \log \Lambda_{ii} \\ &\quad -\frac{1}{2}(\boldsymbol{x} - \boldsymbol{\mu}^*)^T\boldsymbol{\Sigma}^-(\boldsymbol{x} - \boldsymbol{\mu}^*),\end{aligned}$$

where $\boldsymbol{\Sigma}^- = \boldsymbol{V\Lambda}^-\boldsymbol{V}^T$. Here $(\boldsymbol{\Lambda}^-)_{ii}$ is Λ_{ii}^{-1} if $\Lambda_{ii} > 0$ and zero otherwise. In total, this is a quite computationally demanding procedure, as the algorithm is not able to take advantage of the sparse structure of $\boldsymbol{Q}$.

There is an alternative procedure that *corrects* for the constraint, at nearly no cost if $k \ll n$. In the geostatistics literature this is called *conditioning by Kriging*. The result is the following: If we sample from the unconstrained GMRF $\boldsymbol{x} \sim \mathcal{N}(\boldsymbol{\mu}, \boldsymbol{Q}^{-1})$ and then compute

$$\boldsymbol{x}^* = \boldsymbol{x} - \boldsymbol{Q}^{-1}\boldsymbol{A}^T(\boldsymbol{AQ}^{-1}\boldsymbol{A}^T)^{-1}(\boldsymbol{Ax} - \boldsymbol{e}), \tag{2.30}$$

then $\boldsymbol{x}^*$ has the correct conditional distribution. This is clear after comparing the mean and covariance of $\boldsymbol{x}^*$ with (2.28) and (2.29). Note that $\boldsymbol{AQ}^{-1}\boldsymbol{A}^T$ is a dense $k \times k$ matrix, hence its factorization is fast to compute for small k. Algorithm 2.6 generates such a constrained sample, where we denote the dimension of some of the matrices by subscripts.

Algorithm 2.6 Sampling $\boldsymbol{x}|\boldsymbol{A}\boldsymbol{x} = \boldsymbol{e}$ where $\boldsymbol{x} \sim \mathcal{N}(\boldsymbol{\mu}, \boldsymbol{Q}^{-1})$

1: Compute the Cholesky factorization, $\boldsymbol{Q} = \boldsymbol{L}\boldsymbol{L}^T$
2: Sample $\boldsymbol{z} \sim \mathcal{N}(\boldsymbol{0}, \boldsymbol{I})$
3: Solve $\boldsymbol{L}^T\boldsymbol{v} = \boldsymbol{z}$
4: Compute $\boldsymbol{x} = \boldsymbol{\mu} + \boldsymbol{v}$
5: Compute $\boldsymbol{V}_{n\times k} = \boldsymbol{Q}^{-1}\boldsymbol{A}^T$ using Algorithm 2.2 using $\boldsymbol{L}$ from step 1
6: Compute $\boldsymbol{W}_{k\times k} = \boldsymbol{A}\boldsymbol{V}$
7: Compute $\boldsymbol{U}_{k\times n} = \boldsymbol{W}^{-1}\boldsymbol{V}^T$ using Algorithm 2.2
8: Compute $\boldsymbol{c} = \boldsymbol{A}\boldsymbol{x} - \boldsymbol{e}$
9: Compute $\boldsymbol{x}^* = \boldsymbol{x} - \boldsymbol{U}^T\boldsymbol{c}$
10: **Return** $\boldsymbol{x}^*$

For repeated samples we do step 1 and steps 5–7 only once. Note that if $\boldsymbol{z} = \boldsymbol{0}$ then $\boldsymbol{x}^*$ is the conditional mean (2.28). The following trivial but very useful example illustrates the use of (2.30).

Example 2.3 *Let $x_1, \ldots, x_n$ be independent normal variables with mean μ_i and variance σ_i^2. To sample $\boldsymbol{x}$ conditional on $\sum_i x_i = 0$, we first sample $x_i \sim \mathcal{N}(\mu_i, \sigma_i^2)$ for $i = 1, \ldots, n$ and compute the constrained sample $\boldsymbol{x}^*$ via*

$$x_i^* = x_i - c\,\sigma_i^2,$$

where $c = \sum_j x_j / \sum_j \sigma_j^2$.

The log density $\pi(\boldsymbol{x}|\boldsymbol{A}\boldsymbol{x})$ can be rapidly evaluated at $\boldsymbol{x}^*$ using the identity

$$\pi(\boldsymbol{x} \mid \boldsymbol{A}\boldsymbol{x}) = \frac{\pi(\boldsymbol{x})\pi(\boldsymbol{A}\boldsymbol{x} \mid \boldsymbol{x})}{\pi(\boldsymbol{A}\boldsymbol{x})}. \tag{2.31}$$

Note that we can compute each term on the right-hand side easier than the term on the left-hand side: The unconstrained density $\pi(\boldsymbol{x})$ is a GMRF and the log density is computed using (2.26) and $\boldsymbol{L}$ computed in Algorithm 2.6, step 1. The degenerate density $\pi(\boldsymbol{A}\boldsymbol{x}|\boldsymbol{x})$ is either zero or a constant, which must be one for $\boldsymbol{A} = \boldsymbol{I}$. A change of variables gives us

$$\log \pi(\boldsymbol{A}\boldsymbol{x} \mid \boldsymbol{x}) = -\frac{1}{2}\log|\boldsymbol{A}\boldsymbol{A}^T|,$$

i.e., we need to compute the determinant of a $k \times k$ matrix, which is found from its Cholesky factorization. Finally, the denominator $\pi(\boldsymbol{A}\boldsymbol{x})$ in (2.31) is normal with mean $\boldsymbol{A}\boldsymbol{\mu}$ and covariance matrix $\boldsymbol{A}\boldsymbol{Q}^{-1}\boldsymbol{A}^T$. The corresponding Cholesky triangle $\widetilde{\boldsymbol{L}}$ is available from Algorithm 2.6, step 7. The log density can then be computed from (2.25).

Sampling from $\pi(\boldsymbol{x}|\boldsymbol{e})$ where $\boldsymbol{x} \sim \mathcal{N}(\boldsymbol{\mu}, \boldsymbol{Q}^{-1})$ and $\boldsymbol{e}|\boldsymbol{x} \sim \mathcal{N}(\boldsymbol{A}\boldsymbol{x}, \boldsymbol{\Sigma}_\epsilon)$

Let $\boldsymbol{x}$ be a GMRF, where some linear transformation $\boldsymbol{A}\boldsymbol{x}$ is observed with additional normal noise:

$$\boldsymbol{e} \mid \boldsymbol{x} \sim \mathcal{N}(\boldsymbol{A}\boldsymbol{x}, \boldsymbol{\Sigma}_\epsilon). \tag{2.32}$$

Here, $\boldsymbol{e}$ is a vector of length $k < n$, $\boldsymbol{A}$ is a $k \times n$ matrix of rank k, and $\boldsymbol{\Sigma}_\epsilon > 0$ is the covariance matrix of $\boldsymbol{e}$. The log density of $\boldsymbol{x}|\boldsymbol{e}$ is then

$$\log \pi(\boldsymbol{x} \mid \boldsymbol{e}) = -\frac{1}{2}(\boldsymbol{x}-\boldsymbol{\mu})^T \boldsymbol{Q}(\boldsymbol{x}-\boldsymbol{\mu}) - \frac{1}{2}(\boldsymbol{e}-\boldsymbol{A}\boldsymbol{x})^T \boldsymbol{\Sigma}_\epsilon^{-1}(\boldsymbol{e}-\boldsymbol{A}\boldsymbol{x}) + \text{const}, \tag{2.33}$$

i.e.,

$$\boldsymbol{x} \mid \boldsymbol{e} \sim \mathcal{N}_C(\boldsymbol{Q}\boldsymbol{\mu} + \boldsymbol{A}^T \boldsymbol{\Sigma}_\epsilon^{-1} \boldsymbol{e}, \boldsymbol{Q} + \boldsymbol{A}^T \boldsymbol{\Sigma}_\epsilon^{-1} \boldsymbol{A}), \tag{2.34}$$

which could also be derived using (2.16). However, the precision matrix in (2.34) is usually a completely dense matrix and the nice sparse structure of $\boldsymbol{Q}$ is lost. For example, if we observe the sum of x_i with unit variance noise, the conditional precision is $\boldsymbol{Q} + \boldsymbol{1}\boldsymbol{1}^T$, which is a completely dense matrix. In general, we have to sample from (2.33) using Algorithm 2.3 or Algorithm 2.4, which is computationally expensive for large n.

There is however an alternative approach that is feasible for $k \ll n$. If we extend (2.30) to

$$\boldsymbol{x}^* = \boldsymbol{x} - \boldsymbol{Q}^{-1}\boldsymbol{A}^T(\boldsymbol{A}\boldsymbol{Q}^{-1}\boldsymbol{A}^T + \boldsymbol{\Sigma}_\epsilon)^{-1}(\boldsymbol{A}\boldsymbol{x} - \boldsymbol{\epsilon}),$$

where

$$\boldsymbol{\epsilon} \sim \mathcal{N}(\boldsymbol{e}, \boldsymbol{\Sigma}_\epsilon),$$

$\boldsymbol{e}$ is the *observed* value, and $\boldsymbol{x} \sim \mathcal{N}(\boldsymbol{\mu}, \boldsymbol{Q}^{-1})$, then it is easy to show that $\boldsymbol{x}^*$ has the correct conditional distribution (2.34). We denote this case sampling under a *soft constraint*. Algorithm 2.7 is similar to Algorithm 2.6.

Note that if $\boldsymbol{z} = \boldsymbol{0}$ and $\boldsymbol{\epsilon} = \boldsymbol{e}$ in Algorithm 2.7, then $\boldsymbol{x}^*$ is the conditional mean. For repeated samples, we do step 1 and steps 5-7 only once.

Also for soft constraints we can evaluate the log density at $\boldsymbol{x}^*$ using (2.31) with $\boldsymbol{A}\boldsymbol{x} = \boldsymbol{e}$:

$$\pi(\boldsymbol{x} \mid \boldsymbol{e}) = \frac{\pi(\boldsymbol{x})\pi(\boldsymbol{e} \mid \boldsymbol{x})}{\pi(\boldsymbol{e})}.$$

The unconstrained density $\pi(\boldsymbol{x})$ is a GMRF and the log density is computed using (2.26). Regarding $\pi(\boldsymbol{e}|\boldsymbol{x})$, we know from (2.32) that $\pi(\boldsymbol{e}|\boldsymbol{x})$ is normal with mean $\boldsymbol{A}\boldsymbol{x}$ and covariance $\boldsymbol{\Sigma}_\epsilon$. We use (2.25) to compute the log density. Finally, $\boldsymbol{e}$ is normal with mean $\boldsymbol{A}\boldsymbol{\mu}$ and covariance matrix $\boldsymbol{A}\boldsymbol{Q}^{-1}\boldsymbol{A}^T + \boldsymbol{\Sigma}_\epsilon$, hence we can use (2.25) to compute

Algorithm 2.7 Sampling from $\pi(\boldsymbol{x}|\boldsymbol{e})$ where $\boldsymbol{x} \sim \mathcal{N}(\boldsymbol{\mu}, \boldsymbol{Q}^{-1})$ and $\boldsymbol{e}|\boldsymbol{x} \sim \mathcal{N}(\boldsymbol{A}\boldsymbol{x}, \boldsymbol{\Sigma}_{\epsilon})$

1: Compute the Cholesky factorization, $\boldsymbol{Q} = \boldsymbol{L}\boldsymbol{L}^T$
2: Sample $\boldsymbol{z} \sim \mathcal{N}(\boldsymbol{0}, \boldsymbol{I})$
3: Solve $\boldsymbol{L}^T\boldsymbol{v} = \boldsymbol{z}$
4: Compute $\boldsymbol{x} = \boldsymbol{\mu} + \boldsymbol{v}$
5: Compute $\boldsymbol{V}_{n\times k} = \boldsymbol{Q}^{-1}\boldsymbol{A}^T$ using Algorithm 2.2 using $\boldsymbol{L}$ from step 1
6: Compute $\boldsymbol{W}_{k\times k} = \boldsymbol{A}\boldsymbol{V} + \boldsymbol{\Sigma}_{\epsilon}$
7: Compute $\boldsymbol{U}_{k\times n} = \boldsymbol{W}^{-1}\boldsymbol{V}^T$ using Algorithm 2.2
8: Sample $\boldsymbol{\epsilon} \sim \mathcal{N}(\boldsymbol{e}, \boldsymbol{\Sigma}_{\epsilon})$ using Algorithm 2.3.
9: Compute $\boldsymbol{c} = \boldsymbol{A}\boldsymbol{x} - \boldsymbol{\epsilon}$
10: Compute $\boldsymbol{x}^* = \boldsymbol{x} - \boldsymbol{U}^T\boldsymbol{c}$
11: **Return** $\boldsymbol{x}^*$

the log density. Note that all Cholesky triangles required to evaluate the log density are already computed in Algorithm 2.7.

The stochastic version of Example 2.3 now follows.

Example 2.4 *Let* $x_1, \ldots, x_n$ *be independent normal variables with variance* σ_i^2 *and mean* μ_i*. We now observe* $e \sim \mathcal{N}(\sum_i x_i, \sigma_\epsilon^2)$*. To sample from* $\pi(\boldsymbol{x}|e)$*, we sample* $x_i \sim \mathcal{N}(\mu_i, \sigma_i^2)$*, unconditionally, for* $i = 1, \ldots, n$ *while we condition on* $\epsilon \sim \mathcal{N}(e, \sigma_\epsilon^2)$*. A conditional sample* $\boldsymbol{x}^*$ *is then*

$$x_i^* = x_i - c\,\sigma_i^2, \quad \textit{where} \quad c = \frac{\sum_j x_j - \epsilon}{\sum_j \sigma_j^2 + \sigma_\epsilon^2}.$$

We can merge soft and hard constraints into one framework if we allow $\boldsymbol{\Sigma}_{\epsilon}$ to be SPSD, but we have chosen not to, as the details are somewhat tedious.

2.4 Numerical methods for sparse matrices

This section will give a brief introduction to numerical methods for sparse matrices. During our discussion of simulation algorithms for GMRFs, we have shown that they all can be expressed such that the main tasks are to

1. Compute the Cholesky factorization of $\boldsymbol{Q} = \boldsymbol{L}\boldsymbol{L}^T$ where $\boldsymbol{Q} > 0$ is sparse, and
2. Solve $\boldsymbol{L}\boldsymbol{v} = \boldsymbol{b}$ and $\boldsymbol{L}^T\boldsymbol{x} = \boldsymbol{z}$

The second task is faster to compute than the first, but sparsity of $\boldsymbol{Q}$ is also advantageous in this case. We restrict the discussion to sparse Cholesky factorizations but the ideas also apply to sparse $\boldsymbol{LU}$ factorizations for non symmetric matrices.

The goal is to explain *why* a sparse $\boldsymbol{Q}$ allows for fast factorization, *how* we can take advantage of it, *why* we permute the nodes before factorizing the matrix, and *how* statisticians can benefit from recent research results in numerical mathematics. At the end, we will report a small case study factorizing some typical matrices for GMRFs, using classical and more recent methods.

2.4.1 Factorizing a sparse matrix

We start with a dense matrix $\boldsymbol{Q} > 0$ with dimension n, and show how to compute the Cholesky triangle $\boldsymbol{L}$, so $\boldsymbol{Q} = \boldsymbol{L}\boldsymbol{L}^T$, which can be written as

$$Q_{ij} = \sum_{k=1}^{j} L_{ik} L_{jk}, \quad i \geq j.$$

We now define

$$v_i = Q_{ij} - \sum_{k=1}^{j-1} L_{ik} L_{jk}, \quad i \geq j,$$

and we immediately see that $L_{jj}^2 = v_j$ and $L_{ij} L_{jj} = v_i$ for $i > j$. If we know $\{v_i\}$ for fixed j, then $L_{jj} = \sqrt{v_j}$ and $L_{ij} = v_i/\sqrt{v_j}$, for $i = j+1$ to n. This gives the jth column in $\boldsymbol{L}$. The algorithm is completed by noting that $\{v_i\}$ for fixed j only depends on elements of $\boldsymbol{L}$ in the first $j-1$ columns of $\boldsymbol{L}$. Algorithm 2.8 gives the pseudocode using vector notation for simplicity: $v_{j:n} = Q_{j:n,j}$ is short for $v_k = Q_{kj}$ for $k = j$ to n and so on. The overall process involves $n^3/3$ flops. If $\boldsymbol{Q}$ is symmetric but not SPD, then $v_j \leq 0$ for some j and the algorithm fails.

Algorithm 2.8 Cholesky factorization of $\boldsymbol{Q} > 0$

for $j = 1$ **to** n **do**
  $v_{j:n} = Q_{j:n,j}$
  for $k = 1$ **to** $j-1$ **do** $v_{j:n} = v_{j:n} - L_{j:n,k} L_{jk}$
  $L_{j:n,j} = v_{j:n}/\sqrt{v_j}$
end for
Return $\boldsymbol{L}$

The Cholesky factorization is computed without pivoting and its numerical stability follows (roughly) from $\sum_{k=1}^{i} L_{ik}^2 = Q_{ii}$ hence $L_{ij}^2 \leq Q_{ii}$, which shows that the entries in the Cholesky triangle are nicely bounded.

Now we explore the possibilities of a sparse $\boldsymbol{Q}$ to speed up Algorithm 2.8. Recall Theorem 2.7 where we showed that

$$\mathrm{E}(x_i \mid \boldsymbol{x}_{(i+1):n}) = \mu_i - \frac{1}{L_{ii}} \sum_{j=i+1}^{n} L_{ji}(x_j - \mu_j)$$

and $\mathrm{Prec}(x_i|\boldsymbol{x}_{(i+1):n}) = L_{ii}^2$. Another interpretation of Theorem 2.7 is the following result.

Theorem 2.8 *Let $\boldsymbol{x}$ be a GMRF wrt $\mathcal{G}$, with mean $\boldsymbol{\mu}$ and precision matrix $\boldsymbol{Q} > 0$. Let $\boldsymbol{L}$ be the Cholesky triangle of $\boldsymbol{Q}$ and define for $1 \leq i < j \leq n$ the set*

$$F(i,j) = \{i+1, \ldots, j-1, j+1, \ldots, n\},$$

which is the future *of i except j. Then*

$$x_i \perp x_j \mid \boldsymbol{x}_{F(i,j)} \iff L_{ji} = 0. \tag{2.35}$$

Proof. [Theorem 2.8] Assume for simplicity that $\boldsymbol{\mu} = \boldsymbol{0}$ and fix $1 \leq i < j \leq n$. Theorem 2.7 gives that

$$\begin{aligned} \pi(\boldsymbol{x}_{i:n}) &\propto \exp\left(-\frac{1}{2}\sum_{k=i}^{n} L_{kk}^2 \left(x_k + \frac{1}{L_{kk}} \sum_{j=k+1}^{n} L_{jk}x_j\right)^2\right) \\ &= \exp\left(-\frac{1}{2}\boldsymbol{x}_{i:n}^T \boldsymbol{Q}^{(i:n)} \boldsymbol{x}_{i:n}\right), \end{aligned}$$

where $Q_{ij}^{(i:n)} = L_{ii}L_{ji}$. Using Theorem 2.2, it then follows that

$$x_i \perp x_j \mid \boldsymbol{x}_{F(i,j)} \iff L_{ii}L_{ji} = 0,$$

which is equivalent to (2.35) since $L_{ii} > 0$ as $\boldsymbol{Q}^{(i:n)} > 0$. □

The implications of Theorem 2.8 are immediate: If we can verify that L_{ji} is zero, we do not have to compute it in Algorithm 2.8, hence we save computations. However, as Theorem 2.8 relates zeros in the lower triangular of $\boldsymbol{L}$ to conditional independence properties of the successive marginals $\{\pi(\boldsymbol{x}_{i:n})\}_{i=1}^{n}$, there is no easy way to use Theorem 2.8 to check if $L_{ji} = 0$, except computing it and see if L_{ji} turned out to be zero!

Theorem 2.4 provides a simple and *sufficient* criteria for checking if $L_{ji} = 0$, making use of the global Markov property. We state this as a Corollary to Theorem 2.8.

Corollary 2.2 *If $F(i,j)$ separates $i < j$ in $\mathcal{G}$, then $L_{ji} = 0$.*

Proof. [Corollary 2.2] The global Markov property (2.12) ensures that if $F(i,j)$ separates $i < j$ in $\mathcal{G}$, then $x_i \perp x_j \mid \boldsymbol{x}_{F(i,j)}$. Hence, $L_{ji} = 0$ using Theorem 2.8. □

Note that Corollary 2.2 *does not* make use of the actual values in $\boldsymbol{Q}$ to decide if $L_{ji} = 0$, but only uses the conditional independence structure represented by $\mathcal{G}$. Hence, if $L_{ji} = 0$ using Corollary 2.2, then it is zero for all $\boldsymbol{Q} > 0$ with the same graph $\mathcal{G}$. The reverse statement in Corollary 2.2 is of course not true in general. A simple counter-example is the following.

Example 2.5 *Let*

$$\boldsymbol{L} = \begin{pmatrix} L_{11} & & & \\ L_{21} & L_{22} & & \\ L_{31} & 0 & L_{33} & \\ L_{41} & L_{42} & L_{43} & L_{44} \end{pmatrix}$$

so that $\boldsymbol{Q} = \boldsymbol{L}\boldsymbol{L}^T$ *with* $Q_{32} = L_{21}L_{31}$. *Here* 2 *and* 3 *are not separated by* $F(2,3) = 4$, *although* $L_{32} = 0$.

The approach is then to make use of Corollary 2.2 to check if $L_{ji} = 0$, for all $1 \leq i < j \leq n$, and to compute only those L_{ji}'s, that are not known to be zero, using Algorithm 2.8. Note that we always need to compute L_{ji} for $i \sim j$ and $j > i$, since $F(i,j)$ does not separate i and j since $Q_{ij} \neq 0$. A simple example illustrates the procedure.

Example 2.6 *Consider the graph*

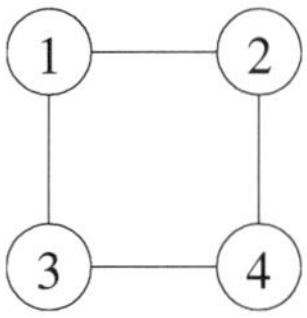

and the corresponding precision matrix $\boldsymbol{Q}$

$$\boldsymbol{Q} = \begin{pmatrix} \times & \times & \times & \\ \times & \times & & \times \\ \times & & \times & \times \\ & \times & \times & \times \end{pmatrix}$$

where the $\times$*'s denote nonzero terms. The only possible zero terms in* $\boldsymbol{L}$ *(in general) are* L_{32} *and* L_{41} *due to Corollary 2.2. Considering* L_{32} *we see that* $F(2,3) = 4$. *This is not a separating subset for* 2 *and* 3 *due to node* 1, *hence* L_{32} *is not known to be zero using Corollary 2.2. For* L_{41} *we see that* $F(1,4) = \{2,3\}$, *which does separate* 1 *and* 4, *hence* $L_{41} = 0$. *In total,* $\boldsymbol{L}$ *has the following structure:*

$$\boldsymbol{L} = \begin{pmatrix} \times & & & \\ \times & \times & & \\ \times & \surd & \times & \\ & \times & \times & \times \end{pmatrix},$$

where the possibly nonzero entry L_{32} *is marked as '*$\surd$*'.*

Applying Corollary 2.2, we know that $\boldsymbol{L}$ is always more or equally dense than the lower triangular part of $\boldsymbol{Q}$, i.e., the number $n_{\boldsymbol{L}}$ of (possible) nonzero elements in $\boldsymbol{L}$ is always larger than the number $n_{\boldsymbol{Q}}$ of nonzero elements in the lower triangular part of $\boldsymbol{Q}$ (including the diagonal). Thus we are concerned about the number of *fill-ins* $n_{\boldsymbol{L}} - n_{\boldsymbol{Q}}$, which we sometimes just simply call *the fill-in.*

Ideally, $n_{\boldsymbol{L}} = n_{\boldsymbol{Q}}$, but there are many other graphs where $n_{\boldsymbol{L}} \gg n_{\boldsymbol{Q}}$. We therefore compare different graphs through the *fill-in ratio* $R = n_{\boldsymbol{L}}/n_{\boldsymbol{Q}}$. Clearly, $R \geq 1$ and the closer R is to unity, the more efficient is the Cholesky factorization of a given precision matrix $\boldsymbol{Q}$. For example, in Example 2.6, $n_Q = 8$, $n_L = 9$ and hence $R = 9/8$.

It will soon become clear that fill-in depends crucially on the ordering of the nodes and is of major concern in numerical linear algebra for sparse matrices. We will return shortly to this issue, but first discuss a simple example and its consequences.

Example 2.7 *Let $\boldsymbol{x}$ be a Gaussian autoregressive process of order* 1, *where*

$$x_t \mid \boldsymbol{x}_{1:(t-1)} \sim \mathcal{N}(\phi x_{t-1}, \sigma^2), \quad |\phi| < 1, \quad t = 2, \ldots, n$$

with $x_1 \sim \mathcal{N}(\mu_1, \sigma_1^2)$, say. Now $x_i \perp x_{i+k} | \boldsymbol{x}_{rest}$ for $k > 1$ and hence

$$\boldsymbol{Q} = \begin{pmatrix} \times & \times & & & & & \\ \times & \times & \times & & & & \\ & \times & \times & \times & & & \\ & & \times & \times & \times & & \\ & & & \times & \times & \times & \\ & & & & \times & \times & \times \\ & & & & & \times & \times \end{pmatrix}$$

is tridiagonal. The $\times$*'s denote the nonzero terms. The (possible) nonzero terms in* $\boldsymbol{L}$ *can now be determined using Corollary 2.2. Since $i \sim i+1$ it follows that $L_{i+1,i}$ is not known to be zero. For $k > 1$, $F(i, i+k)$ separates i and $i+k$, hence all the remaining terms are zero. The consequence is that* $\boldsymbol{L}$ *is (in general) lower tridiagonal,*

$$\boldsymbol{L} = \begin{pmatrix} \times & & & & & & \\ \times & \times & & & & & \\ & \times & \times & & & & \\ & & \times & \times & & & \\ & & & \times & \times & & \\ & & & & \times & \times & \\ & & & & & \times & \times \end{pmatrix}.$$

Note that in this example $R = 1$ and that the bandwidth of both $\boldsymbol{Q}$ *and* $\boldsymbol{L}$ *equals one.*

We can extend Example 2.7 to any autoregressive process of order $p > 1$, where $\boldsymbol{Q}$ will be a band matrix with bandwidth p. For $k > p$, $F(i, i+k)$ separates i and $i+k$, hence $\boldsymbol{L}$ is (in general) lower triangular with the same lower bandwidth p. As a consequence, we have proved that the bandwidth is preserved during Cholesky factorization.

Theorem 2.9 *Let $\boldsymbol{Q} > 0$ be a band matrix with bandwidth p and dimension n, then the Cholesky triangle of $\boldsymbol{Q}$ has (lower) bandwidth p.*

This result is well known and a direct proof is available in Golub and van Loan (1996, Theorem 4.3.1).

We can now do the trivial modification of Algorithm 2.8 to avoid computing L_{ij} and reading Q_{ij} for $|i-j| > p$. The band version of the algorithm is Algorithm 2.9.

Algorithm 2.9 Band-Cholesky factorization of $\boldsymbol{Q}$ with bandwidth p

```
for j = 1 to n do
   λ = min{j + p, n}
   v_{j:λ} = Q_{j:λ,j}
   for k = max{1, j − p} to j − 1 do
      i = min{k + p, n}
      v_{j:i} = v_{j:i} − L_{j:i,k} L_{jk}
   end for
   L_{j:λ,j} = v_{j:λ} / sqrt(v_j)
end for
Return L
```

The overall process involves $n(p^2 + 3p)$ flops assuming $n \gg p$. For an autoregressive process of order p, this is the cost of factorizing $\boldsymbol{Q}$. The costs are linear in n and have been reduced dramatically compared to the general cost $n^3/3$.

Similar efficiency gains also show up if we want to solve $\boldsymbol{L}^T\boldsymbol{x} = \boldsymbol{z}$ via back-substitution:

$$x_i = \frac{1}{L_{ii}}(v_i - \sum_{j=i+1}^{\min\{i+p,n\}} L_{ji}x_j), \quad i = n, \ldots, 1,$$

where the cost is $2np$ flops assuming $n \gg p$. Again, the algorithm is linear in n and we have gained one order of magnitude compared to n^2 flops required in the general case.

2.4.2 Bandwidth reduction

Now we turn to the spatial case where we will demonstrate that the band-Cholesky factorization and the band forward- and back-substitution are

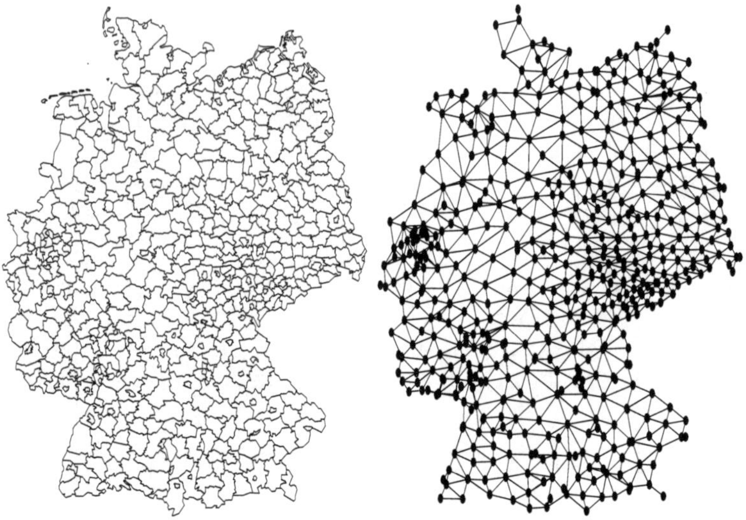

(a) (b)

Figure 2.4 *(a) The map of Germany with $n = 544$ regions, and (b) the corresponding graph for the GMRF where neighboring regions share a common border.*

also applicable for spatial GMRFs. We illustrate this by considering the map of Germany in Figure 2.4(a), where we assume that a GMRF is defined on the regions such that regions sharing a common border are neighbors. The graph for the GMRF is shown in Figure 2.4(b).

If we want to apply the band-Cholesky algorithm, we need to make sure that $\boldsymbol{Q}$ is a band matrix. The ordering of the regions is typically arbitrary (here they are determined through administrative rules), so we cannot expect $\boldsymbol{Q}$ to have a pattern that makes band-Cholesky factorization particularly useful. This is illustrated later in Figure 2.6(a) displaying $\boldsymbol{Q}$ in the original ordering with bandwidth 542. It is, however, easy to permute the nodes: Select one of the $n!$ possible permutations and define the corresponding permutation matrix $\boldsymbol{P}$ such that $\boldsymbol{i}^P = \boldsymbol{P}\boldsymbol{i}$, where $\boldsymbol{i} = (1, \ldots, n)^T$, and $\boldsymbol{i}^P$ is the new ordering of the vertices. This means that node 5, say, is renamed to node i_5^P. We can then try to choose $\boldsymbol{P}$ such that the corresponding precision matrix

$$\boldsymbol{Q}^P = \boldsymbol{P}\boldsymbol{Q}\boldsymbol{P}^T \tag{2.36}$$

is a band matrix with a small bandwidth. Typically it will be impossible to obtain the optimal permutation from all $n!$ possible ones, but a sub-optimal ordering will do as well. For a given ordering, we solve $\boldsymbol{Q}\boldsymbol{\mu} = \boldsymbol{b}$ as follows. Compute the reordered problem, $\boldsymbol{Q}^P$ as in (2.36), $\boldsymbol{b}^P = \boldsymbol{P}\boldsymbol{b}$.

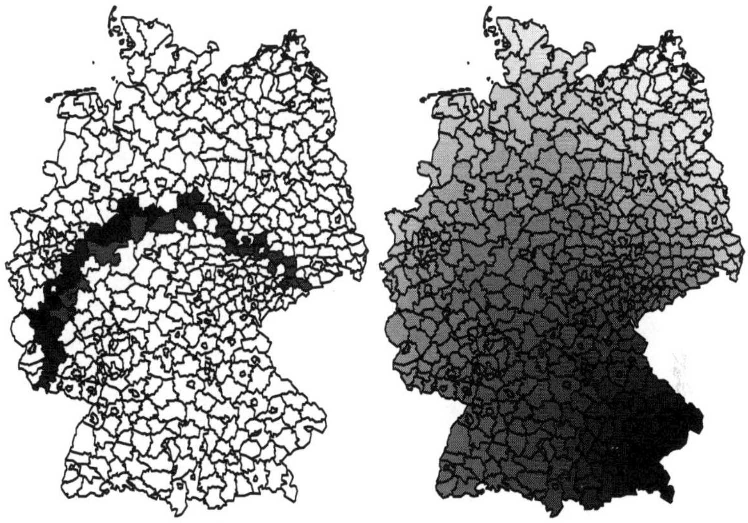

(a) (b)

Figure 2.5 *(a) The black regions make north and south conditionally independent, and (b) displays the automatically computed reordering starting from the white region ending at the black region. This reordering produces the precision matrix in Figure 2.6(b).*

Solve $\boldsymbol{Q}^P \boldsymbol{\mu}^P = \boldsymbol{b}^P$ and then map the solution back, $\boldsymbol{\mu} = \boldsymbol{P}^T \boldsymbol{\mu}^P$.

Return now to Corollary 2.2, which states a sufficient condition for making $L_{ji} = 0$ for $|i-j| > p$. We need to ensure that, after reordering, x_i and x_j are conditional independent given $\boldsymbol{x}_{F(i,j)}$ if $|i-j| > p$. Suppose now we separate the south and the north of Germany through a third subset of regions, the black regions in Figure 2.5(a). Then the north and the south are conditionally independent given the black regions. Clearly we can obtain a similar separation, by sliding the black line from top to bottom. The bandwidth turns out to be the maximal number of (black) regions needed to divide the north and the south. With this reordering, the precision matrix will have a small bandwidth. Section 2.5 gives details about the algorithms used for computing the reordering.

One automatically computed ordering is shown in Figure 2.5(b), from white to black. The reordered $\boldsymbol{Q}$ shown in Figure 2.6(b) has bandwidth 43. As the bandwidth is about $\mathcal{O}(\sqrt{n})$, the costs will be $\mathcal{O}(n^2)$ for the factorization for this and similar type of graphs.

To give an idea of the (amazing) speed of such algorithms, the factorization of $\boldsymbol{Q}^P$ required about 0.0018 seconds on a 1200-MHz CPU. Solving $\boldsymbol{Q}\boldsymbol{\mu} = \boldsymbol{b}$ required about 0.0006 seconds. However, the fill-in ratio for this graph is $R = 5.3$, which suggests that we may do even better

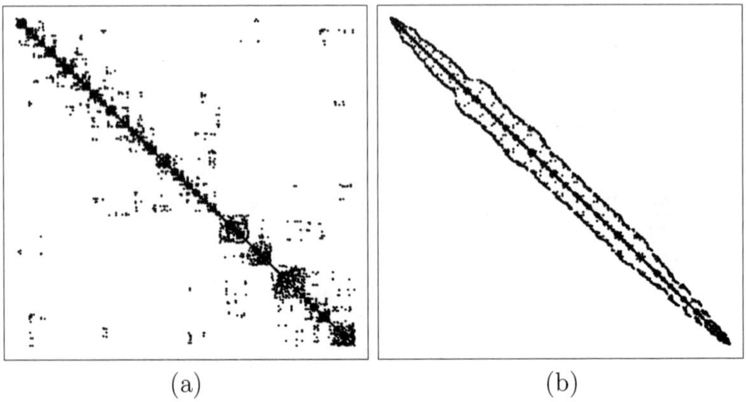

Figure 2.6 *(a) The precision matrix* $\boldsymbol{Q}$ *in the original ordering, and (b) the precision matrix after appropriate reordering to obtain a band matrix with small bandwidth. Only the nonzero terms are shown and those are indicated by a dot.*

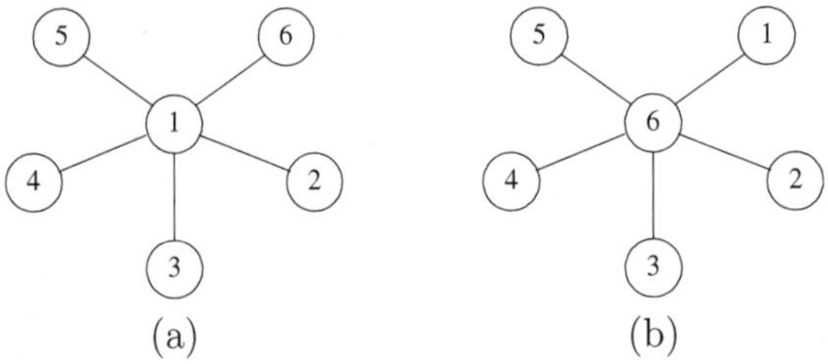

Figure 2.7 *Two graphs with a slight change in the ordering. Graph (a) requires* $\mathcal{O}(n^3)$ *flops to factorize, while graph (b) only requires* $\mathcal{O}(n)$. *Here,* n *represent the number of nodes being neighbors to the center node. The fill-in is maximal in (a) and minimal in (b).*

with other kinds of orderings.

2.4.3 Nested dissection

There has been much work in the numerical and computer science literature on other reordering schemes focusing on reducing the number of fill-ins rather than focusing on reducing the bandwidth. Why reordering schemes that reduce the number of fill-ins may be better can be seen from the following example. For a GMRF with the graph shown in Figure

2.7(a), the precision matrix and its Cholesky triangle are

$$\boldsymbol{Q} = \begin{pmatrix} \times & \times & \times & \times & \times & \times \\ \times & \times & & & & \\ \times & & \times & & & \\ \times & & & \times & & \\ \times & & & & \times & \\ \times & & & & & \times \end{pmatrix}, \quad \boldsymbol{L} = \begin{pmatrix} \times & & & & & \\ \times & \times & & & & \\ \times & \surd & \times & & & \\ \times & \surd & \surd & \times & & \\ \times & \surd & \surd & \surd & \times & \\ \times & \surd & \surd & \surd & \surd & \times \end{pmatrix}. \tag{2.37}$$

In this case the fill-in is maximal and, for general n, the cost is $\mathcal{O}(n^3)$ flops to compute the factorization, where n is the number of nodes in the circle. The reason is that all nodes depend on 1, hence $F(i, j)$ is *never* a separating subset for $i < j$.

However, if we switch the numbers for 1 and $n = 7$ as in Figure 2.7(b), we obtain the following precision matrix and its corresponding Cholesky triangle:

$$\boldsymbol{Q} = \begin{pmatrix} \times & & & & & \times \\ & \times & & & & \times \\ & & \times & & & \times \\ & & & \times & & \times \\ & & & & \times & \times \\ \times & \times & \times & \times & \times & \times \end{pmatrix}, \quad \boldsymbol{L} = \begin{pmatrix} \times & & & & & \\ & \times & & & & \\ & & \times & & & \\ & & & \times & & \\ & & & & \times & \\ \times & \times & \times & \times & \times & \times \end{pmatrix}. \tag{2.38}$$

The situation is now quite different, the fill-in is zero, and we can factorize $\boldsymbol{Q}$ in only $\mathcal{O}(n)$ flops. The remarkable difference to (2.37) is that conditioning on node 7 in Figure 2.7(b) makes all other nodes conditionally independent.

The idea in this example generalizes as follows to determine a good ordering with less fill-in:

- Select a (small) set of nodes whose removal divides the graph into two disconnected subgraphs of almost equal size.
- Order the nodes chosen *after* ordering all the nodes in both subgraphs.
- Apply this procedure recursively to the nodes in each subgraph.

This is the idea of reordering based on *nested dissection*. To demonstrate how this applies to our current example with the graph in Figure 2.4(b), we computed such a (slightly modified) ordering (from white to black) as shown in Figure 2.8. Section 2.5 give details of what algorithms are used for computing the reordering. We see here the idea in practice, first the map is divided into two, then these two parts are divided further, etc. At some stage the recursion is stopped. The reordered precision matrix and its Cholesky triangle are shown in Figure 2.8(b) and (c), where the fill-in is 2866. This corresponds to a fill-in ratio of $R = 2.5$. This is to be compared to $R = 5.3$ for the band reordering shown in Figure 2.5.

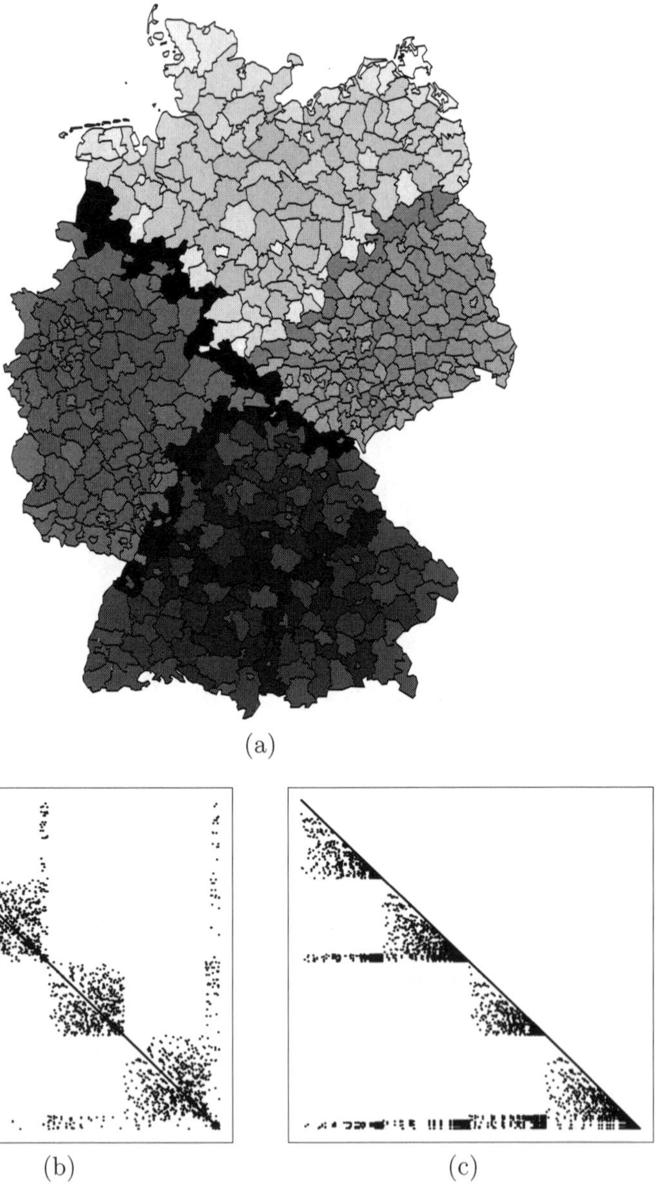

Figure 2.8 *Figure (a) displays the ordering found using a nested dissection algorithm where the ordering is from white to black. (b) displays the reordered precision matrix and (c) its Cholesky triangle. In (b) and (c) only the nonzero elements are shown and those are indicated by a dot.*

Algorithms based on nested dissection ordering require $\mathcal{O}(n^{3/2})$ flops in such and similar cases and gives $\mathcal{O}(n \log n)$ fill-ins, see George and Liu (1981, Ch. 8) for details. This is $\sqrt{n}$ faster than the band-Cholesky approach, but the difference is not that large for many problems of reasonable size. The next section contains an empirical comparison for various 'typical' graphs.

George and Liu (1981, Ch. 8) proves also that *any* reordering would require at least $\mathcal{O}(n^{3/2})$ flops for the factorization and produce at least $\mathcal{O}(n \log n)$ fill-ins for a $\sqrt{n} \times \sqrt{n}$ lattice with a local neighborhood. Hence, the nested dissection reordering is optimal in the order of magnitude sense.

The band-Cholesky factorization approach is quite simple, intuitive, and requires only trivial changes to Algorithm 2.9. Implementation thus requires only a few (hundreds) lines of code, apart from the reordering itself that is somewhat more demanding. More general factorizations like factorizing Figure 2.8(b) to get Figure 2.8(c) efficiently, require substantial knowledge of numerical algorithms, data structures and high-performance computing. The corresponding libraries easily require 10,000 lines of code. The factorizing is usually performed in several steps:

1. A *reordering phase*, where the sparse matrix $\boldsymbol{Q}$ is analyzed to produce a suitable ordering with reduced fill-in.
2. A *symbolical factorization* phase, where (informally) the (possible) nonzero pattern of $\boldsymbol{L}$ is determined and data structures to compute the factorization are constructed.
3. A *numerical factorization* phase, where the numerical values of $\boldsymbol{L}$ are computed.
4. A *solve phase* in which $\boldsymbol{L}^T\boldsymbol{x} = \boldsymbol{z}$ and/or $\boldsymbol{L}\boldsymbol{v} = \boldsymbol{b}$ is solved.

The results from step 1 and 2 can be reused if we factorize several $\boldsymbol{Q}$'s with the same nonzero pattern; This is typical for applications of GMRFs in MCMC algorithms. The data handling problem in such algorithms is significant and it is important to implement this well (step 2) to gain efficiency. We can illustrate this as follows. In the band-Cholesky factorization, we only need to store the $(p+1) \times n$ rectangle, as it contains all nonzeros in $\boldsymbol{L}$, and all loops in Algorithm 2.9 run over this rectangle. If the nonzeros terms of $\boldsymbol{L}$ are spread out, we need to use indirect addressing and loops like

```
for i=1, M
      x(indx(i)) = x(indx(i)) + a*w(i)
endfor
```

There is severe potential of loss of performance using indirect addressing, but clever data handling and data structures can prevent or reduce degradation loss.

In summary, we recommend leaving the issue of constructing and implementing algorithms for factorizing sparse matrices to the numerical and computer science experts. However, statisticians should use their results and libraries for efficient statistical computing. We end this section quoting from Gupta (2002) who summarizes his findings on recent advances for sparse linear solvers:

> *...recent sparse solvers have significantly improved the state of the art of the direct solution of general sparse systems.*

> *...recent years have seen some remarkable advances in the general sparse direct-solver algorithms and software.*

2.5 A numerical case study of typical GMRFs

We will now present a small case study using the algorithms from Section 2.4 on a set of typical GMRFs. The aim is to verify empirically the computational requirements of various algorithms, and to gain some experience in choosing algorithms for different kinds of problems.

As it will become clear in later sections, applications of GMRF models can often be divided into three categories.

1. GMRF models in time or on a line. This includes autoregressive models and models for smooth functions. Neighbors to x_t are then those $\{x_s\}$ such that $|s-t| \leq p$.
2. Spatial GMRF models. Here, the graph is either a regular lattice, or irregular induced by a tessellation or by regions. Neighbors to x_i (spatial index i), are those j spatially 'close' to i, where 'close' is defined from its context.
3. Spatiotemporal GMRF models. These are often appropriate extensions of temporal or spatial models.

We also include the case where additional nodes depend on all other nodes. This occurs in many situations like the following. Let $\boldsymbol{x}$ be a GMRF with a common mean μ, then

$$\boldsymbol{x} \mid \mu \sim \mathcal{N}(\mu\mathbf{1}, \boldsymbol{Q}^{-1}),$$

where $\boldsymbol{Q}$ is sparse. Assume $\mu \sim \mathcal{N}(0, \sigma^2)$, then $(\boldsymbol{x}, \mu)$ is also a GMRF where the node μ is a neighbor of all x_i's.

We apply two different algorithms in our test:

1. The band-Cholesky factorization (BCF) as in Algorithm 2.9. Here we use the LAPACK routines `DPBTRF` and `DTBSV` for the factorization and the forward- or back-substitution, respectively, and the Gibbs-Poole-Stockmeyer algorithm for bandwidth reduction as implemented in Lewis (1982).

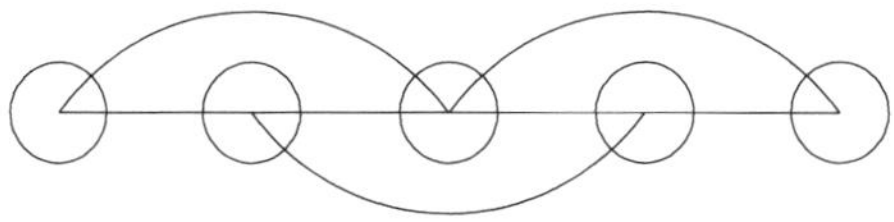

Figure 2.9 *The graph for an autoregressive process with $n = 5$ and $p = 2$.*

2. The multifrontal supernodal Cholesky factorization (MSCF) implementation in the library `TAUCS` (version 2.0) (Toledo et al., 2002) using the nested dissection reordering from the library `METIS` (Karypis and Kumar, 1998).

Both algorithms are available in the `GMRFLib` library described in Appendix B used throughout the book. We use plain LAPACK and BLAS libraries compiled from scratch. Improved performance can be obtained by replacing these with either vendor-supplied BLAS libraries (if available) or libraries from the ATLAS (Automatically Tuned Linear Algebra Software) project, see `http://math-atlas.sourceforge.net/`.

The tasks we want to investigate are:

1. Factorizing $\boldsymbol{Q}$ into $\boldsymbol{L}\boldsymbol{L}^T$
2. Solving $\boldsymbol{L}\boldsymbol{L}^T\boldsymbol{\mu} = \boldsymbol{b}$, i.e., first solving $\boldsymbol{L}\boldsymbol{w} = \boldsymbol{b}$ and then $\boldsymbol{L}^T\boldsymbol{\mu} = \boldsymbol{w}$.

To produce a random sample from a GMRF we need to solve $\boldsymbol{L}^T\boldsymbol{x} = \boldsymbol{z}$, so the cost is half the cost of solving the linear system in step 2 if the factorizing is known. All tests reported here have been conducted on a 1200-MHz CPU.

2.5.1 GMRF models in time

Let $\boldsymbol{Q}$ be a band matrix with bandwidth p and dimension n. This corresponds to an autoregressive process of order p as discussed in Example 2.7 for $p = 1$. The graph of an autoregressive process with $n = 5$ and $p = 2$ is shown in Figure 2.9. For such a problem, using BCF will be (theoretically) optimal with zero fill-in.

Table 2.1 reports the average CPU time (in seconds) used (using 10 replications) for n equals 10^3, 10^4, 10^5, and p equals 5 and 25. The results obtained are quite impressive, which is often the case for 'long and thin' problems. Computing the factorization and solving the system requires np^2 and $2np$ flops, respectively. This theoretical behavior is approximately supported from the results, as small bandwidth makes loops shorter, which gives a reduction in performance. The MSCF is less optimal for band matrices. For $p = 5$ the factorization is about 25 times slower, and the solving step is about 2 to 3 times slower, compared to Table 2.1. This is due to a fill-in of about $n/2$ and because more complicated data structures than needed are used.

	$n = 10^3$		$n = 10^4$		$n = 10^5$	
CPU time	$p = 5$	$p = 25$	$p = 5$	$p = 25$	$p = 5$	$p = 25$
Factorize	0.0005	0.0019	0.0044	0.0271	0.0443	0.2705
Solve	0.0000	0.0004	0.0031	0.0109	0.0509	0.1052

Table 2.1 *The average CPU time for (in seconds) factorizing* $\boldsymbol{Q}$ *into* $\boldsymbol{L}\boldsymbol{L}^T$ *and solving* $\boldsymbol{L}\boldsymbol{L}^T\boldsymbol{\mu} = \boldsymbol{b}$, *for a band matrix of order* n *and bandwidth* p, *using band-Cholesky factorization and band forward- or back-substitution.*

We now add 10 additional nodes, which are neighbors with all others. This makes the bandwidth maximal so the BCF is not a good choice. Using MSCF we have obtained the results shown in Table 2.2. The fill-in is now approximately pn, which is due to the nested dissection ordering used. In this particular case we can compare the result with the optimal reordering giving no fill-ins. This is obtained by placing the 10 global nodes after the n others so we obtain a nonzero structure as in (2.38). With this optimal ordering we obtain a speedup up to about 1.5 for $p = 25$ and slightly less for $p = 5$. However, the effect of not choosing the optimal ordering is not dramatic if the ordering chosen is 'reasonable'. The nested dissection ordering gives good results in all cases considered so far, which will also become clear from the spatial examples shown next.

	$n = 10^3$		$n = 10^4$		$n = 10^5$	
CPU time	$p = 5$	$p = 25$	$p = 5$	$p = 25$	$p = 5$	$p = 25$
Factorize	0.0119	0.0335	0.1394	0.4085	1.6396	4.1679
Solve	0.0007	0.0035	0.0138	0.0306	0.1541	0.3078

Table 2.2 *The average CPU time (in seconds) for factorizing* $\boldsymbol{Q}$ *into* $\boldsymbol{L}\boldsymbol{L}^T$ *and solving* $\boldsymbol{L}\boldsymbol{L}^T\boldsymbol{\mu} = \boldsymbol{b}$, *for a band matrix of order* n, *bandwidth* p *with additional* 10 *nodes that are neighbors to all others. The factorization routine is MSCF.*

2.5.2 Spatial GMRF models

Spatial applications of GMRF models have graphs that are typically either a regular or irregular lattice. Two such examples are provided in Figure 2.10. Figure (a) shows a realization of a GMRF on a regular lattice used as an approximation to a Gaussian field with given correlation function (here the exponential). The neighbors to pixel i are those 24 pixels in a 5×5 window centered at i, illustrated in Figure 2.11(b). We will discuss such approximations in Section 5.1. Figure (b) shows the Dirichlet tessellation found from randomly distributed points on the

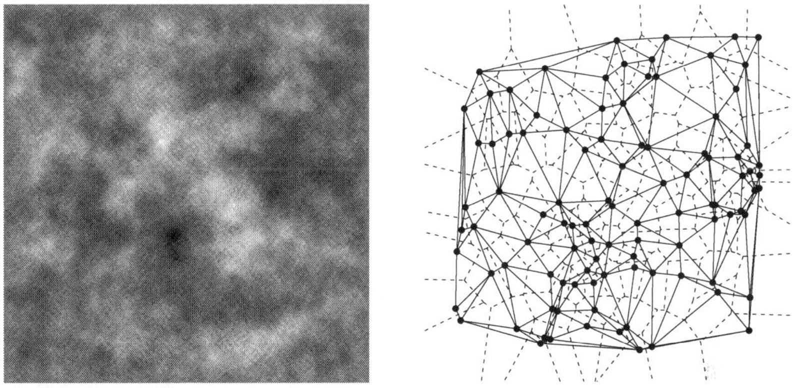

(a) (b)

Figure 2.10 *Two examples of spatial GMRF models; (a) shows a GMRF on a lattice used as an approximation to a Gaussian field with an exponential correlation function, (b) the graph found from Delaunay triangulation of a planar point set.*

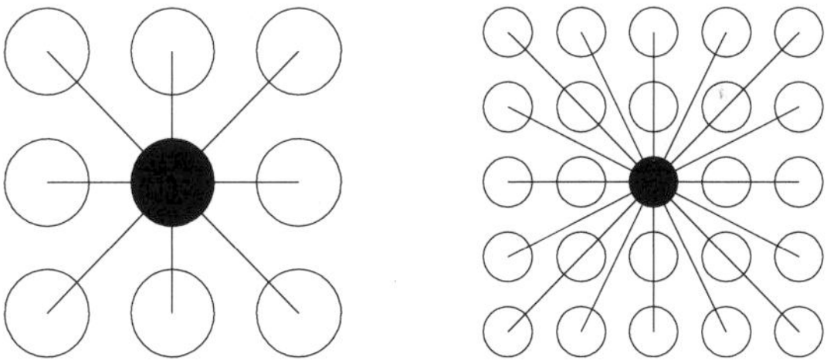

(a) (b)

Figure 2.11 *The neighbors to the black pixel; (a) the* 3×3 *neighborhood system and (b) the* 5×5 *neighborhood system.*

unit square. If adjacent tiles are neighbors, then we obtain the graph found by the Delaunay triangulation of the points. This graph is similar to the one defined by regions of Germany in Figure 2.4, although here, the outline and the configuration of the regions are random as well.

We will report some timing results for lattices only, as they are similar for irregular lattices. The neighbors to a pixel i, will be those 8 (24) in the 3×3 (5×5) window centered at i, and the dimension of the lattice will be 100×100, 150×150, and 200×200. Table 2.3 summarizes our results. The speed of the algorithms is again impressive. The performance in the solve part is quite similar, but for the largest lattice the MSCF really outperform the BCF. The reason is the $\mathcal{O}(n^{3/2})$ cost for MSCF compared to $\mathcal{O}(n^2)$ for the BCF, which is of clear importance for large lattices. For

CPU time	Method	$n = 100^2$ 3×3	$n = 100^2$ 5×5	$n = 150^2$ 3×3	$n = 150^2$ 5×5	$n = 200^2$ 3×3	$n = 200^2$ 5×5
Factorize	BCF	0.51	1.02	2.60	4.93	13.30	38.12
	MSCF	0.17	0.62	0.55	1.92	1.91	4.90
Solve	BCF	0.03	0.05	0.10	0.16	0.24	0.43
	MSCF	0.01	0.04	0.04	0.11	0.08	0.21

Table 2.3 *The average CPU time (in seconds) for factorizing $\boldsymbol{Q}$ into $\boldsymbol{L}\boldsymbol{L}^T$ and solving $\boldsymbol{L}\boldsymbol{L}^T\boldsymbol{\mu} = \boldsymbol{b}$, for a 100^2, 150^2 and 200^2 square lattice with 3×3 and 5×5 neighborhood, using the BCF and MSCF method.*

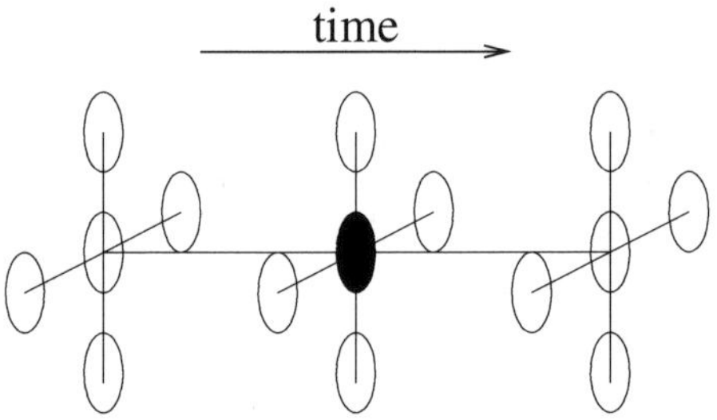

Figure 2.12 *A common neighborhood structure in spatiotemporal GMRF models. In addition to spatial neighbors, also the same node in next and previous time-step can be neighbors.*

large lattices, we need to consider also the memory requirement. MSCF has lower memory requirement than BCF, about $\mathcal{O}(n \log n)$ compared to $\mathcal{O}(n^{3/2})$. The consequence is that BCF runs into memory problems just over $n = 200^2$ while the MSCF runs smoothly until $n = 350^2$ or so, for our machine with 512 Mb memory.

We now add additional 10 nodes that are neighbors to all others, and repeat the test for MSCF and 5×5 neighborhood. For the factorization, we obtained 0.73, 2.40, and 5.38. The solve-part was nearly unchanged. There is not that much extra computational costs adding global nodes.

2.5.3 Spatiotemporal GMRF models

Spatiotemporal GMRF models are extensions of spatial GMRF models to account for additional temporal variation. A typical situation is shown in Figure 2.12. Think of a sequence of T graphs in time, like the graph in Figure 2.4(b). Fix one slice and one node in it. Let this node, x_{it}, say, be the black node in Figure 2.12. Its spatial neighbors are those spatially close $\{x_{jt} \ : \ j \sim i\}$, here shown schematically using 4 neighbors. A common extension to a spatiotemporal GMRF is to take additional neighbors in time into account, like the same node in the next and previous slices; that is, $x_{i,t-1}$ and $x_{i,t+1}$. The results presented in Table 2.4 use this model and the graph of Germany in Figure 2.4(b), for $T = 10$ and $T = 100$, with and without additional 10 nodes that are neighbors to all others. The results show a quite heavy dependency on T. This is

	Without global nodes		With 10 global nodes	
CPU time	$T = 10$	$T = 100$	$T = 10$	$T = 100$
Factorize	0.25	39.96	0.31	39.22
Solve	0.02	0.42	0.02	0.42

Table 2.4 *The average CPU time (in seconds) using the MSCF algorithm for factorizing* $\boldsymbol{Q}$ *into* $\boldsymbol{L}\boldsymbol{L}^T$ *and solving* $\boldsymbol{L}\boldsymbol{L}^T\boldsymbol{\mu} = \boldsymbol{b}$, *for the spatiotemporal GMRF using* T *time steps, and with and without* 10 *global nodes. The dimension of the graph is* $544 \times T$.

due to the denser structure of $\boldsymbol{Q}$ due to the dependency both in space and time. For a cube with n nodes, the MSCF requires $\mathcal{O}(n^2)$ flops to factorize, with neighbors similar to Figure 2.12.

2.6 Stationary GMRFs*

Stationary GMRFs are a special class of GMRFs obtained under rather strong assumptions on both the graph $\mathcal{G}$ and the elements in $\boldsymbol{Q}$. The graph is most often a torus $\mathcal{T}_{\boldsymbol{n}}$ (see Section 2.1.2) and the full conditionals $\{\pi(x_i|\boldsymbol{x}_{-i})\}$ have constant parameters not depending on i. This makes $\boldsymbol{Q}$ a (block) *circulant matrix*, for which nice analytical results about the eigenstructure are available. The practical advantage is that typical operations on such GMRFs can be done using the discrete Fourier transform (DFT). The computational complexity of typical operations is then $\mathcal{O}(n \log n)$ and does not depend on the number of neighbors. Circulant matrices are also well adapted for theoretical studies and we will later use them in Section 2.7.

We will first discuss circulant matrices in general and then apply the results obtained to GMRFs. At the end of this section we will extend these results to scenarios under slightly less strict assumptions. In particular, we will discuss matrices of Toeplitz form and show that they can be approximated by circulant matrices.

2.6.1 Circulant matrices

We will now present some analytical results for (real) *circulant matrices*. For notational convenience, we will change our notation in this section slightly and denote the elements of a vector of length n by the indices $0, \ldots, n-1$, and the elements of an $n \times n$ matrix by the indices $(0,0), \ldots, (n-1, n-1)$.

Circulant matrices have the property that its eigenvalues and eigenvectors are related to the discrete Fourier transform. This allows for fast algorithms operating on circulant matrices, obtaining their inverse, computing the product of two circulant matrices, and so on.

Definition 2.4 (Circulant matrix) *An $n \times n$ matrix $\boldsymbol{C}$ is* circulant *iff it has the form*

$$\boldsymbol{C} = \begin{pmatrix} c_0 & c_1 & c_2 & \cdots & c_{n-1} \\ c_{n-1} & c_0 & c_1 & \cdots & c_{n-2} \\ c_{n-2} & c_{n-1} & c_0 & \cdots & c_{n-3} \\ \vdots & \vdots & \vdots & & \vdots \\ c_1 & c_2 & c_3 & \cdots & c_0 \end{pmatrix} = \left(c_{j-i \bmod n}\right)$$

for some vector $\mathbf{c} = (c_0, c_1, \ldots, c_{n-1})^T$. We call $\mathbf{c}$ the base *of $\boldsymbol{C}$.*

A circulant matrix is hence fully specified by only one column or one row.

The eigenvalues and eigenvectors of a circulant matrix $\boldsymbol{C}$ play a central role. Any eigenvalue λ and eigenvector $\boldsymbol{e}$ of $\boldsymbol{C}$ is a solution of the equation $\boldsymbol{Ce} = \lambda \boldsymbol{e}$. This can be written row by row as n difference equations,

$$\sum_{i=0}^{j-1} c_{n-j+i} e_i + \sum_{i=j}^{n-1} c_{i-j} e_i = \lambda e_j$$

for $j = 0, \ldots, n-1$ and is equivalent to

$$\sum_{i=0}^{n-1-j} c_i e_{i+j} + \sum_{i=n-j}^{n-1} c_i e_{i-(n-j)} = \lambda e_j. \tag{2.39}$$

These linear difference equations have constant coefficients, so we 'guess' that a solution has the form $e_j \propto \rho^j$ for some complex scalar ρ. We will now verify that this is indeed the case.

For $e_j \propto \rho^j$, equation (2.39) reduces to

$$\sum_{i=0}^{n-1-j} c_i \rho^i + \rho^{-n} \sum_{i=n-j}^{n-1} c_i \rho^i = \lambda.$$

If we choose ρ such that $\rho^{-n} = 1$, then

$$\lambda = \sum_{i=0}^{n-1} c_i \rho^i \tag{2.40}$$

and

$$e = \frac{1}{\sqrt{n}}(1, \rho, \rho^2, \ldots, \rho^{n-1})^T. \tag{2.41}$$

The factor $\sqrt{n}$ appears in (2.41) because we require that $e^T e = 1$. The n nth roots of 1 are $\{\exp(2\pi\iota\, j/n), j = 0, \ldots, n-1\}$ where $\iota = \sqrt{-1}$. The jth eigenvalue is found using (2.40) for each nth root of 1,

$$\lambda_j = \sum_{i=0}^{n-1} c_i \exp(-2\pi\iota\, ij/n)$$

and the corresponding jth eigenvector is

$$e_j = \frac{1}{\sqrt{n}}(1, \exp(-2\pi\iota\, j/n), \exp(-2\pi\iota\, j2/n), \ldots, \exp(-2\pi\iota\, j(n-1)/n))^T$$

for $j = 0, \ldots, n-1$.

We now define the eigenvector matrix,

$$\begin{aligned} F &= \left(e_0 \mid e_1 \mid \ldots \mid e_{n-1}\right) \\ &= \frac{1}{\sqrt{n}} \begin{pmatrix} 1 & 1 & 1 & \ldots & 1 \\ 1 & \omega^1 & \omega^2 & \ldots & \omega^{n-1} \\ 1 & \omega^2 & \omega^4 & \ldots & \omega^{2(n-1)} \\ \vdots & \vdots & \vdots & & \vdots \\ 1 & \omega^{n-1} & \omega^{2(n-1)} & \ldots & \omega^{(n-1)(n-1)} \end{pmatrix} \end{aligned} \tag{2.42}$$

where $\omega = \exp(-2\pi\iota/n)$. Note that F does not depend on c. Furthermore, let Λ be a diagonal matrix containing the eigenvalues,

$$\Lambda = \operatorname{diag}(\lambda_0, \lambda_1, \ldots, \lambda_{n-1}).$$

Note that F is unitary, i.e., $F^{-1} = F^H$, where F^H is the conjugate transpose of F and

$$\Lambda = \sqrt{n}\, \operatorname{diag}(Fc).$$

We can verify that $C = F\Lambda F^H$ by a direct calculation:

$$\begin{aligned} C_{ij} &= \frac{1}{n}\sum_{k=0}^{n-1} \exp(2\pi\iota\, k(j-i)/n)\, \lambda_k \\ &= \frac{1}{n}\sum_{k=0}^{n-1} \exp(2\pi\iota\, k(j-i)/n) \sum_{l=0}^{n-1} c_l \exp(-2\pi\iota\, kl/n) \\ &= \frac{1}{n}\sum_{l=0}^{n-1} c_l \sum_{k=0}^{n-1} \exp(2\pi\iota\, k(j-i-l)/n). \end{aligned} \tag{2.43}$$

Using

$$\sum_{k=0}^{n-1} \exp(2\pi\iota\, k(j-i-l)/n) = \begin{cases} n & \text{if } i-j=-l \bmod n \\ 0 & \text{otherwise} \end{cases}$$

we obtain $C_{ij} = c_{j-i \bmod n}$, i.e., all circulant matrices can be expressed as $\boldsymbol{F}\boldsymbol{\Lambda}\boldsymbol{F}^H$ for some diagonal matrix $\boldsymbol{\Lambda}$.

The following theorem now states that the class of circulant matrices is closed under some matrix operations.

Theorem 2.10 *Let $\boldsymbol{C}$ and $\boldsymbol{D}$ be $n \times n$ circulant matrices. Then*

1. *$\boldsymbol{C}$ and $\boldsymbol{D}$ commute, i.e., $\boldsymbol{CD} = \boldsymbol{DC}$, and $\boldsymbol{CD}$ is circulant*
2. *$\boldsymbol{C} \pm \boldsymbol{D}$ is circulant*
3. *$\boldsymbol{C}^p$ is circulant, $p = 1, 2, \ldots$*
4. *if $\boldsymbol{C}$ is non singular then $\boldsymbol{C}^p$ is circulant, $p = -1, -2, \ldots$*

Proof. Recall that a circulant matrix is uniquely described by its n eigenvalues as they all share the same eigenvectors. Let $\boldsymbol{\Lambda}_{\boldsymbol{C}}$ and $\boldsymbol{\Lambda}_{\boldsymbol{D}}$ denote the diagonal matrices with the eigenvalues of $\boldsymbol{C}$ and $\boldsymbol{D}$, respectively, on the diagonal. Then

$$\begin{aligned} \boldsymbol{CD} &= \boldsymbol{F}\boldsymbol{\Lambda}_{\boldsymbol{C}}\boldsymbol{F}^H\, \boldsymbol{F}\boldsymbol{\Lambda}_{\boldsymbol{D}}\boldsymbol{F}^H \\ &= \boldsymbol{F}(\boldsymbol{\Lambda}_{\boldsymbol{C}}\boldsymbol{\Lambda}_{\boldsymbol{D}})\boldsymbol{F}^H; \end{aligned}$$

hence $\boldsymbol{CD}$ is circulant with eigenvalues $\{\lambda_{\boldsymbol{C}i}\lambda_{\boldsymbol{D}i}\}$. The matrices commute since $\boldsymbol{\Lambda}_{\boldsymbol{C}}\boldsymbol{\Lambda}_{\boldsymbol{D}} = \boldsymbol{\Lambda}_{\boldsymbol{D}}\boldsymbol{\Lambda}_{\boldsymbol{C}}$ for diagonal matrices. Using the same argument,

$$\boldsymbol{C} \pm \boldsymbol{D} = \boldsymbol{F}(\boldsymbol{\Lambda}_{\boldsymbol{D}} \pm \boldsymbol{\Lambda}_{\boldsymbol{C}})\boldsymbol{F}^H;$$

hence $\boldsymbol{C} \pm \boldsymbol{D}$ is circulant. Similarly,

$$\boldsymbol{C}^p = \boldsymbol{F}\boldsymbol{\Lambda}_{\boldsymbol{C}}^p\boldsymbol{F}^H, \quad p = \pm 1, \pm 2, \ldots$$

as $\boldsymbol{\Lambda}_{\boldsymbol{C}}^p$ is a diagonal matrix. □

The matrix $\boldsymbol{F}$ in (2.42) is well know as the discrete Fourier transform (DFT) matrix, so computing $\boldsymbol{Fv}$ for a vector $\boldsymbol{v}$ is the same as computing the DFT of $\boldsymbol{v}$. Taking the inverse DFT (IDFT) of $\boldsymbol{v}$ is the same as calculating $\boldsymbol{F}^H\boldsymbol{v}$. Note that if n can be factorized as a product of small primes, the computation of $\boldsymbol{Fv}$ requires only $\mathcal{O}(n \log n)$ flops. 'Small primes' is the 'traditional' requirement, but the (superb) library FFTW, which is a comprehensive collection of fast `C` routines for computing the discrete Fourier transform (`http://www.fftw.org`), allows arbitrary size and employs $\mathcal{O}(n \log n)$ algorithms for all sizes. Small primes are still computational most efficient.

The link to the DFT is useful for computing with circulant matrices. Define the DFT and IDFT of $\boldsymbol{v}$ as

$$\mathrm{DFT}(\boldsymbol{v}) = \boldsymbol{F}\boldsymbol{v} = \frac{1}{\sqrt{n}} \begin{pmatrix} \sum_{j=0}^{n-1} v_j \omega^{j0} \\ \sum_{j=0}^{n-1} v_j \omega^{j1} \\ \vdots \\ \sum_{j=0}^{n-1} v_j \omega^{j(n-1)} \end{pmatrix}$$

and

$$\mathrm{IDFT}(\boldsymbol{v}) = \boldsymbol{F}^H \boldsymbol{v} = \frac{1}{\sqrt{n}} \begin{pmatrix} \sum_{j=0}^{n-1} v_j \omega^{-j0} \\ \sum_{j=0}^{n-1} v_j \omega^{-j1} \\ \vdots \\ \sum_{j=0}^{n-1} v_j \omega^{-j(n-1)} \end{pmatrix}.$$

Recall that '$\odot$' denotes elementwise multiplication, '$\oslash$' denotes elementwise division, and '$\circledwedge$' is elementwise power, see Section 2.1.1. Let $\boldsymbol{C}$ be a circulant matrix with base $\boldsymbol{c}$, then the matrix-vector product $\boldsymbol{C}\boldsymbol{v}$ can be computed as

$$\begin{aligned} \boldsymbol{C}\boldsymbol{v} &= \boldsymbol{F}\boldsymbol{\Lambda}\boldsymbol{F}^H\boldsymbol{v} \\ &= \boldsymbol{F}\,\sqrt{n}\,\mathrm{diag}(\boldsymbol{F}\boldsymbol{c})\,\boldsymbol{F}^H\boldsymbol{v} \\ &= \sqrt{n}\,\mathrm{DFT}(\mathrm{DFT}(\boldsymbol{c}) \odot \mathrm{IDFT}(\boldsymbol{v})). \end{aligned}$$

The product of two circulant matrices $\boldsymbol{C}$ and $\boldsymbol{D}$, with base $\boldsymbol{c}$ and $\boldsymbol{d}$, respectively, can be written as

$$\begin{aligned} \boldsymbol{C}\boldsymbol{D} &= \boldsymbol{F}\,(\boldsymbol{\Lambda}_{\boldsymbol{C}}\boldsymbol{\Lambda}_{\boldsymbol{D}})\,\boldsymbol{F}^H & (2.44) \\ &= \boldsymbol{F}\left(\sqrt{n}\,\mathrm{diag}(\boldsymbol{F}\boldsymbol{c})\sqrt{n}\,\mathrm{diag}(\boldsymbol{F}\boldsymbol{d})\right)\boldsymbol{F}^H. & (2.45) \end{aligned}$$

Since $\boldsymbol{C}\boldsymbol{D}$ is a circulant matrix with (unknown) base $\boldsymbol{p}$, say, then

$$\boldsymbol{C}\boldsymbol{D} = \boldsymbol{F}\left(\sqrt{n}\,\mathrm{diag}(\boldsymbol{F}\boldsymbol{p})\right)\boldsymbol{F}^H. \tag{2.46}$$

Comparing (2.46) and (2.44), we see that

$$\sqrt{n}\,\mathrm{diag}(\boldsymbol{F}\boldsymbol{p}) = \sqrt{n}\,\mathrm{diag}(\boldsymbol{F}\boldsymbol{c})\sqrt{n}\,\mathrm{diag}(\boldsymbol{F}\boldsymbol{d});$$

hence

$$\boldsymbol{p} = \sqrt{n}\,\mathrm{IDFT}\left(\mathrm{DFT}(\boldsymbol{c}) \odot \mathrm{DFT}(\boldsymbol{d})\right).$$

Solving $\boldsymbol{C}\boldsymbol{x} = \boldsymbol{b}$ can be done similarly, since

$$\begin{aligned} \boldsymbol{x} &= \boldsymbol{C}^{-1}\boldsymbol{b} \\ &= \boldsymbol{F}\boldsymbol{\Lambda}^{-1}\boldsymbol{F}^H\boldsymbol{b} \\ &= \frac{1}{\sqrt{n}}\,\mathrm{DFT}(\mathrm{IDFT}(\boldsymbol{b})) \oslash \mathrm{DFT}(\boldsymbol{c})). \end{aligned}$$

The inverse of $\boldsymbol{C}$ is

$$\boldsymbol{C}^{-1} = \boldsymbol{F}\boldsymbol{\Lambda}^{-1}\boldsymbol{F}^{H};$$

hence the base of $\boldsymbol{C}^{-1}$ is

$$\frac{1}{n}\mathrm{IDFT}(\mathrm{DFT}(\boldsymbol{c}) \oslash (-1)).$$

2.6.2 Block-circulant matrices

A natural generalization of circulant matrices are block-circulant matrices. These matrices share the same properties as circulant matrices. Algorithms for block-circulant matrices extend easily from those for circulant matrices by, loosely speaking, replacing the discrete Fourier transform with the two-dimensional discrete Fourier transform. Block-circulant matrices are central for stationary GMRFs defined on a torus, as we will see later.

Definition 2.5 (Block-circulant matrix) *An $Nn \times Nn$ matrix $\boldsymbol{C}$ is* block circulant *with $N \times N$ blocks, iff it can be written as*

$$\boldsymbol{C} = \begin{pmatrix} \boldsymbol{C}_0 & \boldsymbol{C}_1 & \boldsymbol{C}_2 & \dots & \boldsymbol{C}_{N-1} \\ \boldsymbol{C}_{N-1} & \boldsymbol{C}_0 & \boldsymbol{C}_1 & \dots & \boldsymbol{C}_{N-2} \\ \boldsymbol{C}_{N-2} & \boldsymbol{C}_{N-1} & \boldsymbol{C}_0 & \dots & \boldsymbol{C}_{N-3} \\ \vdots & \vdots & \vdots & & \vdots \\ \boldsymbol{C}_1 & \boldsymbol{C}_2 & \boldsymbol{C}_3 & \dots & \boldsymbol{C}_0 \end{pmatrix} = \left(\boldsymbol{C}_{j-i \bmod N}\right)$$

where $\boldsymbol{C}_i$ is a circulant $n \times n$ matrix with base $\boldsymbol{c}_i$. The base of $\boldsymbol{C}$ is the $n \times N$ matrix

$$\boldsymbol{c} = \begin{pmatrix} \boldsymbol{c}_0 & \boldsymbol{c}_1 & \dots & \boldsymbol{c}_{N-1} \end{pmatrix}.$$

A block-circulant matrix is fully specified by one block column, one block row, or the base. The elements of $\boldsymbol{C}$ are defined by the base $\boldsymbol{c}$; element (k, l) in block (i, j) of $\boldsymbol{C}$, is element $l - k \bmod n$ of base $\boldsymbol{c}_{j-i \bmod N}$.

To compute the eigenvalues and eigenvectors of $\boldsymbol{C}$, we will use results from Section 2.6.1. Let $\boldsymbol{F}_n$ and $\boldsymbol{F}_N$ be the eigenvector matrix as defined in (2.42) where the subscript denotes the dimension. As each circulant matrix is diagonalized by $\boldsymbol{F}_n$ (i.e. $\boldsymbol{C}_i = \boldsymbol{F}_n\boldsymbol{\Lambda}_i\boldsymbol{F}_n^H$, where $\boldsymbol{\Lambda}_i = \sqrt{n}\,\mathrm{diag}(\boldsymbol{F}_n\boldsymbol{c}_i)$), we see that

$$\begin{aligned} \boldsymbol{C} &= \begin{pmatrix} \boldsymbol{F}_n & & \\ & \ddots & \\ & & \boldsymbol{F}_n \end{pmatrix} \begin{pmatrix} \boldsymbol{\Lambda}_0 & \dots & \boldsymbol{\Lambda}_{N-1} \\ \vdots & \ddots & \vdots \\ \boldsymbol{\Lambda}_1 & \dots & \boldsymbol{\Lambda}_0 \end{pmatrix} \begin{pmatrix} \boldsymbol{F}_n^H & & \\ & \ddots & \\ & & \boldsymbol{F}_n^H \end{pmatrix} \\ &= \boldsymbol{F}_n^N \boldsymbol{\Lambda} \left(\boldsymbol{F}_n^N\right)^H \end{aligned}$$

with obvious notation. Each $\boldsymbol{\Lambda}_i$ is diagonal so a symmetric permutation of rows and columns in $\boldsymbol{\Lambda}$ will result in a block-diagonal matrix with circulant blocks. Let $\boldsymbol{P}$ be the permutation matrix that takes the ith row of the block row j to the jth row of the block row i. For example, for $n = N = 3$,

$$\boldsymbol{P} = \left(\begin{array}{ccccccccc} 1 & & & & & & & & \\ & & & 1 & & & & & \\ & & & & & & 1 & & \\ \hline & 1 & & & & & & & \\ & & & & 1 & & & & \\ & & & & & & & 1 & \\ \hline & & 1 & & & & & & \\ & & & & & 1 & & & \\ & & & & & & & & 1 \end{array}\right).$$

Then $\boldsymbol{P}\,\boldsymbol{P} = \boldsymbol{I}$ and

$$\boldsymbol{P}\,\boldsymbol{\Lambda}\,\boldsymbol{P} = \begin{pmatrix} \boldsymbol{D}_0 & & & \\ & \boldsymbol{D}_1 & & \\ & & \ddots & \\ & & & \boldsymbol{D}_n \end{pmatrix} = \boldsymbol{D},$$

where $\boldsymbol{D}_i$ is a circulant matrix. The jth element of $\boldsymbol{d}_i$, the base of $\boldsymbol{D}_i$, is the ith diagonal element of $\boldsymbol{\Lambda}_j$, so

$$\begin{pmatrix} \boldsymbol{d}_0 \\ \boldsymbol{d}_1 \\ \vdots \\ \boldsymbol{d}_{n-1} \end{pmatrix} = \sqrt{n}\,\boldsymbol{P} \begin{pmatrix} \boldsymbol{F}_n\boldsymbol{c}_0 \\ \boldsymbol{F}_n\boldsymbol{c}_1 \\ \vdots \\ \boldsymbol{F}_n\boldsymbol{c}_{N-1} \end{pmatrix}.$$

Since $\boldsymbol{D}_i$ is diagonalized by $\boldsymbol{F}_N$ (i.e. $\boldsymbol{D}_i = \boldsymbol{F}_N\boldsymbol{\Gamma}_i\boldsymbol{F}_N^H$, where $\boldsymbol{\Gamma}_i = \sqrt{N}\,\mathrm{diag}(\boldsymbol{F}_N\boldsymbol{d}_i)$), we obtain

$$\boldsymbol{C} = \left(\boldsymbol{F}_n^N\,\boldsymbol{P}\,\boldsymbol{F}_N^n\right)\,\boldsymbol{\Gamma}\,\left((\boldsymbol{F}_N^n)^H\,\boldsymbol{P}\,(\boldsymbol{F}_n^N)^H\right), \tag{2.47}$$

where $\boldsymbol{F}_N^n = \mathrm{diag}(\boldsymbol{F}_N)$ and $\boldsymbol{\Gamma} = \mathrm{diag}(\boldsymbol{\Gamma}_0, \ldots, \boldsymbol{\Gamma}_{n-1})$. (Here, 'diag' operates on matrices instead of scalars.)

We have demonstrated by (2.47) that the nice factorization result obtained for circulant matrices extends also to block-circulant matrices and so does Theorem 2.10. It will also extend to higher dimensions, i.e., a block-circulant matrix where each block is a block-circulant matrix and so on, by following the same route.

Although Equation (2.47) gives the recipe of how to compute the eigenvalues, we typically do not want to use this expression directly.

Instead, we can use the relation between the two-dimensional discrete Fourier transform, and the eigenvectors and eigenvalues of a block-circulant matrix. Let the block-diagonal matrix $\boldsymbol{\Gamma}$ contain all nN eigenvalues on the diagonal. Store these eigenvalues in an $n \times N$ matrix $\boldsymbol{\Pi}$, where row i of $\boldsymbol{\Pi}$ is the diagonal of $\boldsymbol{\Gamma}_i$. Since $\boldsymbol{F}$ is the discrete Fourier transform matrix, we can compute $\boldsymbol{\Pi}$ as follows: compute the DFT of each row of the base $\boldsymbol{c}$, compute the DFT of each column and scale both with $\sqrt{Nn}$. The result is that $\boldsymbol{\Pi}$ is the two-dimensional discrete Fourier transform of the base $\boldsymbol{c}$. Similarly, the block matrix $\boldsymbol{F}_n^N \ \boldsymbol{P} \ \boldsymbol{F}_N^n$ is the two-dimensional discrete Fourier transform matrix.

Computations for block-circulant matrices are as easy as for circulant matrices, if we extend the notation to two-dimensional discrete Fourier transforms. Let $\boldsymbol{a}$ be an $n \times N$ matrix. The two-dimensional DFT of $\boldsymbol{a}$, $\text{DFT2}(\boldsymbol{a})$ is an $n \times N$ matrix with elements

$$\frac{1}{\sqrt{nN}} \sum_{i'=0}^{n-1} \sum_{j'=0}^{N-1} a_{i'j'} \exp\left(-2\pi\iota \left(\frac{ii'}{n} + \frac{jj'}{N}\right)\right),$$

$i = 0, \ldots, n-1$, $j = 0, \ldots, N-1$, and the inverse DFT of $\boldsymbol{a}$, $\text{IDFT2}(\boldsymbol{a})$, is an $n \times N$ matrix with elements

$$\frac{1}{\sqrt{nN}} \sum_{i'=0}^{n-1} \sum_{j'=0}^{N-1} a_{i'j'} \exp\left(2\pi\iota \left(\frac{ii'}{n} + \frac{jj'}{N}\right)\right).$$

Using this notation, the $n \times N$ matrix

$$\boldsymbol{\Pi} = \sqrt{nN}\ \text{DFT2}(\boldsymbol{c})$$

contains all eigenvalues of $\boldsymbol{C}$; a block-circulant matrix with base $\boldsymbol{c}$. Let $\boldsymbol{v}$ be an $n \times N$ matrix and $\text{vec}(\boldsymbol{v})$ its vector representation obtained by stacking the columns one above the other, see Section 2.1.1. The matrix-vector product $\text{vec}(\boldsymbol{u}) = \boldsymbol{C}\text{vec}(\boldsymbol{v})$ can then be computed as

$$\boldsymbol{u} = \sqrt{nN}\ \text{DFT2}(\text{DFT2}(\boldsymbol{c}) \odot \text{IDFT2}(\boldsymbol{v})). \tag{2.48}$$

The product of two block-circulant matrices $\boldsymbol{C}$ and $\boldsymbol{D}$, with base $\boldsymbol{c}$ and $\boldsymbol{d}$, respectively, is a block-circulant matrix with base

$$\sqrt{nN}\ \text{IDFT2}\left(\text{DFT2}(\boldsymbol{c}) \odot \text{DFT2}(\boldsymbol{d})\right). \tag{2.49}$$

The solution of $\boldsymbol{C}\text{vec}(\boldsymbol{x}) = \text{vec}(\boldsymbol{b})$ is

$$\boldsymbol{x} = \frac{1}{\sqrt{nN}}\ \text{DFT2}(\text{IDFT2}(\boldsymbol{b}) \oslash \text{DFT2}(\boldsymbol{c}))$$

while the inverse of $\boldsymbol{C}$ has base

$$\frac{1}{nN}\ \text{IDFT2}(\text{DFT2}(\boldsymbol{c}) \circledast (-1)). \tag{2.50}$$

2.6.3 GMRFs with circulant precision matrices

The relevance of block-circulant matrices regarding GMRFs appears when we study a *stationary* GMRF on a torus.

This puts strong assumptions on both the graph and $\boldsymbol{Q}$, *but* it is an important special case. For illustration, a torus is shown in Figure 2.1(a).

Let a zero mean GMRF be defined on $\mathcal{T}_{\boldsymbol{n}}$ through the conditional moments

$$\begin{aligned} \mathrm{E}(x_{ij} \mid \boldsymbol{x}_{-ij}) &= -\frac{1}{\theta_{00}} \sum_{i'j' \neq 00} \theta_{i'j'} x_{i+i',j+j'}, \\ \mathrm{Prec}(x_{ij} \mid \boldsymbol{x}_{-ij}) &= \theta_{00}, \end{aligned} \tag{2.51}$$

where usually only a few of the $\theta_{i'j'}$'s are nonzero, for example, $|i'| \leq 1$ and $|j'| \leq 1$, or $|i'| \leq 2$ and $|j'| \leq 2$. Let $\mathcal{G}$ be the graph induced by the torus $\mathcal{T}_{\boldsymbol{n}}$ and the nonzero $\{\theta_{ij}\}$. The precision matrix $\boldsymbol{Q}$ is

$$\boldsymbol{Q}_{(i,j),(i',j')} = \theta_{i-i',j-j'}$$

and $\theta_{i'j'} = \theta_{-i',-j'}$ due to symmetry. Here we assume that the elements are stored by row, i.e., $(i,j) = i + jn_1$, so $\boldsymbol{Q}$ is a block-cyclic matrix with base $\boldsymbol{\theta}$. We assume further that $\boldsymbol{Q}$ is SPD.

The so-defined GMRF is called *stationary* if the mean vector is constant and if

$$\mathrm{Cov}(x_{ij}, x_{i'j'}) = c(i-i', j-j')$$

for some function $c(\cdot,\cdot)$, i.e., the covariance matrix is a block-cyclic matrix with base $\boldsymbol{c}$. Often $c(i-i', j-j')$ depends on $i-i'$ and $j-j'$ only through the Euclidean distance (on the torus) between (i,j) and (i',j'). The precision matrix for a stationary GMRF is then block-cyclic by the generalization of Theorem 2.10, with the consequence that a stationary GMRF has the full conditionals (2.51). Similarly, a GMRF with full conditionals (2.51) and constant mean is stationary.

Fast algorithms can now be derived based on operations on block-circulant matrices using the discrete Fourier transform. As $\boldsymbol{Q}$ is SPD, all eigenvalues are real, positive, and all eigenvectors are real.

To describe how to sample a zero mean GMRF $\boldsymbol{x}$ defined in (2.51), let $\boldsymbol{x}$ be stored as an $n \times N$ matrix and similar with $\boldsymbol{z}$. The spectral decomposition of $\boldsymbol{Q}$ is $\boldsymbol{Q} = \boldsymbol{V}\boldsymbol{\Lambda}\boldsymbol{V}^T$, so solving

$$\boldsymbol{\Lambda}^{1/2}\boldsymbol{V}^T \mathrm{vec}(\boldsymbol{x}) = \mathrm{vec}(\boldsymbol{z}),$$

where $\mathrm{vec}(\boldsymbol{z}) \sim \mathcal{N}_{nN}(\boldsymbol{0}, \boldsymbol{I})$, gives

$$\mathrm{vec}(\boldsymbol{x}) = \boldsymbol{V}\boldsymbol{\Lambda}^{-1/2}\mathrm{vec}(\boldsymbol{z}).$$

This can be computed using the DFT2 as illustrated in Algorithm 2.10 where $\boldsymbol{\Lambda}$ is an $n \times N$ matrix. The imaginary part of $\boldsymbol{v}$ is not used in this

Algorithm 2.10 Sampling a zero mean GMRF with block-circulant precision

Sample $\boldsymbol{z}$, where $\mathrm{Re}(z_{ij}) \overset{\text{iid}}{\sim} \mathcal{N}(0,1)$ and $\mathrm{Im}(z_{ij}) \overset{\text{iid}}{\sim} \mathcal{N}(0,1)$
Compute the (real) eigenvalues, $\boldsymbol{\Lambda} = \sqrt{nN}\ \mathrm{DFT2}(\boldsymbol{\theta})$
$\boldsymbol{v} = \mathrm{DFT2}((\boldsymbol{\Lambda} \oslash (-\frac{1}{2})) \odot \boldsymbol{z})$
$\boldsymbol{x} = \mathrm{Re}(\boldsymbol{v})$
Return $\boldsymbol{x}$

algorithm. We can make use of it since $\mathrm{Im}(\boldsymbol{v})$ has the same distribution as $\mathrm{Re}(\boldsymbol{v})$, and $\mathrm{Im}(\boldsymbol{v})$ and $\mathrm{Re}(\boldsymbol{v})$ are independent.

The log density can be evaluated as

$$-\frac{Nn}{2}\log 2\pi + \frac{1}{2}\log|\boldsymbol{Q}| - \frac{1}{2}\mathrm{vec}(\boldsymbol{x})^T \boldsymbol{Q}\ \mathrm{vec}(\boldsymbol{x}),$$

where

$$\log|\boldsymbol{Q}| = \sum_{ij} \log \Lambda_{ij}$$

with $\boldsymbol{\Lambda}$ as computed in Algorithm 2.10. To obtain the quadratic term $q = \mathrm{vec}(\boldsymbol{x})^T \boldsymbol{Q}\ \mathrm{vec}(\boldsymbol{x})$, we use (2.48) to obtain $\mathrm{vec}(\boldsymbol{u}) = \boldsymbol{Q}\ \mathrm{vec}(\boldsymbol{x})$, and then $q = \mathrm{vec}(\boldsymbol{x})^T\ \mathrm{vec}(\boldsymbol{u})$.

Example 2.8 *Let $\mathcal{T}_n$ be a* 128×128 *torus, where*

$$\begin{aligned} E(x_{ij} \mid \boldsymbol{x}_{-ij}) &= \frac{1}{4+\delta}\left(x_{i+1,j} + x_{i-1,j} + x_{i,j+1} + x_{i,j-1}\right) \\ Prec(x_{ij} \mid \boldsymbol{x}_{-ij}) &= 4+\delta, \quad \delta > 0. \end{aligned}$$

The precision matrix $\boldsymbol{Q}$ is then block-circulant with base (128×128 *matrix)*

$$\boldsymbol{c} = \begin{pmatrix} 4+\delta & -1 & & & -1 \\ -1 & & & & \\ & & & & \\ & & & & \\ & & & & \\ -1 & & & & \end{pmatrix},$$

where we display only the nonzero terms. Note that $\boldsymbol{Q}$ is symmetric and diagonal dominant hence $\boldsymbol{Q} > 0$. A sample using Algorithm 2.10 is displayed in Figure 2.13(a) using $\delta = 0.1$. We also compute the correlation matrix, i.e., the scaled covariance matrix $\boldsymbol{Q}^{-1}$, which has

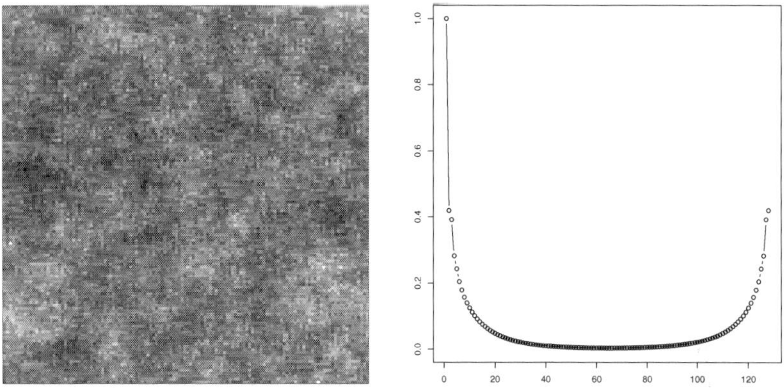

(a) (b)

Figure 2.13 *Illustrations to Example 2.8, the sample (a) and the first column of the base of $\boldsymbol{Q}^{-1}$ in (b).*

base equal to

$$\begin{pmatrix} 1.00 & 0.42 & 0.39 & 0.28 & 0.24 & 0.20 & 0.18 & 0.16 & 0.14 & \cdots \\ 0.42 & 0.33 & 0.31 & 0.26 & 0.23 & 0.20 & 0.18 & 0.16 & 0.14 & \cdots \\ 0.39 & 0.31 & 0.29 & 0.25 & 0.22 & 0.19 & 0.17 & 0.15 & 0.14 & \cdots \\ 0.28 & 0.26 & 0.25 & 0.22 & 0.20 & 0.18 & 0.16 & 0.15 & 0.13 & \cdots \\ 0.24 & 0.23 & 0.22 & 0.20 & 0.18 & 0.17 & 0.15 & 0.14 & 0.13 & \cdots \\ 0.20 & 0.20 & 0.19 & 0.18 & 0.17 & 0.15 & 0.14 & 0.13 & 0.12 & \cdots \\ 0.18 & 0.18 & 0.17 & 0.16 & 0.15 & 0.14 & 0.13 & 0.12 & 0.11 & \cdots \\ 0.16 & 0.16 & 0.15 & 0.15 & 0.14 & 0.13 & 0.12 & 0.11 & 0.10 & \cdots \\ 0.14 & 0.14 & 0.14 & 0.13 & 0.13 & 0.12 & 0.11 & 0.10 & 0.09 & \cdots \\ \vdots & \vdots & \vdots & \vdots & \vdots & \vdots & \vdots & \vdots & \vdots & \ddots \end{pmatrix}.$$

The first column of the base is displayed in Figure 2.13(b).

2.6.4 Toeplitz matrices and their approximations

The toroidal assumption for a torus, where opposite sides of a regular lattice in d dimensions are adjacent, may for many applications be somewhat artificial.

However, the nice analytical properties of circulant matrices and their superior computational properties through the connection to the DFT raises the question whether it is possible to approximate nontoroidal GMRFs with toroidal GMRFs. We will now look at approximations of so-called Toeplitz matrices through circulant matrices.

Definition 2.6 (Toeplitz matrix) *An* $n \times n$ *matrix* $\boldsymbol{T}$ *is called* Toeplitz *iff it has the form*

$$\boldsymbol{T} = \begin{pmatrix} t_0 & t_1 & t_2 & \dots & t_{n-1} \\ t_{-1} & t_0 & t_1 & \dots & t_{n-2} \\ t_{-2} & t_{-1} & t_0 & \dots & t_{n-3} \\ \vdots & \vdots & \vdots & & \vdots \\ t_{-(n-1)} & t_{-(n-2)} & t_{-(n-3)} & \dots & t_0 \end{pmatrix} = (t_{j-i})$$

for a vector $\boldsymbol{t} = (t_{-(n-1)}, \dots, t_{-1}, t_0, t_1, \dots, t_{n-1})^T$*, called the* base *of* $\boldsymbol{T}$*. If* $\boldsymbol{T}$ *is symmetric, then* $t_k = t_{-k}$ *and the base is* $\boldsymbol{t} = (t_0, t_1, \dots, t_{n-1})^T$*.*

A Toeplitz matrix is fully specified by one column and one row. A symmetric Toeplitz matrix is fully specified by either one column or one row.

Example 2.9 *Let* x_t*,* $t = \dots, -1, 0, 1, \dots$*, be a zero mean stationary Gaussian autoregressive process of order* K *where* t *denotes time. The precision matrix of* $\boldsymbol{x}_n = (x_0, \dots, x_{n-1})^T$ *where* $n > 2p$*, conditioned on* $x_t = 0$ *for* $t \notin \{0, \dots, n-1\}$*, is then a symmetric Toeplitz matrix with base*

$$\boldsymbol{t} = (\theta_0, \theta_1, \dots, \theta_p, 0, \dots, 0)^T, \tag{2.52}$$

say. The (conditional) log density of $\boldsymbol{x}_n$ *is*

$$\log \pi(\boldsymbol{x}_n) = -\frac{n}{2} \log 2\pi + \frac{1}{2} \log |\boldsymbol{T}_n| - \frac{1}{2} \boldsymbol{x}_n^T \boldsymbol{T}_n \boldsymbol{x}_n. \tag{2.53}$$

If n *is large, then we might approximate* $\boldsymbol{T}_n$ *with a circulant matrix* $\boldsymbol{C}_n$ *with base*

$$\boldsymbol{c} = (\theta_0, \dots, \theta_p, 0, \dots, 0, \theta_p, \dots, \theta_1)^T. \tag{2.54}$$

to obtain a more computational feasible log density,

$$\log \pi_c(\boldsymbol{x}_n) = -\frac{n}{2} \log 2\pi + \frac{1}{2} \log |\boldsymbol{C}_n| - \frac{1}{2} \boldsymbol{x}_n^T \boldsymbol{C}_n \boldsymbol{x}_n, \tag{2.55}$$

which is an approximation to (2.53).

The rationale for the approximation in Example 2.9 is that $\boldsymbol{T}_n$ and $\boldsymbol{C}_n$ are *asymptotically equivalent* (to be defined) as $n \to \infty$. This will enable us to prove rather easily that

$$\left| \frac{1}{n} \log \pi(\boldsymbol{x}_n) - \frac{1}{n} \log \pi_c(\boldsymbol{x}_n) \right| \to 0,$$

almost surely as $n \to \infty$. As a consequence, the (conditional) maximum likelihood estimator (MLE) of some unknown parameters $\boldsymbol{\theta}$ tends to the same limit as the one obtained using the circulant approximation.

To define asymptotically equivalent matrices, we need to define the strong and weak matrix norm.

Definition 2.7 (Weak and strong norm) *Let* $\boldsymbol{A}$ *be a real* $n \times n$ *matrix. The* strong *and* weak norm *is defined as*

$$\|\boldsymbol{A}\|_s = \max_{\boldsymbol{x}\,:\,\boldsymbol{x}^T\boldsymbol{x}=1} \left(\boldsymbol{x}^T(\boldsymbol{A}^T\boldsymbol{A})\boldsymbol{x}\right)^{1/2} \quad \textit{and}$$

$$\|\boldsymbol{A}\|_w = \left(\frac{1}{n}\sum_{ij} A_{ij}^2\right)^{1/2},$$

respectively.

Both norms can be expressed in terms of the eigenvalues $\{\lambda_k\}$ of $\boldsymbol{A}^T\boldsymbol{A}$:

$$\|\boldsymbol{A}\|_s^2 = \max_k \lambda_k, \quad \text{and} \quad \|\boldsymbol{A}\|_w^2 = \frac{1}{n}\text{trace}(\boldsymbol{A}^T\boldsymbol{A}) = \frac{1}{n}\sum_k \lambda_k.$$

If $\boldsymbol{A}$ is symmetric with eigenvalues $\{\alpha_k\}$ then $\lambda_k = \alpha_k^2$.

Definition 2.8 (Asymptotically equivalent matrices) *Two sequences of* $n \times n$ *matrices* $\boldsymbol{A}_n$ *and* $\boldsymbol{B}_n$ *are said to be* asymptotically equivalent *if*

1. *There exists* $M < \infty$ *such that* $\|\boldsymbol{A}_n\|_s < M$ *and* $\|\boldsymbol{B}_n\|_s < M$, *and*
2. $\|\boldsymbol{A}_n - \boldsymbol{B}_n\|_w \to 0$ *as* $n \to \infty$.

Asymptotically equivalent matrices have nice properties, for example, certain functions of the eigenvalues converge to the same limit, see for example Gray (2002, Thm. 2.4).

Theorem 2.11 *Let* $\boldsymbol{A}_n$ *and* $\boldsymbol{B}_n$ *be asymptotically equivalent matrices with eigenvalues* $\alpha_{n,k}$ *and* $\beta_{n,k}$ *and suppose there exist* $m > 0$ *and a finite* M *such that*

$$m < |\alpha_{n,k}| < M \quad \textit{and} \quad m < |\beta_{n,k}| < M.$$

Let $f(\cdot)$ *be a continuous function on* $[m, M]$, *then*

$$\lim_{n\to\infty} \frac{1}{n}\sum_k f(\alpha_{n,k}) = \lim_{n\to\infty} \frac{1}{n}\sum_k f(\beta_{n,k})$$

if one of the limits exists.

The proof of Theorem 2.11 is too long to give here, but the idea is to prove that the mean of powers of the eigenvalues converges to the same limit as a remedy to show convergence for any polynomial $f(\cdot)$. Then the Stone-Weierstrass approximation theorem is used to obtain the result for any continuous function $f(\cdot)$.

We are now able to prove that the error we do by approximating $\boldsymbol{T}_n$ with $\boldsymbol{C}_n$ is asymptotically negligible, hence we can approximate the conditional log density by (2.55) and evaluate it utilizing the connection to the DFT.

Theorem 2.12 *Let $\boldsymbol{x}_n \sim \mathcal{N}(\boldsymbol{0}, \boldsymbol{T}_n^{-1})$, where the precision matrix $\boldsymbol{T}_n$ is Toeplitz with base (2.52) and let $\boldsymbol{C}_n$ be the corresponding circulant approximation with base (2.54). Assume the eigenvalues of $\boldsymbol{T}_n$ and $\boldsymbol{C}_n$ are bounded away from 0, then as $n \to \infty$,*

$$\frac{1}{n}\left|\frac{1}{2}\log|\boldsymbol{T}_n| - \frac{1}{2}\boldsymbol{x}_n\boldsymbol{T}_n\boldsymbol{x}_n - \frac{1}{2}\log|\boldsymbol{C}_n| + \frac{1}{2}\boldsymbol{x}_n\boldsymbol{C}_n\boldsymbol{x}_n\right| \to 0 \qquad (2.56)$$

almost surely.

Proof. First we show that $\boldsymbol{T}_n$ and $\boldsymbol{C}_n$ are asymptotically equivalent. Note that $\boldsymbol{T}_n$ and $\boldsymbol{C}_n$ only differ in $\mathcal{O}(p^2)$ terms. This ensures that $\|\boldsymbol{T}_n - \boldsymbol{C}_n\|_w \to 0$. The eigenvalues of $\boldsymbol{T}_n$ and $\boldsymbol{C}_n$ are bounded from above since the elements of $\boldsymbol{T}_n$ and $\boldsymbol{C}_n$ are bounded and so are the eigenvalues of $\boldsymbol{T}_n^T\boldsymbol{T}_n$ and $\boldsymbol{C}_n^T\boldsymbol{C}_n$. The strong norm is then bounded. By Definition 2.8, $\boldsymbol{T}_n$ and $\boldsymbol{C}_n$ are asymptotically equivalent.

The triangle inequality applied to (2.56) gives the following upper bound:

$$\underbrace{\frac{1}{2n}\big|\log|\boldsymbol{T}_n| - \log|\boldsymbol{C}_n|\big|}_{\text{term 1}} + \underbrace{\frac{1}{2n}\left|\boldsymbol{x}_n^T\boldsymbol{T}_n\boldsymbol{x}_n - \boldsymbol{x}_n^T\boldsymbol{C}_n\boldsymbol{x}_n\right|}_{\text{term 2}}.$$

Consider first term 1. Using Theorem 2.11 with $f(\cdot) = \log(\cdot)$ and that $\boldsymbol{T}_n$ and $\boldsymbol{C}_n$ are asymptotically equivalent matrices with bounded eigenvalues from above and from below (by assumption), then

$$\lim_{n\to\infty}\frac{1}{n}\log|\boldsymbol{T}_n| = \lim_{n\to\infty}\frac{1}{n}\log|\boldsymbol{C}_n|.$$

Regarding term 2, note that only $\mathcal{O}(p^2)$ terms in $\boldsymbol{T}_n - \boldsymbol{C}_n$ are nonzero. Further, each x_i is bounded in probability since the eigenvalues of $\boldsymbol{T}_n$ and $\boldsymbol{C}_n$ are bounded from below (by assumption) and above. This ensures that term 2 tends to zero almost surely. □

In applications we may use (2.55) to approximate the MLE of unknown parameters $\boldsymbol{\theta}$. However, two issues arise.

1. We want the MLE from the (exact) log likelihood and not the conditional log likelihood.

2. We also need to consider the rate of convergence of the circulant approximation to the log likelihood *and* its partial derivatives wrt parameters that govern $\gamma(\cdot)$, to compare the bias caused by a circulant approximation with the bias and random error of the MLE.

Intuitively, the boundary conditions are increasingly important for increasing dimension. For the d-dimensional sphere with radius r, the volume is $\mathcal{O}(r^d)$ while the surface is $\mathcal{O}(r^{d-1})$. Hence the appropriateness of the circulant approximation may depend on the dimension.

We will now report both more precise and more general results by Kent and Mardia (1996) which also generalize results of Guyon (1982). Let $\boldsymbol{x}$ be a zero mean stationary Gaussian process on a d-dimensional lattice with size $N = n \times n \times \cdots \times n$. Let $\boldsymbol{\Sigma}_N$ be the (block) Toeplitz covariance matrix and $\boldsymbol{S}_N$ its (block) circulant approximation. Under some regularity conditions on

$$\mathrm{Cov}(x_{\boldsymbol{i}}, x_{\boldsymbol{j}}) = \gamma(\boldsymbol{i} - \boldsymbol{j}),$$

where $\boldsymbol{i}$ and $\boldsymbol{j}$ are points in the d-dimensional space and $\gamma(\cdot)$ is the so-called covariance function (see Definition 5.1 in Section 5.1), they proved that

$$\begin{aligned} -\frac{1}{2}\log|\boldsymbol{\Sigma}_N| + \frac{1}{2}\log|\boldsymbol{S}_N| &= \mathcal{O}(n^{d-1}) \\ -\frac{1}{2}\boldsymbol{x}^T\boldsymbol{\Sigma}_N^{-1}\boldsymbol{x} + \frac{1}{2}\boldsymbol{x}^T\boldsymbol{S}_N^{-1}\boldsymbol{x} &= \mathcal{O}_p(n^{d-1}). \end{aligned} \tag{2.57}$$

The result (2.57) also holds for partial derivatives wrt parameters that govern $\gamma(\cdot)$. The error in the deterministic and stochastic part of (2.57) is of the same order. The consequence is that the log likelihood and its circulant approximation differ by $\mathcal{O}_p(n^{d-1})$. We can also use the results of Kent and Mardia (1996) (their Lemma 4.3 in particular) to give the same bound on the difference of the conditional log likelihood (or its circulant approximation) and the log likelihood.

Let $\hat{\boldsymbol{\theta}}$ be the true MLE estimator and $\widetilde{\boldsymbol{\theta}}$ be the MLE estimator computed using the circulant approximation. Maximum likelihood theory states that, under some mild regularity conditions, $\hat{\boldsymbol{\theta}}$ is asymptotically normal with

$$N^{1/2}(\hat{\boldsymbol{\theta}} - \boldsymbol{\theta}) \sim \mathcal{N}(\boldsymbol{0}, \boldsymbol{H})$$

where $\boldsymbol{H} > 0$. The standard deviation for a component of $\boldsymbol{\theta}$ is then $\mathcal{O}(N^{-1/2})$. The bias in the MLE is for this problem $\mathcal{O}(N^{-1})$ (Mardia, 1990). Kent and Mardia (1996) show that $\widetilde{\boldsymbol{\theta}}$ has bias of $\mathcal{O}(1/n)$. From this we can conclude, that for $d = 1$ the bias caused by the circulant approximation is of smaller order than the standard deviation. The circulant approximation is harmless and $\widetilde{\boldsymbol{\theta}}$ has the same asymptotic properties as $\hat{\boldsymbol{\theta}}$. For $d = 2$ the bias caused by the circulant approximation is of the same order as the standard deviation, so the error we make is of the same order as the random error. The circulant approximation is then tolerable, bearing in mind this bias. For $d \geq 3$ the bias is of larger order than the standard deviation so the error due to the circulant approximation dominates completely. An alternative is then to use the modified Whittle approximation to the log likelihood that is discussed in Section 2.6.5.

2.6.5 Stationary GMRFs on infinite lattices

As an alternative for a zero mean stationary GMRF on the torus we may consider a zero mean GMRFs on an infinite lattice $\mathcal{I}_\infty$. Such a process does exist on $\mathcal{I}_\infty$ (Rosanov, 1967) and can be represented and defined using its spectral density function (SDF). We will use the common term *conditional autoregression* for this process. An approximation to the log likelihood of a finite restriction of the process to $\mathcal{I}_n$ can be constructed using the SDF. This section defines conditional autoregression on $\mathcal{I}_\infty$, the SDF and the log likelihood approximation using the SDF.

Let $\boldsymbol{x}$ (shorthand for $\{x_i\}$) be a zero mean and stationary Gaussian process on $\mathcal{I}_\infty$, and define the covariances $\{\gamma_{ij}\}$,

$$\gamma_{ij} = \mathrm{E}(x_{kl}x_{k+i,l+j}),$$

which does not depend on k and l due to stationarity. We assume throughout that $\sum_{ij\in\mathcal{I}_\infty} |\gamma_{ij}|$ is finite.

The covariances $\{\gamma_{ij}\}$ define the *spectral density function* of $\boldsymbol{x}$.

Definition 2.9 (Spectral density function) *The spectral density function of $\boldsymbol{x}$ is*

$$f(\omega_1, \omega_2) = \frac{1}{4\pi^2} \sum_{ij\in\mathcal{I}_\infty} \gamma_{ij} \exp\left(-\iota(i\omega_1 + j\omega_2)\right),$$

where $\iota = \sqrt{-1}$ and $(\omega_1, \omega_2) \in (-\pi, \pi]^2$.

The SDF is the Fourier transform of $\{\gamma_{ij}\}$, hence we can express γ_{ij} from the SDF using the inverse transform:

$$\gamma_{ij} = \int_{-\pi}^{\pi}\int_{-\pi}^{\pi} f(\omega_1, \omega_2)\, \exp\left(\iota(i\omega_1 + j\omega_2)\right)\, d\omega_1\, d\omega_2.$$

Since $\gamma_{ij} = \gamma_{-i,-j}$, the SDF is real.

A *conditional autoregression* on $\mathcal{I}_\infty$ is a Gaussian process with a SDF of a specific form.

Definition 2.10 (Conditional autoregression) *A zero mean stationary Gaussian process on $\mathcal{I}_\infty$ is called a* conditional autoregression *if the SDF is*

$$f(\omega_1, \omega_2) = \frac{1}{4\pi^2} \frac{1}{\sum_{ij\in\mathcal{I}_\infty} \theta_{ij} \exp\left(-\iota(i\omega_1 + j\omega_2)\right)}, \tag{2.58}$$

where

1. *the number of nonzero coefficients θ_{ij} is finite*
2. *$\theta_{ij} = \theta_{-i,-j}$*
3. *$\theta_{00} > 0$*
4. *$\{\theta_{ij}\}$ is so that $f(\omega_1, \omega_2) > 0$ for all $(\omega_1, \omega_2) \in (-\pi, \pi]^2$*

Site ij and kl are called neighbors iff $\theta_{i-k,j-l} \neq 0$ and $ij \neq kl$.

Conditions 2, 3, and 4 correspond to $\boldsymbol{Q}$ being SPD. The definition of the conditional autoregression is consistent with the definition of a finite GMRF and the term 'neighbor' has also the same meaning.

Theorem 2.13 *The full conditionals of a conditional autoregression are normal with moments*

$$E(x_{ij} \mid \boldsymbol{x}_{-ij}) = -\frac{1}{\theta_{00}} \sum_{kl \in \mathcal{I}_\infty \setminus 00} \theta_{kl} x_{i+k,j+k} \tag{2.59}$$

$$Prec(x_{ij} \mid \boldsymbol{x}_{-ij}) = \theta_{00}. \tag{2.60}$$

To prove Theorem 2.13 we proceed as follows. First, define the *covariance generating function* (CGF) that compactly represents $\{\gamma_{ij}\}$.

Definition 2.11 (Covariance generating function) *The covariance generating function is defined by*

$$\Gamma(z_1, z_2) = \sum_{ij \in \mathcal{I}_\infty} \gamma_{ij} z_1^i z_2^j,$$

where it exists.

The covariance γ_{ij} can be extracted from the CGF using

$$\left. \frac{\partial^{i+j}}{\partial z_1^i \partial z_2^j} \Gamma(z_1, z_2) \right|_{(z_1,z_2)=(0,0)} = \gamma_{ij}$$

for $i \geq 0$ and $j \geq 0$, and the SDF can be expressed using the CGF as

$$f(\omega_1, \omega_2) = \frac{1}{4\pi^2} \Gamma(\exp(-\iota\omega_1), \exp(-\iota\omega_2)). \tag{2.61}$$

We need the following result.

Lemma 2.4 *The covariances of the conditional autoregression satisfy the recurrence equations*

$$\sum_{kl \in \mathcal{I}_\infty} \theta_{kl} \gamma_{i+k,j+l} = \begin{cases} 1, & ij = 00 \\ 0, & \textit{otherwise.} \end{cases} \tag{2.62}$$

Proof. [Lemma 2.4] Let δ_{ij} be 1 if $ij = 00$ and zero otherwise. Then (2.62) is equivalent to

$$\gamma_{ij} = \frac{1}{\theta_{00}} (\delta_{ij} - \sum_{kl \in \mathcal{I}_\infty \setminus 00} \theta_{kl} \gamma_{i+k,j+l}).$$

Define

$$d_{ij} = \gamma_{ij} - \frac{1}{\theta_{00}} (\delta_{ij} - \sum_{kl \in \mathcal{I}_\infty \setminus 00} \theta_{kl} \gamma_{i+k,j+l}).$$

By showing that $\sum_{ij\in\mathcal{I}_\infty} d_{ij} z_1^i z_2^j = 0$ for all (z_1, z_2) where $\Gamma(z_1, z_2)$ exists, we can conclude that $d_{ij} = 0$ for all ij. We obtain

$$
\begin{aligned}
\sum_{ij\in\mathcal{I}_\infty} d_{ij} z_1^i z_2^j &= \Gamma(z_1, z_2) - \frac{1}{\theta_{00}} + \frac{1}{\theta_{00}} \sum_{ij\in\mathcal{I}_\infty} \sum_{kl\in\mathcal{I}_\infty\setminus 00} \theta_{kl} z_1^i z_2^j \gamma_{i+k,j+l} \\
&= \Gamma(z_1, z_2) \left(1 + \frac{1}{\theta_{00}} \sum_{kl\in\mathcal{I}_\infty\setminus 00} \theta_{kl} z_1^{-k} z_2^{-l}\right) - \frac{1}{\theta_{00}} \\
&= \frac{1}{\theta_{00}} - \frac{1}{\theta_{00}} = 0,
\end{aligned}
$$

using (2.58) expressed by the CGF (2.61). □

Proof. [Theorem 2.13] We can verify (2.59) by showing that

$$
\mathrm{E}\left(\left(x_{ij} + \frac{1}{\theta_{00}} \sum_{kl\in\mathcal{I}_\infty\setminus 00} \theta_{kl} x_{i+k,j+l}\right) \sum_{k'l'\in\mathcal{I}_\infty\setminus 00} \theta_{k'l'} x_{i+k',j+l'}\right) = 0,
$$

which follows by expanding the terms and then using Lemma 2.4. To show (2.60) we start with

$$
\mathrm{Var}(x_{ij}) = \mathrm{Var}(\mathrm{E}(x_{ij} \mid \boldsymbol{x}_{-ij})) + \mathrm{E}(\mathrm{Var}(x_{ij} \mid \boldsymbol{x}_{-ij}))
$$

to compute

$$
\begin{aligned}
\mathrm{E}(\mathrm{Var}(x_{ij} \mid \boldsymbol{x}_{-ij})) &= \gamma_{00} - \mathrm{Var}(\mathrm{E}(x_{ij} \mid \boldsymbol{x}_{-ij})) \\
&= \gamma_{00} - \mathrm{E}\left(\frac{1}{\theta_{00}^2} \sum_{kl\in\mathcal{I}_\infty\setminus 00} \sum_{k'l'\in\mathcal{I}_\infty\setminus 00} \theta_{kl}\theta_{k'l'} x_{i+k,j+l} x_{i+k',j+l'}\right) \\
&= \gamma_{00} - \frac{1}{\theta_{00}^2} \sum_{kl\in\mathcal{I}_\infty\setminus 00} \theta_{kl} \sum_{k'l'\in\mathcal{I}_\infty\setminus 00} \theta_{k'l'} \gamma_{k'-k,l'-l} \\
&= \gamma_{00} - \frac{1}{\theta_{00}^2} \sum_{kl\in\mathcal{I}_\infty\setminus 00} \theta_{kl} (-\theta_{00}) \gamma_{kl} \\
&= \frac{1}{\theta_{00}} \left(\theta_{00}\gamma_{00} + \sum_{kl\in\mathcal{I}_\infty\setminus 00} \theta_{kl}\gamma_{kl}\right) \\
&= \frac{1}{\theta_{00}},
\end{aligned}
$$

where we have used Lemma 2.4 twice. From this (2.60) follows since $\mathrm{Var}(x_{ij}|\boldsymbol{x}_{-ij})$ is a constant. □

We end by presenting Whittle's approximation, which uses the SDF of the process on $\mathcal{I}_\infty$ to approximate the log likelihood if $\boldsymbol{x}$ is observed at a finite lattice $\mathcal{I}_n$ (Whittle, 1954). The empirical covariances of $\boldsymbol{x}$ are

$$\hat{\gamma}_{ij} = \frac{1}{n} \sum_{kl \in \mathcal{I}_\infty} x_{kl} x_{k+i,l+j}$$

and the empirical SDF is

$$\hat{f}(\omega_1, \omega_2) = \frac{1}{4\pi^2} \sum_{ij \in \mathcal{I}_\infty} \hat{\gamma}_{ij} \exp(-\iota(i\omega_1 + j\omega_2)).$$

Then Whittle's approximation is

$$\begin{aligned} \log \pi(\boldsymbol{x}) \quad \approx \quad & -\frac{n}{2} \log 2\pi \\ & -\frac{n}{8\pi^2} \int_{-\pi}^{\pi} \int_{-\pi}^{\pi} \log(4\pi^2 f(\omega_1, \omega_2)) \, d\omega_1 \, d\omega_2 \\ & -\frac{n}{8\pi^2} \int_{-\pi}^{\pi} \int_{-\pi}^{\pi} \frac{\hat{f}(\omega_1, \omega_2)}{f(\omega_1, \omega_2)} \, d\omega_1 \, d\omega_2. \end{aligned} \tag{2.63}$$

The properties of this approximation have been studied by Guyon (1982). The approximation shares the same property as the circulant approximation in Section 2.6.4 (Kent and Mardia, 1996), but modifications to (2.63) can be made to correct for the bias either using unbiased estimates for $\hat{\gamma}_{ij}$ (Guyon, 1982), or better, using data tapers (Dahlhaus and Künsch, 1987).

2.7 Parameterization of GMRFs★

In this section we will consider the case, where $\boldsymbol{Q}$ is a function of some parameters $\boldsymbol{\theta}$. In this case it is important to know the set of values of $\boldsymbol{\theta}$ for which $\boldsymbol{Q}$ is SPD, hence $\boldsymbol{x}$ is a GMRF. Suppose the precision matrix $\boldsymbol{Q}$ is parameterized by some parameter vector $\boldsymbol{\theta} \in \boldsymbol{\Theta}$. We assume throughout that $\boldsymbol{Q}(\boldsymbol{\theta})$ is symmetric and has strictly positive diagonal entries for all $\boldsymbol{\theta} \in \boldsymbol{\Theta}$. In some cases the parameterization is such that $\boldsymbol{Q}(\boldsymbol{\theta})$ is by definition positive definite for all $\boldsymbol{\theta} \in \boldsymbol{\Theta}$. One such example is

$$\pi(\boldsymbol{x} \mid \boldsymbol{\theta}) \propto \exp\left(-\frac{\theta_1}{2} \sum_{i \sim j} (x_i - x_j)^2 - \frac{\theta_2}{2} \sum_i x_i^2 \right)$$

which is a GMRF with a SPD precision matrix if $\theta_1 > 0$ and $\theta_2 > 0$.

In this section we are concerned with the case where the parameter space $\boldsymbol{\Theta}$ has to be restricted to ensure $\boldsymbol{Q}(\boldsymbol{\theta}) > 0$, hence we need to know the *valid parameter space*

$$\boldsymbol{\Theta}^+ = \{\boldsymbol{\theta} \in \boldsymbol{\Theta} : \boldsymbol{Q}(\boldsymbol{\theta}) > 0\}.$$

Unfortunately, it is hard to determine $\boldsymbol{\Theta}^+$ in general. However, it is always possible to check if $\boldsymbol{\theta} \in \boldsymbol{\Theta}^+$ by a direct verification if $\boldsymbol{Q}(\boldsymbol{\theta}) > 0$ or not. This is most easily done by trying to compute the Cholesky factorization, which will be successful iff $\boldsymbol{Q}(\boldsymbol{\theta}) > 0$, see Section 2.4.

Although this 'brute force' method is possible, it is often computationally expensive, intellectually not satisfying and gives little insight in properties of $\boldsymbol{\Theta}^+$. To obtain some knowledge we will in Section 2.7.1 study $\boldsymbol{\Theta}^+$ for a stationary GMRF on a torus (see Section 2.6). By using the properties of (block) circulant matrices some analytical results are possible to obtain. These analytical results are also useful for precision matrices that are Toeplitz, a relationship we comment on at the end of Section 2.7.1.

Since the characterization of $\boldsymbol{\Theta}^+$ is difficult, a frequently used approach is to use a sufficient condition, *diagonal dominance*, to ensure that $\boldsymbol{Q}(\boldsymbol{\theta})$ is SPD. This is the theme for Section 2.7.2. This sufficient condition restricts $\boldsymbol{\Theta}$ to a subset $\boldsymbol{\Theta}^{++}$, say, where $\boldsymbol{\Theta}^{++} \subseteq \boldsymbol{\Theta}^+$. The parameter space $\boldsymbol{\Theta}^{++}$ can in most cases be determined analytically without much effort. We will compare $\boldsymbol{\Theta}^{++}$ with $\boldsymbol{\Theta}^+$ using the exact results obtained in Section 2.7.1.

2.7.1 The valid parameter space

Let a zero mean GMRF be defined on the torus $\mathcal{T}_{\boldsymbol{n}}$ through the conditional moments

$$\begin{aligned} \mathrm{E}(x_{ij} \mid \boldsymbol{x}_{-ij}) &= -\sum_{i'j' \neq 00} \theta_{i'j'} x_{i+i',j+j'} \qquad \text{and} \\ \mathrm{Prec}(x_{ij} \mid \boldsymbol{x}_{-ij}) &= 1, \end{aligned} \tag{2.64}$$

where the sum is over a few terms only, for example, $|i'| \leq 1$ and $|j'| \leq 1$. The elements in $\boldsymbol{Q}$ are found from (2.64) using Theorem 2.6:

$$Q_{(i,j),(i',j')} = \theta_{i-i',j-j'}, \tag{2.65}$$

where $\theta_{i'j'} = \theta_{-i',-j'}$ due to the symmetry of $\boldsymbol{Q}$. Here (i,j) is short for the index $i + jn_1$.

We now specify the parameters $\boldsymbol{\theta}$ as

$$\{\theta_{i'j'}, i' = 0, \pm 1, \ldots, \pm m_1,\ j' = 0, \pm 1, \ldots, \pm m_2\} \tag{2.66}$$

for some fixed $\boldsymbol{m} = (m_1, m_2)^T$, where the remaining terms are zero and $\theta_{00} = 1$. We assume in the following that $\boldsymbol{n} > 2\boldsymbol{m}$, to simplify the discussion.

The precision matrix (2.65) is a block-circulant matrix and its properties were discussed in Section 2.6. For this class of matrices, the eigenvalues are known to be

$$\lambda_{ij}(\boldsymbol{\theta}) = \sum_{i'j'} \theta_{i'j'} \cos 2\pi \left(\frac{ii'}{n_1} + \frac{jj'}{n_2} \right) \tag{2.67}$$

for $i = 0, \ldots, n_1 - 1$ and $j = 0, \ldots, n_2 - 1$. Recall that the eigenvalues can be computed using the two-dimensional discrete Fourier transform of the matrix $(\theta_{i'j'})$, see Section 2.6.

Some properties about $\boldsymbol{\Theta}^+$ can be derived from (2.67). We need the notion of a *polyhedron*, which is defined as a space that can be built from line segments, triangles, tetrahedra, and their higher-dimensional analogues by gluing them together along their faces. Alternatively, a polyhedron can be viewed as an intersection of half spaces.

Theorem 2.14 *Let $\boldsymbol{x}$ be a GMRF on $\mathcal{T}_{\boldsymbol{n}}$ with dimension $\boldsymbol{n} > 2\boldsymbol{m}$ with full conditionals as in (2.64) and $\boldsymbol{m}$ as defined in (2.66). Then the valid parameter space $\boldsymbol{\Theta}^+$ is a bounded and convex polyhedron.*

A bounded polyhedron is also called a *polytope*.

Proof. From (2.67) it is clear that the eigenvalues are linear in $\boldsymbol{\theta}$. Since $\boldsymbol{Q}(\boldsymbol{\theta})$ is SPD iff all eigenvalues are strictly positive, it follows that

$$\boldsymbol{\Theta}^+ = \{\boldsymbol{\theta} \,:\, \boldsymbol{C}\boldsymbol{\theta} > \mathbf{0}\}$$

for some matrix $\boldsymbol{C}$, hence $\boldsymbol{\Theta}^+$ is a polyhedron. Let $\boldsymbol{\theta}'$ and $\boldsymbol{\theta}'$ be two configurations in $\boldsymbol{\Theta}^+$, then consider

$$\boldsymbol{\theta}(\alpha) = \alpha\boldsymbol{\theta}' + (1-\alpha)\boldsymbol{\theta}'$$

for $0 \le \alpha \le 1$. As

$$\boldsymbol{C}(\boldsymbol{\theta}(\alpha)) = \boldsymbol{C}(\alpha\boldsymbol{\theta}' + (1-\alpha)\boldsymbol{\theta}') = \alpha\boldsymbol{C}\boldsymbol{\theta}' + (1-\alpha)\boldsymbol{C}\boldsymbol{\theta}' > 0,$$

it follows that $\boldsymbol{\theta}(\alpha) \in \boldsymbol{\Theta}^+$ and that $\boldsymbol{\Theta}^+$ is convex. Furthermore, $\boldsymbol{\Theta}^+$ is *bounded* as $\max_{i \ne j} |Q_{ij}| < 1$ (Section 2.1.6) as $Q_{ii} = 1$ for all i. □

A further complication is that $\boldsymbol{\Theta}^+$ also depends on $\boldsymbol{n}$. To investigate this issue, we now write $\boldsymbol{\Theta}^+_{\boldsymbol{n}}$ to make this dependency explicit.

Assume $\boldsymbol{\theta} \in \boldsymbol{\Theta}^+_{\boldsymbol{n}}$ and change the dimension to $\boldsymbol{n}'$ keeping $\boldsymbol{\theta}$ fixed. Can we conclude that $\boldsymbol{\theta} \in \boldsymbol{\Theta}^+_{\boldsymbol{n}'}$? A simple counterexample demonstrates that this is not true in general. Let $\boldsymbol{n} = (10, 1)^T$ with

$$\theta_{00} = 1, \quad \theta_{\pm1,0} = 0.745, \quad \theta_{\pm2,0} = 0.333,$$

then all eigenvalues are positive. If we change the dimension to $\boldsymbol{n}' = (11, 1)^T$ or $\boldsymbol{n}' = (9, 1)^T$, then the smallest eigenvalue is negative. This is rather disappointing; if we estimate $\boldsymbol{\theta}$ for one grid size, we cannot necessarily use the estimates for a different grid size.

However, if we reduce the dimension from $\boldsymbol{n}$ to $\boldsymbol{n}' > 2\boldsymbol{m}$, where $\boldsymbol{n}/\boldsymbol{n}'$ is a positive integer (or a vector of positive integers), then the following result states that $\boldsymbol{\Theta}^+_{\boldsymbol{n}} \subseteq \boldsymbol{\Theta}^+_{\boldsymbol{n}'}$.

Theorem 2.15 *Let $\boldsymbol{A}$ and $\boldsymbol{B}$ be block-circulant matrices of dimension $(k_1 n_1\ k_2 n_2)^2 \times (k_1 n_1\ k_2 n_2)^2$ and $(n_1\ n_2)^2 \times (n_1\ n_2)^2$, respectively, with entries $A_{(i,j),(i',j')} = \theta_{i-i',j-j'}$ and $B_{(i,j),(i',j')} = \theta_{i-i',j-j'}$ where k_1 and k_2 are positive integers. Here, $\boldsymbol{\theta}$ is defined in (2.66) with $\boldsymbol{n} > 2\boldsymbol{m}$ and $\theta_{i'j'} = \theta_{-i',-j'}$. If $\boldsymbol{A}$ is SPD then $\boldsymbol{B}$ is SPD.*

Proof. Both $\boldsymbol{A}$ and $\boldsymbol{B}$ are symmetric as $\theta_{i'j'} = \theta_{-i',-j'}$. The eigenvalues for $\boldsymbol{A}$ are

$$\lambda_{ij}^{\boldsymbol{A}}(\boldsymbol{\theta}) = \sum_{i'j'} \theta_{i'j'} \cos 2\pi \left(\frac{ii'}{k_1 n_1} + \frac{jj'}{k_2 n_2} \right) \tag{2.68}$$

for $i = 0, \ldots, k_1 n_1 - 1$, $j = 0, \ldots, k_2 n_2 - 1$. The eigenvalues for $\boldsymbol{B}$ are

$$\lambda_{ij}^{\boldsymbol{B}}(\boldsymbol{\theta}) = \sum_{i'j'} \theta_{i'j'} \cos 2\pi \left(\frac{ii'}{n_1} + \frac{jj'}{n_2} \right) \tag{2.69}$$

for $i = 0, \ldots, n_1 - 1$, $j = 0, \ldots, n_2 - 1$. Comparing (2.68) and (2.69), we see that

$$\lambda_{ij}^{\boldsymbol{B}}(\boldsymbol{\theta}) = \lambda_{k_0 i, k_1 j}^{\boldsymbol{A}}(\boldsymbol{\theta})$$

for $i = 0, \ldots, n_1 - 1$, $j = 0, \ldots, n_2 - 1$, as $\boldsymbol{n} > 2\boldsymbol{m}$. Since $\boldsymbol{A}$ is SPD then all $\lambda_{ij}^{\boldsymbol{A}}(\boldsymbol{\theta})$ are strictly positive $\forall ij$, hence $\boldsymbol{B}$ is SPD. $\square$

One consequence of Theorem 2.15 is, for example, that

$$\boldsymbol{\Theta}_{\boldsymbol{n}}^{+} \supseteq \boldsymbol{\Theta}_{2\boldsymbol{n}}^{+} \supseteq \boldsymbol{\Theta}_{4\boldsymbol{n}}^{+} \supseteq \boldsymbol{\Theta}_{8\boldsymbol{n}}^{+},$$

but as we have shown by a counterexample,

$$\boldsymbol{\Theta}_{\boldsymbol{n}}^{+} \not\supseteq \boldsymbol{\Theta}_{\boldsymbol{n}+\boldsymbol{1}}^{+}$$

in general. Since the size of $\boldsymbol{\Theta}_{\boldsymbol{n}}^{+}, \boldsymbol{\Theta}_{2\boldsymbol{n}}^{+}, \boldsymbol{\Theta}_{4\boldsymbol{n}}^{+}, \ldots$ is nonincreasing, we may hope that the intersection of all $\boldsymbol{\Theta}_{\boldsymbol{n}}^{+}$'s is nonempty. We define

$$\boldsymbol{\Theta}_{\infty}^{+} = \bigcap_{\boldsymbol{n} > 2\boldsymbol{m}} \boldsymbol{\Theta}_{\boldsymbol{n}}^{+}. \tag{2.70}$$

If we can determine $\boldsymbol{\Theta}_{\infty}^{+}$, then any configuration in this set is valid for all $\boldsymbol{n} > 2\boldsymbol{m}$.

Theorem 2.16 *The set $\boldsymbol{\Theta}_{\infty}^{+}$ as defined in (2.70) is nonempty, bounded, convex and*

$$\boldsymbol{\Theta}_{\infty}^{+} = \left\{ \boldsymbol{\theta} : \sum_{ij} \theta_{ij} \cos(i\omega_1 + j\omega_2) > 0,\ (\omega_1, \omega_2) \in [0, 2\pi)^2 \right\}. \tag{2.71}$$

Note that (2.71) corresponds to the SDF (2.58) being strictly positive.

Proof. The diagonal dominance criterion, see Section 2.1.6, states that if $\sum_{ij} |\theta_{ij}| < 1$ then $\boldsymbol{Q}(\boldsymbol{\theta})$ is SPD for any $\boldsymbol{n} > 2\boldsymbol{m}$, hence $\boldsymbol{\Theta}_{\infty}^{+}$ is nonempty. Further, $\boldsymbol{\Theta}_{\infty}^{+}$ is bounded as $|\theta_{ij}| < 1$ (see Section 2.1.6), and convex by direct verification. For the remaining part of the proof, it is sufficient to show that $\boldsymbol{\theta} \in \boldsymbol{\Theta}_{\infty}^{+}$ implies that $\boldsymbol{\theta} \in \boldsymbol{\Theta}_{\boldsymbol{n}}^{+}$ for any $\boldsymbol{n} > 2\boldsymbol{m}$, and if $\boldsymbol{\theta} \notin \boldsymbol{\Theta}_{\infty}^{+}$ then $\boldsymbol{\theta} \notin \boldsymbol{\Theta}_{\boldsymbol{n}}^{+}$ for at least one $\boldsymbol{n} > 2\boldsymbol{m}$. The first part follows by the definition of $\boldsymbol{\Theta}_{\infty}^{+}$. For the latter part, fix $\boldsymbol{\theta} \notin \boldsymbol{\Theta}_{\infty}^{+}$ so that

$$\sum_{ij} \theta_{ij} \cos(i\omega_1^* + j\omega_2^*) < 0 \tag{2.72}$$

for some irrational numbers (ω_1^*, ω_2^*). By continuity of (2.72), we can find rational numbers i'/n_1 and j'/n_2 such that

$$\sum_{ij} \theta_{ij} \cos(2\pi \frac{ii'}{n_1} + 2\pi \frac{jj'}{n_2}) < 0;$$

hence $\boldsymbol{\theta} \notin \boldsymbol{\Theta}_{\boldsymbol{n}}^{+}$. □

Lakshmanan and Derin (1993) showed that (2.71) is equivalent to a bivariate reciprocal polynomial not having any zeros inside the unit bicircle. They use some classical results concerning the geometry of zero sets of reciprocal polynomials and obtain a complete procedure for verifying the validity of any $\boldsymbol{\theta}$ and for identifying $\boldsymbol{\Theta}_{\infty}^{+}$. The approach taken is still somewhat complex but explicit results for $\boldsymbol{m} = (1, 1)^T$ are known. We will now report these results.

Let $\boldsymbol{m} = (1, 1)^T$ so $\boldsymbol{\theta}$ can be represented as

$$\begin{bmatrix} \text{sym} & \theta_{01} & \theta_{11} \\ \text{sym} & 1 & \theta_{10} \\ \text{sym} & \text{sym} & \theta_{1-1} \end{bmatrix},$$

where the entries marked with 'sym' follow from $\theta_{ij} = \theta_{-i,-j}$. Then $\boldsymbol{\theta} \in \boldsymbol{\Theta}_{\infty}^{+}$ iff the following four conditions are satisfied:

$$\rho = 4(\theta_{11}^2 + \theta_{01}^2 + \theta_{1-1}^2 - \frac{1}{2}\theta_{10}^2) - 1 < 0, \tag{2.73}$$

$$2\left(4\theta_{11}\theta_{1-1} - \theta_{10}^2\right) + 2\left(4\theta_{01}(\theta_{11} + \theta_{1-1}) - 2\theta_{10}\right) + \rho < 0,$$
$$2\left(4\theta_{11}\theta_{1-1} - \theta_{10}^2\right) - 2\left(4\theta_{01}(\theta_{11} + \theta_{1-1}) - 2\theta_{10}\right) + \rho < 0,$$

and either

$$R = 16\left(4\theta_{11}\theta_{1-1} - \theta_{10}^2\right)^2 - \left(4\theta_{01}(\theta_{11} + \theta_{1-1}) - 2\theta_{10}\right)^2 < 0$$

or

$$R \geq 0, \text{ and } R^2 - \Big[-8\rho\left(4\theta_{11}\theta_{1-1} - \theta_{10}^2\right) + 3\left(4\theta_{01}\left(\theta_{11} + \theta_{1-1}\right) - 2\theta_{10}\right)^2\Big]^2 < 0. \tag{2.74}$$

These inequalities are somewhat involved but special cases are of great interest. First consider the case where $\theta_{01} = \theta_{10}$ and $\theta_{11} = \theta_{1-1} = 0$, which gives the requirement $|\theta_{01}| < 1/4$.

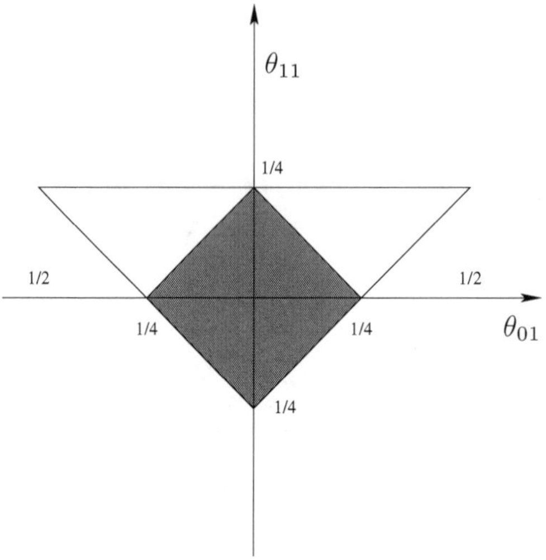

Figure 2.14 *The valid parameter space* $\boldsymbol{\Theta}_\infty^+$ *where* $\theta_{01} = \theta_{10}$ *and* $\theta_{11} = \theta_{1-1}$, *where restriction to diagonal dominance* $\boldsymbol{\Theta}^{++}$ *defined in (2.76), is shown as gray.*

Suppose now we include the diagonal terms $\theta_{01} = \theta_{10}$ and $\theta_{11} = \theta_{1-1}$, the inequalities (2.73) to (2.74) reduce to $|\theta_{01}| - \theta_{11} < 1/4$, and $\theta_{11} < 1/4$. Figure 2.14 shows $\boldsymbol{\Theta}^+$ in this case. The smaller gray area in Figure 2.14 is found from using a sufficient condition only, the diagonal dominant criterion, which we discuss in Section 2.7.2.

The analytical results obtained for a stationary GMRF on $\mathcal{T}_{\boldsymbol{n}}$ are also informative for a nonstationary GMRF on the lattice $\mathcal{I}_{\boldsymbol{n}'}$ with full conditionals

$$\mathrm{E}(x_{ij} \mid \boldsymbol{x}_{-ij}) = - \sum_{\substack{i'j' \neq 00 \\ (i+i', j+j') \in \mathcal{I}_{\boldsymbol{n}'}}} \theta_{i'j'} x_{i+i', j+j'} \qquad \text{and}$$
$$\mathrm{Prec}(x_{ij} \mid \boldsymbol{x}_{-ij}) = 1.$$

The full conditionals equal those in (2.64) in the interior of the lattice, but differ at the boundary. Let $\boldsymbol{\Theta}_{\boldsymbol{n}'}^{+,\mathcal{I}}$ be the valid parameter space for the GMRF on $\mathcal{I}_{\boldsymbol{n}'}$. We now observe that the (block Toeplitz) precision matrix for the GMRF on the lattice is a principal submatrix of the (block-circulant) precision matrix for the GMRF on the torus, if $\boldsymbol{n}' \leq \boldsymbol{n} - \boldsymbol{m}$. The consequence is that

$$\boldsymbol{\Theta}_{\boldsymbol{n}}^+ \subseteq \boldsymbol{\Theta}_{\boldsymbol{n}'}^{+,\mathcal{I}} \quad \text{for } \boldsymbol{n}' \leq \boldsymbol{n} - \boldsymbol{m}. \tag{2.75}$$

Further details are provided in Section 5.1.4. The block-Toeplitz and the block-circulant precision matrix are asymptotically equivalent matrices so $\boldsymbol{\Theta}_{\infty}^{+}$ is also the intersection of $\{\boldsymbol{\Theta}_{\boldsymbol{n}'}^{+;\mathcal{I}}\}$ for all $\boldsymbol{n}' > 2\boldsymbol{m}$. This follows from Grenander and Szegö (1984, p. 65) provided there exists $\delta > 0$ such that the SDF $f(\omega_1, \omega_2) > \delta > 0$, see also Gray (2002, Sec. 4.4).

2.7.2 Diagonal dominance

A frequently used approach in practice is to impose a sufficient condition, *diagonal dominance*, to ensure that $\boldsymbol{Q}(\boldsymbol{\theta})$ is SPD. The diagonal dominance criterion is most often easy to treat analytically. On the downside, the criterion is known to be conservative. After presenting this criterion we will compare $\boldsymbol{\Theta}^{++}$ with the exact results for $\boldsymbol{\Theta}^{+}$ obtained in Section 2.7.1.

Recall from Section 2.1.6 that a matrix $\boldsymbol{A}$ is called *diagonal dominant*, if $|A_{ii}| > \sum_{j:j\neq i} |A_{ij}|$, $\forall i$. If a precision matrix is diagonal dominant, then

$$Q_{ii} > \sum_{j:j\neq i} |Q_{ij}|, \quad \forall i$$

as the diagonal is always strictly positive. It is easy to show that this is a sufficient condition for $\boldsymbol{Q} > 0$:

Theorem 2.17 *Let $\boldsymbol{A}$ be an $n \times n$ diagonal-dominant symmetric matrix with strictly positive diagonal entries, then $\boldsymbol{A}$ is SPD.*

The converse of Theorem 2.17 is not true in general.

Proof. Let λ be an eigenvalue of $\boldsymbol{A}$ with eigenvector $\boldsymbol{v}$ and define $i = \arg\max_j |v_j|$, break ties arbitrarily. Since $\boldsymbol{A}\boldsymbol{v} = \lambda\boldsymbol{v}$ it follows that $\lambda v_i - A_{ii}v_i = \sum_{j:j\neq i} A_{ij}v_j$. Using the triangle inequality we see that

$$|\lambda v_i - A_{ii}v_i| \leq \sum_{j:j\neq i} |A_{ij}v_j| \leq |v_i| \sum_{j:j\neq i} |A_{ij}|$$

so

$$\lambda \geq A_{ii} - \sum_{j:j\neq i} |A_{ij}|.$$

The lower bound is strictly positive as $\boldsymbol{A}$ is diagonal dominant and $A_{ii} > 0$. As λ was any eigenvalue, all n eigenvalues of $\boldsymbol{A}$ are strictly positive and $\boldsymbol{A}$ has full rank. Let $\boldsymbol{\Lambda}$ be a diagonal matrix with the eigenvalues on the diagonal and the corresponding eigenvectors in the corresponding column of $\boldsymbol{V}$, so $\boldsymbol{A} = \boldsymbol{V}\boldsymbol{\Lambda}\boldsymbol{V}^T$. For $\boldsymbol{x} \neq \boldsymbol{0}$,

$$\boldsymbol{x}^T\boldsymbol{A}\boldsymbol{x} = \boldsymbol{x}^T(\boldsymbol{V}\boldsymbol{\Lambda}\boldsymbol{V}^T)\boldsymbol{x} = (\boldsymbol{V}^T\boldsymbol{x})^T\boldsymbol{\Lambda}(\boldsymbol{V}^T\boldsymbol{x}) > 0,$$

hence $\boldsymbol{A}$ is SPD. □

Using the diagonal dominance criterion on the stationary GMRF in Section 2.7.1, we obtain

$$\boldsymbol{\Theta}^{++} = \left\{ \boldsymbol{\theta} : \sum_{i'j' \neq 00} |\theta_{i'j'}| < 1 \right\}. \tag{2.76}$$

Compared to the complexity of locating $\boldsymbol{\Theta}^{+}$ given by the inequalities (2.73) to (2.74) for $\boldsymbol{m} = (1,1)^T$, the simplicity of (2.76) is striking. However, if $\boldsymbol{\Theta}^{+}_{\infty}$ is too conservative we might lose more than we gain.

Let us reconsider the special cases of the inequalities (2.73) to (2.74). First consider the case where $\theta_{01} = \theta_{10}$ and $\theta_{11} = \theta_{1-1} = 0$. Diagonal dominance gives the requirement $|\theta_{01}| < 1/4$. This is the same as the diagonal dominance criterion, hence $\boldsymbol{\Theta}^{++} = \boldsymbol{\Theta}^{+}_{\infty}$. If we include the diagonal terms $\theta_{01} = \theta_{10}$ and $\theta_{11} = \theta_{1-1}$, we obtain the constraints: $|\theta_{01}| + |\theta_{11}| < 1/4$. Both $\boldsymbol{\Theta}^{+}_{\infty}$ and $\boldsymbol{\Theta}^{++}$ are shown in Figure 2.14, where $\boldsymbol{\Theta}^{++}$ is the gray area. We see that $\boldsymbol{\Theta}^{+}_{\infty}$ is twice the size of $\boldsymbol{\Theta}^{++}$, hence using a diagonal-dominant parameterization is much more restrictive than necessary.

Extrapolating these results suggests that diagonal dominance as a criterion for positive definiteness becomes more restrictive for an increasing number of parameters. Although we do not have a formal proof of this claim, we have verified it through simulation in the one-dimensional case using an autoregressive process of order M. We can define $\boldsymbol{\Theta}^{+}$ implicit using the partial correlation function defined on $(-1,1)^M$ and compute the parameters $\boldsymbol{\theta}$ using the Levinson algorithm (Brockwell and Davis, 1987).

2.8 Bibliographic notes

The definitions of matrices and SPD matrices are extracted from Searle (1982) and Harville (1997). The definitions of graph related terms and conditional independence are extracted from Whittaker (1990) and Lauritzen (1996).

Guyon (1995), Mardia (1988), Whittaker (1990) and Lauritzen (1996) are alternative sources for the results in Section 2.2. Brook's lemma (Brook, 1964) is discussed by Besag (1974). Multivariate GMRFs are discussed in Mardia (1988).

Circulant matrices are discussed by Davis (1979) and Gray (2002). The derivation of the eigenvalues and eigenvectors in Section 2.6.1 follow Gray (2002).

Algorithm 2.10 is the FFT algorithm to sample stationary Gaussian fields on toruses and has been reinvented several times. An early reference is Woods (1972), see also Dietrich and Newsam (1997), Hunt

(1973), Krogstad (1989) and Wood and Chan (1994). Dietrich and Newsam (1996) discuss a nice extension to conditional simulation by cyclic embedding.

Gray (2002) discuss Toeplitz matrices and their circulant approximations, which is our source for Section 2.6.4, see also Grenander and Szegö (1984) for a more technical discussion. The proofs in Section 2.6.5 are based on some unpublished notes by J. Besag. Box and Tiao (1973) also make use of (2.31) to evaluate a log density under constraints.

The statistical aspects in Section 2.3 and Section 2.4 are from Rue (2001) and Rue and Follestad (2003). Numerical methods for sparse matrices are discussed in Dongarra et al. (1998), Duff et al. (1989), George and Liu (1981) and Gupta (2002) gives a review and comparison. Section 2.3.1 follows any standard reference in numerical linear algebra, for example, Golub and van Loan (1996). Wilkinson and Yeung (2002, 2004) discuss propagation algorithms and their connection to the sparse matrix approach. Rue (2001) presents a divide-and-conquer strategy for the simulation of large GMRFs using iterative numerical techniques for SPD linear systems. Steinsland (2003) discusses sampling from GMRFs using parallel algorithms for sparse matrices while Wilkinson (2004) discuss parallel computing relevant for statistics in general.

A similar result to Theorem 2.7 can also be derived for the Cholesky triangle of the covariance matrix. For details regarding conditioning by kriging, see, for example, Chilés and Delfiner (1999), Cressie (1993) and Lantuéjoul (2002).

Gaussian random fields with a continuous parameter obeying the Markov property are not discussed in this chapter. Refer to Künsch (1979) for a well-written survey of the subject, see also Pitt (1971), Wong (1969), and Pitt and Robeva (2003).

CHAPTER 3

Intrinsic Gaussian Markov random fields

This chapter will introduce a special type of GMRFs, so-called *intrinsic GMRFs* (IGMRF). IGMRFs are *improper*, i.e., they have precision matrices not of full rank. We will use these quite extensively later on as prior distributions in various applications. Of particular importance are IGMRFs that are invariant to any trend that is a polynomial of the locations of the nodes up to a specific order.

IGMRFs have been studied in some depth on regular lattices. There have been different attempts to define the *order* of an IGMRF, Besag and Kooperberg (1995) adopt a definition for GMRFs to IGMRFs, where the order is defined through the chosen neighborhood structure. However, Matheron (1973) uses the term to describe the level of degeneracy of continuum intrinsic processes, and subsequently Künsch (1987) does the same for IGMRFs. This chapter also describes IGMRFs on irregular lattices and nonpolynomial IGMRFs, so a more general definition is needed. Inspired by Matheron (1973) and Künsch (1987), we define the order of an IGMRF as the rank deficiency of its precision matrix. This seems to be a natural choice, in particular since we only discuss IGMRF's on finite graphs. Note, however, that any autoregression that merely has an indeterminate mean is zero order according to Künsch (1987), but first order according to our definition.

To prepare the forthcoming construction of IGMRF's, we will start with a section where we first introduce some additional notation and describe forward differences and polynomials. We then discuss the conditional properties of $\boldsymbol{x}|\boldsymbol{Ax}$, where $\boldsymbol{x}$ is a GMRF. This will be valuable to understand and construct IGMRFs on finite graphs.

3.1 Preliminaries

3.1.1 Some additional definitions

The *null space* of a matrix $\boldsymbol{A}$ is the set of all vectors $\boldsymbol{x}$ such that $\boldsymbol{Ax} = \boldsymbol{0}$. The *nullity* of $\boldsymbol{A}$ is the dimension of the null space. For an $n \times m$ matrix the rank is $\min\{n, m\} - k$ where k is the nullity. For a singular $n \times n$ matrix $\boldsymbol{A}$ with nullity k, we denote by $|\boldsymbol{A}|^*$ the product of the $n-k$ nonzero eigenvalues of $\boldsymbol{A}$. We label this term the *generalized determinant* due to lack of any standard terminology.

The *Kronecker product* of two matrices $\boldsymbol{A}$ and $\boldsymbol{B}$ is denoted by $\boldsymbol{A} \otimes \boldsymbol{B}$ and produces a larger matrix with a special block structure. Let $\boldsymbol{A}$ be a $n \times m$ matrix and $\boldsymbol{B}$ a $p \times q$ matrix, then their Kronecker product

$$\boldsymbol{A} \otimes \boldsymbol{B} = \begin{pmatrix} A_{11}\boldsymbol{B} & \dots & A_{1m}\boldsymbol{B} \\ \vdots & \ddots & \vdots \\ A_{n1}\boldsymbol{B} & \dots & A_{nm}\boldsymbol{B} \end{pmatrix}$$

is an $np \times mq$ matrix. For example,

$$\begin{pmatrix} 1 & 2 \\ 0 & -1 \end{pmatrix} \otimes \begin{pmatrix} 1 & 2 & 3 \\ 4 & 5 & 6 \end{pmatrix} = \begin{pmatrix} 1 & 2 & 3 & 2 & 4 & 6 \\ 4 & 5 & 6 & 8 & 10 & 12 \\ 0 & 0 & 0 & -1 & -2 & -3 \\ 0 & 0 & 0 & -4 & -5 & -6 \end{pmatrix}.$$

Let $\boldsymbol{A}$, $\boldsymbol{B}$, $\boldsymbol{C}$, and $\boldsymbol{D}$ be matrices of appropriate dimensions. The basic properties of the Kronecker product are the following:

1. For a scalar a

$$\begin{aligned} \boldsymbol{A} \otimes (a\boldsymbol{B}) &= a(\boldsymbol{A} \otimes \boldsymbol{B}) \\ (a\boldsymbol{A}) \otimes \boldsymbol{B} &= a(\boldsymbol{A} \otimes \boldsymbol{B}) \end{aligned}$$

2. Kronecker product distributes over addition

$$\begin{aligned} (\boldsymbol{A} + \boldsymbol{B}) \otimes \boldsymbol{C} &= (\boldsymbol{A} \otimes \boldsymbol{C}) + (\boldsymbol{B} \otimes \boldsymbol{C}) \\ \boldsymbol{A} \otimes (\boldsymbol{B} + \boldsymbol{C}) &= (\boldsymbol{A} \otimes \boldsymbol{B}) + (\boldsymbol{A} \otimes \boldsymbol{C}) \end{aligned}$$

3. The Kronecker product is associative

$$(\boldsymbol{A} \otimes \boldsymbol{B}) \otimes C = \boldsymbol{A} \otimes (\boldsymbol{B} \otimes \boldsymbol{C})$$

4. The Kronecker product is in general not commutative

$$\boldsymbol{A} \otimes \boldsymbol{B} \neq \boldsymbol{B} \otimes \boldsymbol{A}$$

5. Transpose does not invert order

$$(\boldsymbol{A} \otimes \boldsymbol{B})^T = \boldsymbol{A}^T \otimes \boldsymbol{B}^T$$

6. Matrix multiplication

$$(\boldsymbol{A} \otimes \boldsymbol{B})(\boldsymbol{C} \otimes \boldsymbol{D}) = \boldsymbol{AC} \otimes \boldsymbol{BD}$$

7. For invertible matrices $\boldsymbol{A}$ and $\boldsymbol{B}$

$$(\boldsymbol{A} \otimes \boldsymbol{B})^{-1} = \boldsymbol{A}^{-1} \otimes \boldsymbol{B}^{-1}$$

8. For an $n \times n$ matrix $\boldsymbol{A}$ and $m \times m$ matrix $\boldsymbol{B}$

$$|\boldsymbol{A} \otimes \boldsymbol{B}| = |\boldsymbol{A}|^m \, |\boldsymbol{B}|^n \tag{3.1}$$

9. Rank of the Kronecker product of two matrices

$$\text{rank}(\boldsymbol{A} \otimes \boldsymbol{B}) = \text{rank}(\boldsymbol{A}) \, \text{rank}(\boldsymbol{B})$$

3.1.2 Forward differences

Intrinsic GMRFs of order k in one dimension are often constructed using forward differences of order k. Here we introduce the necessary notation.

Definition 3.1 (Forward difference) *Define the first-order forward difference of a function* $f(\cdot)$ *as*

$$\Delta f(z) = f(z+1) - f(z).$$

Higher-order forward differences are defined recursively:

$$\Delta^k f(z) = \Delta\Delta^{k-1} f(z)$$

so

$$\Delta^2 f(z) = f(z+2) - 2f(z+1) + f(z) \tag{3.2}$$

and in general for $k = 1, 2, \ldots,$

$$\Delta^k f(z) = (-1)^k \sum_{j=0}^{k} (-1)^j \binom{k}{j} f(z+j).$$

For a vector $\boldsymbol{z} = (z_1, z_2, \ldots, z_n)^T$, $\Delta\boldsymbol{z}$ has elements $\Delta z_i = z_{i+1} - z_i$, $i = 1, \ldots, n-1$.

We can think of the forward difference of kth order as an approximation to the kth derivative of $f(z)$. Consider, for example, the first derivative

$$f'(z) = \lim_{h \to 0} \frac{f(z+h) - f(z)}{h},$$

which for $h = 1$ equals the first-order forward difference.

3.1.3 Polynomials

Many IGMRFs are invariant to the addition of a polynomial of a certain degree. Here we introduce the necessary notation for polynomials, first on a line and then in higher dimensions.

Let $s_1 < s_2 < \cdots < s_n$ denote the ordered locations on the line and define $\boldsymbol{s} = (s_1, \ldots, s_n)^T$. Let $p_k(s_i)$ denote a polynomial of degree k, evaluated at the locations $\boldsymbol{s}$,

$$p_k(s_i) = \beta_0 + \beta_1 s_i + \frac{1}{2}\beta_2 s_i^2 + \cdots + \frac{1}{k!}\beta_k s_i^k, \tag{3.3}$$

with some coefficients $\boldsymbol{\beta}_k = (\beta_0, \ldots, \beta_k)^T$. In matrix notation, this is

$$\begin{pmatrix} p_k(s_1) \\ p_k(s_2) \\ \vdots \\ p_k(s_n) \end{pmatrix} = \begin{pmatrix} 1 & s_1 & \ldots & \frac{1}{k!}s_1^k \\ 1 & s_2 & \ldots & \frac{1}{k!}s_2^k \\ \vdots & \vdots & & \vdots \\ 1 & s_n & \ldots & \frac{1}{k!}s_n^k \end{pmatrix} \begin{pmatrix} \beta_0 \\ \beta_1 \\ \vdots \\ \beta_k \end{pmatrix}, \tag{3.4}$$

which we write compactly as

$$\boldsymbol{p}_k = \boldsymbol{S}_k \boldsymbol{\beta}_k. \tag{3.5}$$

The matrix $\boldsymbol{S}_k$ is called the *polynomial design matrix of degree k*. Throughout we assume that $\boldsymbol{S}_k$ is of full rank $k+1$.

Polynomials can also be defined in higher dimension, but this requires more notation. Let

$$\boldsymbol{s}_i = (s_{i_1}, s_{i_2}, \ldots, s_{i_d})$$

denote the spatial location of node i, where s_{i_j} is the location of node i in the jth dimension, $j = 1, \ldots, d$. We now use a compact notation and define

$$\begin{aligned} \boldsymbol{j} &= (j_1, j_2, \ldots, j_d), \\ \boldsymbol{s}_{\boldsymbol{i}}^{\boldsymbol{j}} &= s_{i_1}^{j_1} s_{i_2}^{j_2} \cdots s_{i_d}^{j_d}, \qquad \text{and} \\ \boldsymbol{j!} &= j_1! j_2! \cdots j_d!. \end{aligned}$$

A polynomial trend of degree k in d dimensions will consist of

$$m_{k,d} = \binom{d+k}{k} \tag{3.6}$$

terms and is expressed as

$$p_{k,d}(\boldsymbol{s}_i) = \sum_{0 \le j_1 + \cdots + j_d \le k} \frac{1}{\boldsymbol{j!}} \beta_{\boldsymbol{j}} \boldsymbol{s}_{\boldsymbol{i}}^{\boldsymbol{j}}, \tag{3.7}$$

where $j_l \in \{0, 1, \ldots, k\}$ for $l \in \{1, 2, \ldots, d\}$.

For illustration, for $d = 2$ the number of terms $m_{k,d}$ is 3, 6 and 10 for $k = 1, 2$ and 3. For $d = k = 2$, (3.7) is

$$p_{2,2}(s_{i_1}, s_{i_2}) = \beta_{00} + \beta_{10} s_{i_1} + \beta_{01} s_{i_2} + \frac{1}{2}\beta_{20} s_{i_1}^2 + \beta_{11} s_{i_1} s_{i_2} + \frac{1}{2}\beta_{02} s_{i_2}^2. \tag{3.8}$$

Similar to (3.5), let $\boldsymbol{p}_{k,d}$ denote the vector of the polynomial evaluated at each of the n locations. We can express this vector as

$$\boldsymbol{p}_{k,d} = \boldsymbol{S}_{k,d} \boldsymbol{\beta}_{k,d}, \tag{3.9}$$

where $\boldsymbol{\beta}_{k,d}$ is a vector of all the coefficients and $\boldsymbol{S}_{k,d}$ is the polynomial designmatrix with elements $\frac{1}{j!}\boldsymbol{s}_i^j$. We do not consider degenerated cases and assume therefore that $\boldsymbol{S}_{k,d}$ is of full rank.

3.2 GMRFs under linear constraints

IGMRFs have much in common with GMRFs conditional on linear constraints. In this section we consider proper (full-rank) GMRFs, and derive their precision matrix, conditional on such linear constraints. We then introduce informally IGMRFs. IGMRFs are always improper, i.e., their precision matrices do not have full rank.

Let $\boldsymbol{x}$ be a zero mean GMRF of dimension n with precision matrix $\boldsymbol{Q} > 0$. Let $\lambda_1, \lambda_2, \ldots, \lambda_n$ be the eigenvalues and $\boldsymbol{e}_1, \ldots, \boldsymbol{e}_n$ the corresponding eigenvectors of $\boldsymbol{Q}$, such that

$$\boldsymbol{Q} = \sum_i \lambda_i \boldsymbol{e}_i \boldsymbol{e}_i^T = \boldsymbol{V}\boldsymbol{\Lambda}\boldsymbol{V}^T,$$

where $\boldsymbol{V} = (\boldsymbol{e}_1, \boldsymbol{e}_2, \ldots, \boldsymbol{e}_n)$, $\boldsymbol{V}^T\boldsymbol{V} = \boldsymbol{I}$ and $\boldsymbol{\Lambda} = \text{diag}(\lambda_1, \lambda_2, \ldots, \lambda_n)$. Consider now the conditional density

$$\pi(\boldsymbol{x} \mid \boldsymbol{A}\boldsymbol{x} = \boldsymbol{a}), \tag{3.10}$$

where the $k \times n$ matrix $\boldsymbol{A}$ has the special form

$$\boldsymbol{A}^T = (\boldsymbol{e}_1, \boldsymbol{e}_2, \ldots, \boldsymbol{e}_k) \tag{3.11}$$

and $\boldsymbol{a} = (a_1, \ldots, a_k)^T$ is arbitrary. The specific form of $\boldsymbol{A}$ is not a restriction as we will explain at the end of this section.

To derive the explicit form of (3.10), it is useful to change variables to $\boldsymbol{y} = \boldsymbol{V}^T\boldsymbol{x}$, which can easily be shown to have mean zero and $\text{Prec}(\boldsymbol{y}) = \boldsymbol{\Lambda}$, i.e., $y_i \overset{\text{iid}}{\sim} \mathcal{N}(0, \lambda_i^{-1})$, $i = 1\ldots, n$. Now $\boldsymbol{A}\boldsymbol{x} = \boldsymbol{y}_{1:k}$ where $\boldsymbol{y}_{i:j} = (y_i, y_{i+1}, \ldots, y_j)^T$ and

$$\pi(\boldsymbol{y} \mid \boldsymbol{A}\boldsymbol{x} = \boldsymbol{a}) = 1_{[\boldsymbol{y}_{1:k}=\boldsymbol{a}]} \prod_{i=k+1}^{n} \pi(y_i).$$

Hence $\text{E}(\boldsymbol{y}|\boldsymbol{A}\boldsymbol{x} = \boldsymbol{a}) = (\boldsymbol{a}^T, \boldsymbol{0}^T)^T$ and $\text{Prec}(\boldsymbol{y}|\boldsymbol{A}\boldsymbol{x} = \boldsymbol{a}) = \widetilde{\boldsymbol{\Lambda}}$, where $\widetilde{\boldsymbol{\Lambda}} = \text{diag}(0, \ldots, 0, \lambda_{k+1}, \ldots, \lambda_n)$, from which it follows that

$$\begin{aligned} \text{E}(\boldsymbol{x} \mid \boldsymbol{A}\boldsymbol{x} = \boldsymbol{a}) &= \boldsymbol{V}\begin{pmatrix}\boldsymbol{a}\\ \boldsymbol{0}\end{pmatrix} = a_1\boldsymbol{e}_1 + \cdots + a_k\boldsymbol{e}_k \qquad \text{and} \\ \text{Prec}(\boldsymbol{x} \mid \boldsymbol{A}\boldsymbol{x} = \boldsymbol{a}) &= \boldsymbol{V}\widetilde{\boldsymbol{\Lambda}}\boldsymbol{V}^T. \end{aligned}$$

Hence (3.10) can be written as

$$\begin{aligned}\log \pi(\boldsymbol{x} \mid \boldsymbol{A}\boldsymbol{x} = \boldsymbol{a}) &= -\frac{n-k}{2}\log 2\pi + \frac{1}{2}\sum_{i=k+1}^{n} \log \lambda_i \\ &\quad - \frac{1}{2}\left(\boldsymbol{x} - \boldsymbol{V}\begin{pmatrix}\boldsymbol{a}\\ \boldsymbol{0}\end{pmatrix}\right)^T \boldsymbol{V}\widetilde{\boldsymbol{\Lambda}}\boldsymbol{V}^T \left(\boldsymbol{x} - \boldsymbol{V}\begin{pmatrix}\boldsymbol{a}\\ \boldsymbol{0}\end{pmatrix}\right) \qquad (3.12)\\ &= -\frac{n-k}{2}\log 2\pi + \frac{1}{2}\sum_{i=k+1}^{n} \log \lambda_i - \frac{1}{2}\boldsymbol{x}^T\widetilde{\boldsymbol{Q}}\boldsymbol{x},\end{aligned}$$

where $\widetilde{\boldsymbol{Q}} = \boldsymbol{V}\widetilde{\boldsymbol{\Lambda}}\boldsymbol{V}^T$. Note that $\boldsymbol{e}_1, \ldots, \boldsymbol{e}_k$ do not contribute to (3.12) explicitly, and that (3.12) does depend on $\boldsymbol{a}$ only implicitly in the sense that $\pi(\boldsymbol{x}|\boldsymbol{A}\boldsymbol{x} = \boldsymbol{a})$ is nonzero only for $\boldsymbol{A}\boldsymbol{x} = \boldsymbol{a}$. Hence, for the specific choice of $\boldsymbol{A}$ in (3.11), (3.10) has a particularly simple form. It can be obtained from the corresponding unconstrained density $\pi(\boldsymbol{x})$ by (a) setting all eigenvalues in $\boldsymbol{\Lambda}$ that correspond to eigenvectors in $\boldsymbol{A}$ to zero and (b) adjusting the normalizing constant accordingly.

Example 3.1 *Assume*

$$\boldsymbol{Q} = \begin{pmatrix} 6 & -1 & 0 & -1 \\ -1 & 6 & -1 & 0 \\ 0 & -1 & 6 & -1 \\ -1 & 0 & -1 & 6 \end{pmatrix},$$

then $\boldsymbol{\Lambda} = diag(4, 6, 6, 8)$ *and* $\boldsymbol{e}_1 = 1/2 \cdot (1, 1, 1, 1)^T$, $\boldsymbol{e}_2 = \sqrt{1/2} \cdot (1, 0, -1, 0)^T$, $\boldsymbol{e}_3 = \sqrt{1/2} \cdot (0, -1, 0, 1)^T$ *and* $\boldsymbol{e}_4 = 1/2 \cdot (1, -1, 1, -1)^T$. *If we now condition on* $\boldsymbol{e}_1^T\boldsymbol{x} = 0$, *we obtain the conditional precision matrix:*

$$\widetilde{\boldsymbol{Q}} = \begin{pmatrix} 5 & -2 & -1 & -2 \\ -2 & 5 & -2 & -1 \\ -1 & -2 & 5 & -2 \\ -2 & -1 & -2 & 5 \end{pmatrix}.$$

Note that each row (and of course each column) sums up to zero.

Of key interest for IGMRFs is an *improper* version of the (log) density (3.12), which we *define* as

$$\log \pi^*(\boldsymbol{x}) = -\frac{n-k}{2}\log 2\pi + \frac{1}{2}\sum_{i=k+1}^{n} \log \lambda_i - \frac{1}{2}\boldsymbol{x}^T\widetilde{\boldsymbol{Q}}\boldsymbol{x} \qquad (3.13)$$

for *any* $\boldsymbol{x} \in \mathbb{R}^n$. The rank of $\widetilde{\boldsymbol{Q}}$ is $n-k$ and appears because $\pi^*(\boldsymbol{x})$ is now defined for all $\boldsymbol{x} \in \mathbb{R}^n$, and not just for those $\boldsymbol{x}$ that satisfy $\boldsymbol{A}\boldsymbol{x} = \boldsymbol{a}$, as was the case for (3.12).

To see more specifically what is going on, note that any $\boldsymbol{x} \in \mathbb{R}^n$ can be decomposed as

$$\begin{aligned} \boldsymbol{x} &= (c_1\boldsymbol{e}_1 + \cdots + c_k\boldsymbol{e}_k) + (d_{k+1}\boldsymbol{e}_{k+1} + \cdots + d_n\boldsymbol{e}_n) \\ &= \boldsymbol{x}^{\parallel} + \boldsymbol{x}^{\perp}, \end{aligned} \tag{3.14}$$

where $\boldsymbol{x}^{\parallel}$ is the part of $\boldsymbol{x}$ in the subspace spanned by the columns of $\boldsymbol{A}^T$, the *null space* of $\widetilde{\boldsymbol{Q}}$, and $\boldsymbol{x}^{\perp}$ is the part of $\boldsymbol{x}$ orthogonal to $\boldsymbol{x}^{\parallel}$, where of course $(\boldsymbol{x}^{\parallel})^T\boldsymbol{x}^{\perp} = 0$. For a given $\boldsymbol{x}$, the coefficients in (3.14) are $c_i = \boldsymbol{e}_i^T\boldsymbol{x}$ and $d_j = \boldsymbol{e}_j^T\boldsymbol{x}$.

Using this decomposition we immediately see that

$$\pi^*(\boldsymbol{x}) = \pi^*(\boldsymbol{x}^{\perp}), \tag{3.15}$$

so $\pi^*(\boldsymbol{x})$ does not depend on $c_1, \ldots, c_k$. Hence, π^* is *invariant* to the addition of any $\boldsymbol{x}^{\parallel}$ and this is *the* important feature of IGMRFs. Also note that $\pi^*(\boldsymbol{x}^{\perp})$ is equal to $\pi(\boldsymbol{x}|\boldsymbol{A}\boldsymbol{x} = \boldsymbol{a})$.

We can interpret $\pi^*(\boldsymbol{x})$ as a limiting form of a proper density $\tilde{\pi}(\boldsymbol{x})$. Let $\breve{\pi}(\boldsymbol{x}^{\parallel})$ be the density of $\boldsymbol{x}^{\parallel}$ and define the proper density

$$\tilde{\pi}(\boldsymbol{x}) = \pi^*(\boldsymbol{x}^{\perp})\, \breve{\pi}(\boldsymbol{x}^{\parallel}).$$

Let $\breve{\pi}(\boldsymbol{x}^{\parallel})$ be a zero mean Gaussian with precision matrix $\gamma\boldsymbol{I}$. Then in the limit as $\gamma \to 0$,

$$\tilde{\pi}(\boldsymbol{x}) \propto \pi^*(\boldsymbol{x}). \tag{3.16}$$

Roughly speaking, $\pi^*(\boldsymbol{x})$ can be decomposed into the proper density for $\boldsymbol{x}^{\perp} \in \mathbb{R}^{n-k}$ times a diffuse improper density for $\boldsymbol{x}^{\parallel} \in \mathbb{R}^k$.

Example 3.2 *Consider again Example 3.1, where we now look at the improper density π^* defined in (3.13) with 'mean' zero and 'precision' $\widetilde{\boldsymbol{Q}}$. Suppose we are interested in the density value π^* of the vector $\boldsymbol{x} = (0, -2, 0, -2)^T$, which can be factorized into $\boldsymbol{x} = 2\boldsymbol{e}_1 + 2\boldsymbol{e}_4$, hence $\boldsymbol{x}^{\parallel} = 2\boldsymbol{e}_1$ and $\boldsymbol{x}^{\perp} = 2\boldsymbol{e}_4$. Since $\boldsymbol{e}_1$ is a constant vector, the density $\pi^*(\boldsymbol{x})$ is invariant to the addition of any arbitrary constant to $\boldsymbol{x}$. This can be interpreted as a diffuse prior on the overall level of $\boldsymbol{x}$, i.e., $\boldsymbol{x}^{\parallel} \sim \mathcal{N}(\boldsymbol{0}, \kappa^{-1}\boldsymbol{I})$ with $\kappa \to 0$.*

Using this interpretation of $\pi^*(\boldsymbol{x})$, we will now define how to generate a 'sample' from $\pi^*(\boldsymbol{x})$, where we use quotes to emphasize that the density is actually improper. Since the rank deficiency is only due to $\boldsymbol{x}^{\parallel}$, we *define* that a sample from $\pi^*(\boldsymbol{x})$ means a sample from the proper part $\pi^*(\boldsymbol{x}^{\perp})$, bearing in mind (3.15). For known eigenvalues and eigenvectors of $\widetilde{\boldsymbol{Q}}$ it is easy to sample from $\pi^*(\boldsymbol{x}^{\perp})$ using Algorithm 3.1.

Example 3.3 *We have generated 1000 samples from π^* defined in Example 3.2, shown in a 'pairs plot' in Figure 3.1. At first sight, these samples seem well-behaved and proper, but note that $\boldsymbol{1}^T\boldsymbol{x} = 0$ for all samples, and that the empirical correlation matrix is singular.*

Algorithm 3.1 Sampling from an improper GMRF with mean zero

for $j = k+1$ to n **do**
  $y_j \sim \mathcal{N}(0, \lambda_j^{-1})$
end for
Return $\boldsymbol{x} = y_{k+1}\boldsymbol{e}_{k+1} + y_{k+2}\boldsymbol{e}_{k+2} + \cdots + y_n\boldsymbol{e}_n$

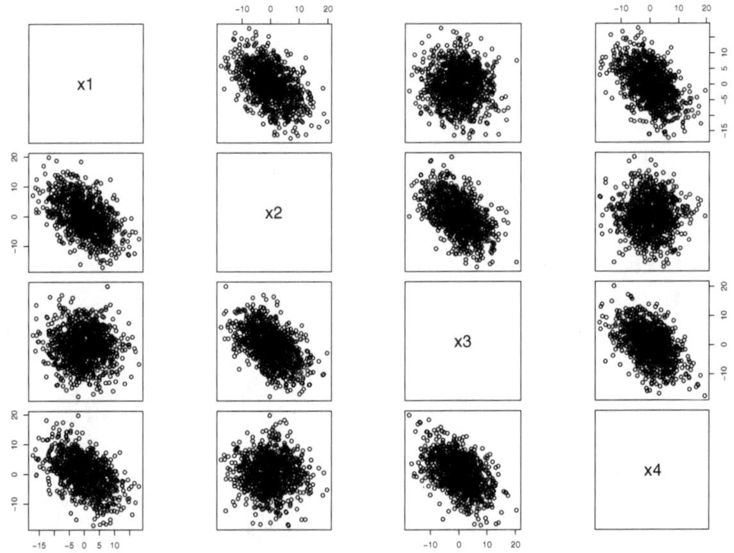

Figure 3.1 *Pairs plot for 1000 samples from an improper GMRF with mean and precision defined in Example 3.2.*

As it will become clear later, in most cases the matrix $\widetilde{\boldsymbol{Q}}$ and the eigenvectors $\boldsymbol{e}_1, \ldots, \boldsymbol{e}_k$ are known explicitly *by construction.* However, the remaining eigenvalues and eigenvectors will typically not be known. Hence an alternative algorithm based on Algorithm 2.6 will be useful. Here we use the fact that $\pi^*(\boldsymbol{x}^\perp)$ equals $\pi(\boldsymbol{x}|\boldsymbol{A}\boldsymbol{x} = \boldsymbol{0})$, from which we can sample in two steps: first sample from the unconstrained density and then correct the obtained sample for the constraint $\boldsymbol{A}\boldsymbol{x} = \boldsymbol{0}$ via Equation (2.30). More specifically, for $\pi(\boldsymbol{x})$ we have to use a zero mean GMRF with SPD precision matrix

$$\boldsymbol{Q} = \widetilde{\boldsymbol{Q}} + \sum_{i=1}^{k} a_i \boldsymbol{e}_i \boldsymbol{e}_i^T.$$

Note that this method works for any strictly positive values of $a_1, \ldots, a_k$;

for simplicity we may use $a_1 = \ldots = a_k = 1$.

Example 3.4 *In the previous example* $\mathbf{e}_1\mathbf{e}_1^T$ *is a matrix with entries equal to* $1/4$*, hence* $\boldsymbol{Q}$ *can be obtained by adding an arbitrary but strictly positive value to* $\widetilde{\boldsymbol{Q}}$*. For any such* $\boldsymbol{Q}$*, the correction step (2.30) now simply corresponds to the subtraction of the (unweighted) mean value of the sample.*

Finally, we will comment on the specific form of $\boldsymbol{A}$ in (3.11). At first sight this choice seems rather restrictive, but we will now explain why this is not the case. Consider the more general situation where we want to compute the conditional density of $\boldsymbol{x}|\boldsymbol{Bx} = \boldsymbol{b}$ for *any* $k \times n$ matrix $\boldsymbol{B}$ with rank $0 < k < n$. If $\text{Cov}(\boldsymbol{x}) = \boldsymbol{\Sigma} > 0$, then

$$\text{Cov}(\boldsymbol{x} \mid \boldsymbol{Bx}) = \boldsymbol{\Sigma} - \boldsymbol{\Sigma}\boldsymbol{B}^T \left(\boldsymbol{B}\boldsymbol{\Sigma}\boldsymbol{B}^T\right)^{-1} \boldsymbol{B}\boldsymbol{\Sigma}. \tag{3.17}$$

The generality of our argument is evident by verifying that the k columns of $\boldsymbol{B}^T$ span the null space of $\text{Cov}(\boldsymbol{x}|\boldsymbol{Bx})$, i.e.,

$$\left(\boldsymbol{\Sigma} - \boldsymbol{\Sigma}\boldsymbol{B}^T \left(\boldsymbol{B}\boldsymbol{\Sigma}\boldsymbol{B}^T\right)^{-1} \boldsymbol{B}\boldsymbol{\Sigma}\right) \boldsymbol{B}^T = \boldsymbol{0}.$$

The condition $\boldsymbol{Bx} = \boldsymbol{b}$ is then equivalent to $\boldsymbol{Ax} = \boldsymbol{a}$ in (3.11) expressed in terms of those eigenvectors of (3.17) that have zero eigenvalues.

3.3 IGMRFs of first order

Using the results from Section 3.2, we will now define (polynomial) intrinsic GMRFs of first order. We start by defining an improper GMRF with rank deficiency k.

Definition 3.2 (Improper GMRF) *Let* $\boldsymbol{Q}$ *be an* $n \times n$ *SPSD matrix with rank* $n - k > 0$*. Then* $\boldsymbol{x} = (x_1, \ldots, x_n)^T$ *is an improper GMRF of rank* $n - k$ *with parameters* $(\boldsymbol{\mu}, \boldsymbol{Q})$*, if its density is*

$$\pi(\boldsymbol{x}) = (2\pi)^{\frac{-(n-k)}{2}} \left(|\boldsymbol{Q}|^*\right)^{1/2} \exp\left(-\frac{1}{2}(\boldsymbol{x} - \boldsymbol{\mu})^T \boldsymbol{Q}(\boldsymbol{x} - \boldsymbol{\mu})\right). \tag{3.18}$$

Further, $\boldsymbol{x}$ *is an improper GMRF wrt to the labelled graph* $\mathcal{G} = (\mathcal{V}, \mathcal{E})$*, where*

$$Q_{ij} \neq 0 \iff \{i, j\} \in \mathcal{E} \quad \textit{for all} \quad i \neq j.$$

Recall that $|\cdot|^*$ denote the generalized determinant as defined in Section 3.1.1. The parameters $(\boldsymbol{\mu}, \boldsymbol{Q})$ do *no longer* represent the mean and the precision since they formally do not exist; however, for convenience we will continue to denote them as the 'mean' and the 'precision', even without the quotes. The Markov properties of an IGMRF are to be interpreted as those obtained from the limit of a proper density. This is similar to the argument leading to (3.16). Let the columns of $\boldsymbol{A}^T$ span the null space of $\boldsymbol{Q}$, then define

$$\boldsymbol{Q}(\gamma) = \boldsymbol{Q} + \gamma \boldsymbol{A}^T \boldsymbol{A}. \tag{3.19}$$

Now each element in $\boldsymbol{Q}(\gamma)$ tends to the corresponding one in $\boldsymbol{Q}$ as $\gamma \to 0$. Similarly, a statement like

$$\mathrm{E}(x_i \mid \boldsymbol{x}_{-i}) = \mu_i - \frac{1}{Q_{ii}} \sum_{j \sim i} Q_{ij}(x_j - \mu_j)$$

(using (2.5)) will be meaningful by the same limiting argument.

An IGMRF of first order is an improper GMRF of rank $n-1$, where the vector $\mathbf{1}$ spans the null space of $\boldsymbol{Q}$.

Definition 3.3 (IGMRF of first order) *An intrinsic GMRF of first order is an improper GMRF of rank $n-1$ where $\boldsymbol{Q}\mathbf{1} = \mathbf{0}$.*

The condition $\boldsymbol{Q}\mathbf{1} = \mathbf{0}$ simply means that $\sum_j Q_{ij} = 0$, for all i.

We can relate this to the discussion in Section 3.2, using $\boldsymbol{A}^T = \mathbf{1}$. It then follows directly that the density for an IGMRF of first order is invariant to the addition of $c_1\mathbf{1}$, for any arbitrary c_1, see (3.14) and (3.15). To illustrate this feature, let $\boldsymbol{\mu} = \mathbf{0}$ so

$$\mathrm{E}(x_i \mid \boldsymbol{x}_{-i}) = -\frac{1}{Q_{ii}} \sum_{j:j\sim i} Q_{ij} x_j,$$

where $-\sum_{j:j\sim i} Q_{ij}/Q_{ii} = 1$. Hence, the conditional mean of x_i is simply a weighted mean of its neighbors, but does not involve an overall level. In applications, this 'local' behavior is often desirable. We can then concentrate on the deviation from *any* overall mean level without having to specify the overall mean level itself. Many IGMRFs are constructed such that the *deviation* from the overall level is a smooth curve in time or a smooth surface in space.

3.3.1 IGMRFs of first order on the line

We will now construct a widely used model known as the first-order random walk. We first assume that the location of the nodes i are all positive integers, i.e., $i = 1, 2, \ldots, n$. It is not uncommon in this case to think of i as time t. The distance between the nodes is constant and equal to 1. We will later discuss a modification for the case where the nodes are nonequally spaced, i.e., the distance between the nodes varies.

The first-order random walk model for regular locations

The first-order random walk model is constructed assuming *independent increments*

$$\Delta x_i \stackrel{\text{iid}}{\sim} \mathcal{N}(0, \kappa^{-1}), \quad i = 1, \ldots, n-1. \tag{3.20}$$

This immediately implies that

$$x_j - x_i \ \sim \ \mathcal{N}(0, (j-i)\kappa^{-1}) \quad \text{for} \quad i < j.$$

Also, if the intersection between $\{i, \ldots, j\}$ and $\{k, \ldots, l\}$ is empty for $i < j$ and $k < l$, then

$$\text{Cov}(x_j - x_i, x_l - x_k) = 0.$$

These properties are well known and coincide with those of a *Wiener process* observed in discrete time. We will define the Wiener process shortly in Definition 3.4.

The density for $\boldsymbol{x}$ is derived from its $n-1$ increments (3.20) as

$$\begin{aligned}
\pi(\boldsymbol{x} \mid \kappa) \quad &\propto \quad \kappa^{(n-1)/2} \exp\left(-\frac{\kappa}{2}\sum_{i=1}^{n-1}(\Delta x_i)^2\right) \\
&= \quad \kappa^{(n-1)/2} \exp\left(-\frac{\kappa}{2}\sum_{i=1}^{n-1}(x_{i+1} - x_i)^2\right) \\
&= \quad \kappa^{(n-1)/2} \exp\left(-\frac{1}{2}\boldsymbol{x}^T\boldsymbol{Q}\boldsymbol{x}\right),
\end{aligned} \tag{3.21}$$

where $\boldsymbol{Q} = \kappa\boldsymbol{R}$ and where $\boldsymbol{R}$ is the so-called *structure matrix*:

$$\boldsymbol{R} = \begin{pmatrix}
1 & -1 & & & & & \\
-1 & 2 & -1 & & & & \\
 & -1 & 2 & -1 & & & \\
 & & \ddots & \ddots & \ddots & & \\
 & & & -1 & 2 & -1 & \\
 & & & & -1 & 2 & -1 \\
 & & & & & -1 & 1
\end{pmatrix}. \tag{3.22}$$

The form of $\boldsymbol{R}$ follows easily from

$$\sum_{i=1}^{n-1}(\Delta x_i)^2 = (\boldsymbol{D}\boldsymbol{x})^T(\boldsymbol{D}\boldsymbol{x}) = \boldsymbol{x}^T\boldsymbol{D}^T\boldsymbol{D}\boldsymbol{x} = \boldsymbol{x}^T\boldsymbol{R}\boldsymbol{x},$$

where the $(n-1) \times n$ matrix $\boldsymbol{D}$ has the form

$$\boldsymbol{D} = \begin{pmatrix} -1 & 1 & & & \\ & -1 & 1 & & \\ & & \ddots & \ddots & \\ & & & -1 & 1 \end{pmatrix}.$$

Note that the eigenvalues of $\boldsymbol{R}$ are equal to

$$\lambda_i = 2 - 2\cos\left(\pi(i-1)/n\right), \quad i = 1, \dots, n, \tag{3.23}$$

which can be used for analytic calculation of the generalized determinant appearing in (3.18).

It is clear that $\boldsymbol{Q}\mathbf{1} = \mathbf{0}$ by either verifying that $\pi(\boldsymbol{x}|\kappa)$ is invariant to the addition of a constant vector $c_1\mathbf{1}$, or that $\sum_j Q_{ij} = 0$ from (3.22). The rank of $\boldsymbol{Q}$ is $n-1$. Hence (3.21) is an IGMRF of first order, by Definition 3.3. We denote this model by RW1(κ) or short RW1 as an abbreviation for a *random walk* of first order.

The invariance to the addition of any constant to the overall mean is evident from the full conditional distributions

$$x_i \mid \boldsymbol{x}_{-i}, \kappa \sim \mathcal{N}(\frac{1}{2}(x_{i-1} + x_{i+1}), 1/(2\kappa)), \quad 1 < i < n, \tag{3.24}$$

because there is no shrinkage toward an overall mean. An alternative interpretation of the conditional mean can be obtained by fitting a first-order polynomial, i.e., a simple line

$$p(j) = \beta_0 + \beta_1 j,$$

locally through the points $(i-1, x_{i-1})$ and $(i+1, x_{i+1})$ using least squares. The conditional mean turns out to be equal to $p(i)$.

If we fix the random walk at locations $1, \dots, i$, future values have the conditional distribution

$$x_{i+k} \mid x_1, \dots, x_i, \kappa \sim \mathcal{N}(x_i, k/\kappa), \quad 0 < i < i+k \le n.$$

Hence, this model gives a constant forecast equal to the last observed x_i with linearly increasing variance.

To give some intuition about the form of realizations of a RW1 model, we have generated 10 samples of $\boldsymbol{x}^{\perp}$ with $n = 99$ and $\kappa = 1$. We set $x^{\parallel} = 0$ as described in Section 3.2. The samples are shown in Figure 3.2(a). The (theoretical) variances $\mathrm{Var}(x_i^{\perp})$ for $i = 1, \dots, n$, are shown in Figure 3.2(b) after being normalized with the average variance. There is more variability at and near the ends compared to the interior. To study the correlation structure, we have also computed $\mathrm{Corr}(x_{n/2}^{\perp}, x_i^{\perp})$ for $i = 1, \dots, n$ and the result is shown in Figure 3.2(c). The behavior is quasiexponential where the negative correlation at the endpoints with

$x_{n/2}$ is due to the sum-to-zero constraint $x^{\|} = 0$. Note that a different κ only involves a change of the scale; the correlation structure is invariant with respect to κ, but not with respect to n.

Alternatively, we could define a random walk of first order on a circle, i.e., we would include the pair $x_1 \sim x_n$ in the graph and correspondingly the term $(x_1 - x_n)^2$ in the sum in the exponent of (3.21). The precision matrix will then be circulant, and so will be the inverse, which implies a constant variance, see Section 2.6.1. Note that such a *circular* first-order random walk is also a IGMRF of first order, the distribution of $\boldsymbol{x}$ will still be invariant to the addition of a constant.

We now have a closer look at the elements in the structure matrix (3.22) of the (noncircular) RW1 model. Each row in $\boldsymbol{R}$ (except for the first and the last one) has coefficients $-1, 2, -1$, which are simply the coefficients in $-\Delta^2$ as defined in (3.2). So, if Δx_i are the increments, then $\boldsymbol{R}$ consists of $-\Delta^2$ terms, apart from corrections at the boundary. We may interpret

$$-x_{i-1} + 2x_i - x_{i+1}$$

as an estimate of the negative second derivative of an underlying continuous function $x(t)$ at $t = i$, making use of the observations at $\{i-1, i, i+1\}$. In Section 3.4 when we consider random walks of second order, $\Delta^2 x_i$ are the increments and the precision matrix will consist of $-\Delta^4$ terms plus boundary corrections. A closer look into the theory of constructing continuous splines will provide further insight in this direction, see, for example, Gu (2002).

The first-order random walk model for irregular locations

We will now discuss the case where x_i is assigned to a location s_i but where the distance between s_{i+1} and s_i is not constant. Without loss of generality we assume that $s_1 < s_2 < \cdots < s_n$, and define the distance

$$\delta_i = s_{i+1} - s_i. \tag{3.25}$$

To obtain the precision matrix in this case, we will consider x_i as the realization of an integrated Wiener process in continuous time, at time s_i.

Definition 3.4 (Wiener process) *A Wiener process with precision κ is a continuous-time stochastic process $W(t)$ for $t \geq 0$ with $W(0) = 0$ and such that the increments $W(t) - W(s)$ are normal with mean 0 and variance $(t-s)/\kappa$ for any $0 \leq s < t$. Furthermore, increments for nonoverlapping time intervals are independent. For $\kappa = 1$, this process is called a standard Wiener process.*

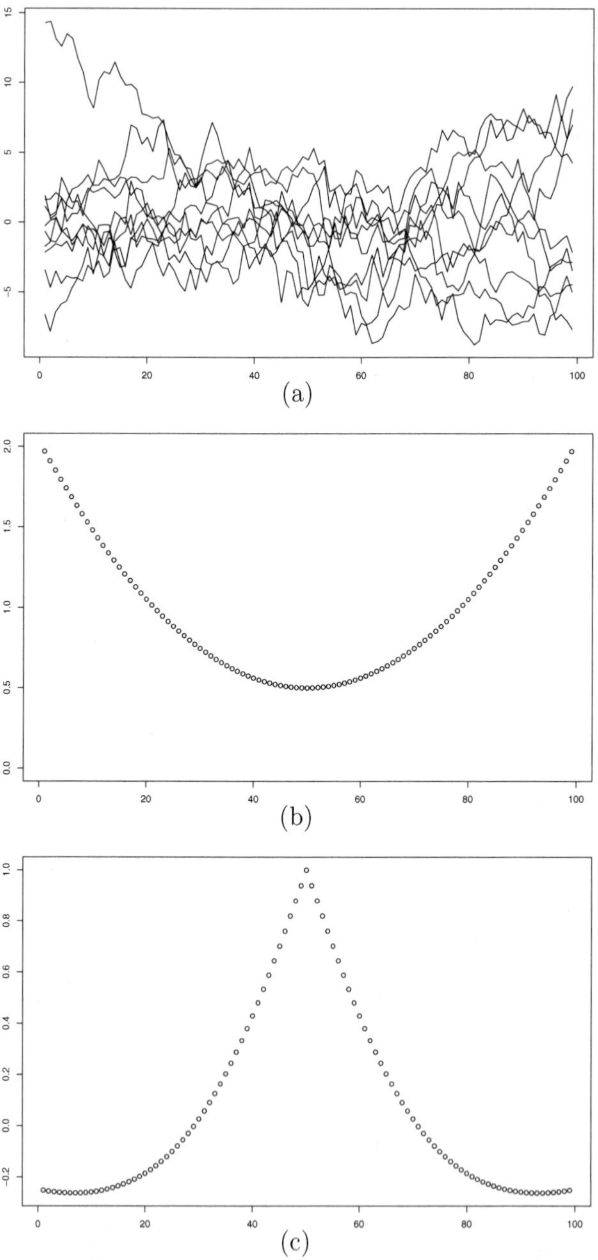

Figure 3.2 *Illustrations of the properties of the RW1 model with* $n = 99$*: (a) displays* 10 *samples of* $\boldsymbol{x}^{\perp}$*, (b) displays* $Var(x_i^{\perp})$ *for* $i = 1, \ldots, n$ *normalized with the average variance, and (c) displays* $Corr(x_{n/2}^{\perp}, x_i^{\perp})$ *for* $i = 1, \ldots, n$.

The full conditional in the above case has moments

$$\mathrm{E}(x_i \mid \boldsymbol{x}_{-i}, \kappa) = \frac{\delta_i}{\delta_{i-1}+\delta_i} x_{i-1} + \frac{\delta_{i-1}}{\delta_{i-1}+\delta_i} x_{i+1}$$

$$\mathrm{Prec}(x_i \mid \boldsymbol{x}_{-i}, \kappa) = \kappa \left(\frac{1}{\delta_{i-1}} + \frac{1}{\delta_i} \right).$$

Here κ is a precision parameter and chosen so that we obtain the same result as in (3.24) if $\delta_i = 1$ for all i.

The precision matrix can now be obtained using Theorem 2.6 as

$$Q_{ij} = \kappa \begin{cases} \frac{1}{\delta_{i-1}} + \frac{1}{\delta_i} & j = i \\ -\frac{1}{\delta_i} & j = i+1 \\ 0 & \text{otherwise} \end{cases}$$

for $1 < i < n$ where the $Q_{i,i-1}$ terms are found via $Q_{i,i-1} = Q_{i-1,i}$. A proper correction at the boundary (implicitly we use a diffuse prior for $W(0)$ rather than the fixed $W(0) = 0$) gives the remaining diagonal terms $Q_{11} = \kappa/\delta_1$, $Q_{nn} = \kappa/\delta_{n-1}$. Clearly, $\boldsymbol{Q}\mathbf{1} = \mathbf{0}$ still holds and the joint density of $\boldsymbol{x}$,

$$\pi(\boldsymbol{x} \mid \kappa) \propto \kappa^{(n-1)/2} \exp\left(-\frac{\kappa}{2} \sum_{i=1}^{n-1} (x_{i+1} - x_i)^2 / \delta_i \right), \tag{3.26}$$

is invariant to the addition of a constant. The scaling with δ_i is because $\mathrm{Var}(x_{i+1} - x_i) = \delta_i/\kappa$ according to the continuous-time model.

The interpretation of a RW1 model as a discretely observed Wiener process (continuous in time) justifies the corrections needed for non-equally spaced locations. Hence, the underlying model *is the same*, it is only observed differently.

To compare this model with its regular counterpart, we reproduced Figure 3.2 with $n = 99$, but now with the locations $s_2, \ldots, s_{49}$ sampled uniformly between $s_1 = 1$ and $s_{50} = 50$. The locations s_i for $i = 51, \ldots, 99$ are obtained requiring symmetry around $s_{50} = 50$: $s_i + s_{100-i} = 100$ for $i = 1, \ldots, n = 99$. The results are shown in Figure 3.3.1 and show a very similar behavior as in the regular case. Note, however, that in the case of nonsymmetric distributed random locations between $s_1 = 1$ and $s_{99} = 99$ the marginal variances and correlations are not exactly symmetric around $s_{50} = 50$ (not shown).

When the mean is only locally constant

The approach to model $\boldsymbol{Q}$ through forward differences of first order as normal increments does more than 'just' being invariant to the addition

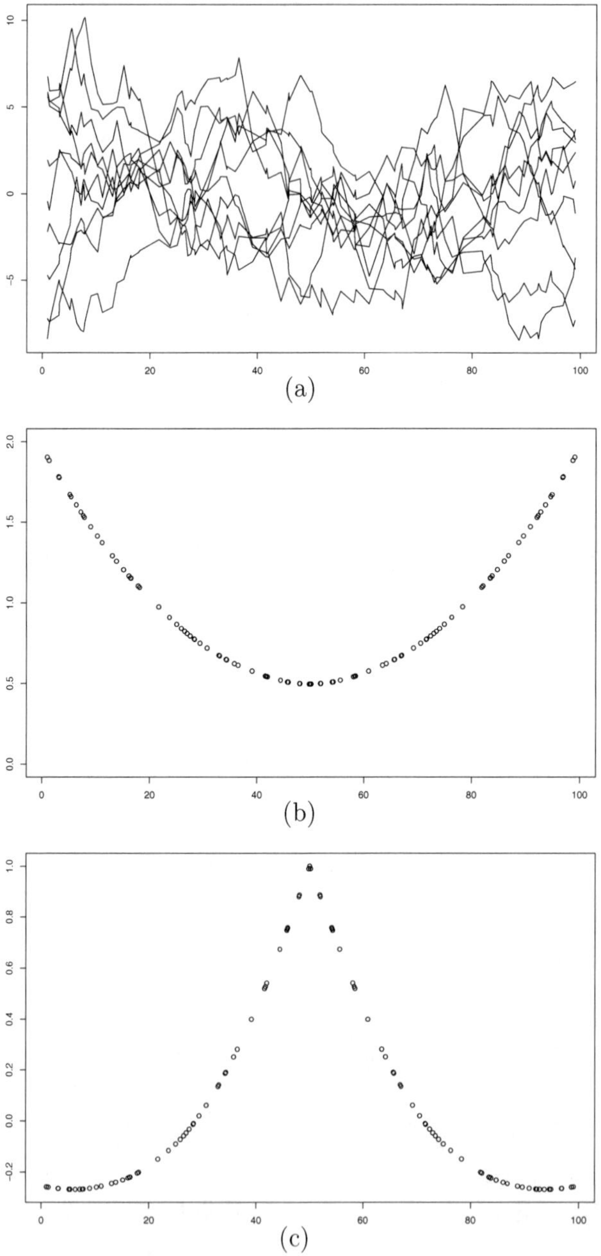

Figure 3.3 *Illustrations of the properties of the RW1 model with* $n = 99$ *irregular locations with* $s_1 = 1$ *and* $s_n = n$*: (a) displays* 10 *samples of* $\boldsymbol{x}^{\perp}$ *, (b) displays* $Var(x_i^{\perp})$ *for* $i = 1, \ldots, n$ *normalized with the average variance, and (c) displays* $Corr(x_{n/2}^{\perp}, x_i^{\perp})$ *for* $i = 1, \ldots, n$*. The horizontal axis relates to the locations* s_i*.*

of a constant. Consider the alternative IGMRF

$$\pi(\boldsymbol{x}) \propto \kappa^{(n-1)/2} \exp\left(-\frac{\kappa}{2}\sum_{i=1}^{n}(x_i - \overline{\boldsymbol{x}})^2\right), \tag{3.27}$$

where $\overline{\boldsymbol{x}}$ is the empirical mean of $\boldsymbol{x}$. Without writing out the precision matrix, we immediately see that $\pi(\boldsymbol{x})$ is invariant to the addition of a constant and that its rank is $n-1$, hence (3.27) defines an IGMRF of first order. Although both (3.27) and the RW1 model (3.21) are maximized at $\boldsymbol{x} \propto \mathbf{1}$, the benefit of the RW1 model is evident when we consider the following vector $\boldsymbol{x}$. Assume n is even and

$$x_i = \begin{cases} 0, & 1 \le i \le n/2 \\ 1, & n/2 < i \le n. \end{cases} \tag{3.28}$$

Thus $\boldsymbol{x}$ is locally constant with two levels. If we evaluate the density of (3.28) under the RW1 model and the alternative (3.27), we obtain

$$\kappa^{(n-1)/2} \exp\left(-\frac{\kappa}{2}\right) \quad \text{and} \quad \kappa^{(n-1)/2} \exp\left(-n\frac{\kappa}{8}\right),$$

respectively. The log ratio of the densities is then of order $\mathcal{O}(n)$. The reason for the drastic difference is that the RW1 model only penalizes the *local* deviation from a constant level (interpret Δx_i as the derivative at location i) whereas (3.27) penalizes the *global* deviation from a constant level. This *local* behavior of the RW1 model is obviously quite advantageous in applications if the mean level of $\boldsymbol{x}$ is approximately or locally constant. A similar argument will also apply to polynomial IGMRFs of higher order; those will be constructed using forward differences of order k as independent normal increments.

3.3.2 IGMRFs of first order on lattices

For regular or irregular lattices, the construction of IGMRFs of first order follows the same concept, but is now based on 'independent' increments writing '·' due to (hidden) linear constraints imposed by the more complicated geometry. We first look at the irregular case.

First-order IGMRFs on irregular lattices

Let us reconsider Figure 2.4(a) and the map of the 544 regions in Germany. Here we may define two regions as *neighbors* if they share a common border. The corresponding graph is shown in Figure 2.4(b). Between neighboring regions i and j, say, we define a normal increment

$$x_i - x_j \sim \mathcal{N}(0, \kappa^{-1}) \tag{3.29}$$

and the assumption of 'independent' increments yields the IGMRF model:

$$\pi(\boldsymbol{x}) \propto \kappa^{(n-1)/2} \exp\left(-\frac{\kappa}{2}\sum_{i\sim j}(x_i - x_j)^2\right). \tag{3.30}$$

Here $i \sim j$ denotes the set of all *unordered* pairs of neighbors. The requirement for the pair to be unordered prevents us from double counting as $i \sim j \Leftrightarrow j \sim i$. Note that the number of increments $|i \sim j|$ is typically larger than n, but the rank of the corresponding precision matrix is still $n-1$. This implies that there are hidden constraints in the increments due to the more complicated geometry on a lattice than on the line, hence the use of the term 'independent'. To see this consider the following simple example.

Example 3.5 *Let $n = 3$ where all nodes are neighbors. Then (3.29) gives $x_1 - x_2 = \epsilon_1$, $x_2 - x_3 = \epsilon_2$, and $x_3 - x_1 = \epsilon_3$, where ϵ_1, ϵ_2 and ϵ_3 are the increments. Adding the two first equations and comparing with the last implies that $\epsilon_1 + \epsilon_2 + \epsilon_3 = 0$, which is the 'hidden' linear constraint.*

However, under the linear constraints the density of $\boldsymbol{x}$ is (3.30) by the following argument. Let $\boldsymbol{\epsilon}$ be $|i \sim j|$ independent increments and $\boldsymbol{A\epsilon}$ the linear constraints saying that the increments sum to zero over all circuits in the graph. Then $\pi(\boldsymbol{\epsilon}|\boldsymbol{A\epsilon}) \propto \pi(\boldsymbol{\epsilon})$ if $\boldsymbol{\epsilon}$ is a configuration satisfying the constraint from (2.31). We now change variables from $\boldsymbol{\epsilon}$ to $\boldsymbol{x}$ and we obtain the exponent in (3.30). See also Besag and Higdon (1999, p. 740). The normalization constant follows from the generalized determinant of $\boldsymbol{Q}$, the precision matrix of $\boldsymbol{x}$.

The density (3.30) is equal to the density of an RW1 model if we are on a line and $i \sim j$ iff $|i - j| = 1$. However, in general there is no longer an underlying continuous stochastic process that we can relate to this density. Hence, if we change the spatial resolution or split a region into two new ones, we really change the model.

Let n_i denote the number of neighbors of region i. The precision matrix $\boldsymbol{Q}$ in (3.30) has elements

$$Q_{ij} = \kappa \begin{cases} n_i & i = j, \\ -1 & i \sim j, \\ 0 & \text{otherwise}, \end{cases} \tag{3.31}$$

from which it follows directly that

$$x_i \mid \boldsymbol{x}_{-i}, \kappa \sim \mathcal{N}\Big(\frac{1}{n_i}\sum_{j:j\sim i} x_j, \frac{1}{n_i\kappa}\Big). \tag{3.32}$$

Similar to the RW1 model for nonequally spaced locations, we can incorporate positive and symmetric weights w_{ij} for each pair of adjacent nodes i and j in (3.30). For example, one could use the inverse Euclidean distance between the centroids of each region.

Assuming the 'independent' increments

$$x_i - x_j \sim \mathcal{N}(0, 1/(w_{ij}\kappa)),$$

the joint density becomes

$$\pi(\boldsymbol{x}) \propto \kappa^{(n-1)/2} \exp\left(-\frac{\kappa}{2} \sum_{i \sim j} w_{ij}(x_i - x_j)^2 \right). \tag{3.33}$$

The corresponding precision matrix $\boldsymbol{Q}$ now has elements

$$Q_{ij} = \kappa \begin{cases} \sum_{k:k\sim i} w_{ik} & i = j, \\ -w_{ij} & i \sim j, \\ 0 & \text{otherwise,} \end{cases} \tag{3.34}$$

and the full conditional $\pi(x_i|\boldsymbol{x}_{-i}, \kappa)$ is normal with mean and precision

$$\frac{\sum_{j:j\sim i} x_j w_{ij}}{\sum_{j:j\sim i} w_{ij}} \quad \text{and} \quad \kappa \sum_{j:j\sim i} w_{ij},$$

respectively. It is straightforward to verify that $\boldsymbol{Q}\mathbf{1} = \mathbf{0}$ both for (3.31) and (3.34).

Example 3.6 *We will now have a closer look at the IGMRFs defined in (3.30) and (3.33) using the graph shown in Figure 2.4(b) found from the map of the administrative regions in Germany, as shown in Figure 2.4(a). Throughout we show samples of $\boldsymbol{x}^{\perp}$ and discuss its properties, as argued for in Section 3.2. Additionally, we fix $\kappa = 1$. Note that the correlation structure for $\boldsymbol{x}^{\perp}$ does not depend on κ. Of interest is how samples from the density look, but also how much the variance of $x_i^{\perp}$ varies through the graph.*

We first consider (3.30) and display in Figure 3.4 two samples of $\boldsymbol{x}^{\perp}$. The gray scale is linear from the minimum value to the maximum value, and the samples shows a relatively smooth variation over the map. The dependence of the conditional precision on the number of neighbors (3.32) is typically inherited to the (marginal) variance, although the actual form of the graph also plays a certain role. In (c) we display the scaled variance in relationship to the number of neighbors. The scaling is so that the average variance is 1. We added some random noise to the number of neighbors to make each dot visible. It can be seen clearly that, on average, the variance increases with decreasing number of neighbors, with a range from 0.7 to 2.0. This is a considerable variation.

In particular, the large values of the variance for regions with only one neighbor are worrisome. A possible (ad hoc) *remedy is to choose all adjacent regions, but if there is only one neighbor, we also add all the neighbors of this neighbor. Unfortunately, the variation of the variance for the others is about the same (not shown).*

It is tempting to make an adjustment for the 'distance' $d(i,j)$ *between region* i *and* j*, which is incorporated in (3.33) using the weights* $w_{ij} = 1/d(i,j)$*. A feasible choice is to use the Euclidean distance between the centroids of region* i *and* j*. When we have access to the outline of each region, this is easy to compute using contour-integrals. Note that the correlation structure only depends on the distances up to a multiplicative constant for the same reason as the correlation structure does not depend on the actual value of* κ*.*

Figure 3.5(a) and (b) displays two samples of $\boldsymbol{x}^{\perp}$ *with the distance correction included and (c) the variance in relation to the number of neighbors. The samples look slightly smoother compared to Figure 3.4, but a closer study is needed to quantify this more objectively. Most regions with only one neighbor are in the interior of its neighbor, typically representing a larger city within its rural suburbs. In this case, the distance between the city region and its surrounding region may be small. One* (ad hoc) *remedy out of this is to apply the same correction as for the unweighted model but increase the number of neighbors for those regions with only one neighbor. The adjustment with distances results in a more homogeneous variance, but the variance still decreases due to the increasing number of neighbors (not shown).*

First-order IGMRFs on regular lattices

For a lattice $\mathcal{I}_{\boldsymbol{n}}$ with $n = n_1 n_2$ nodes, let ij or (i,j) denote the node in the ith row and jth column that also defines its location. In the interior, we can now define the nearest four sites of ij as its neighbors, i.e.,

$$(i+1,j),\ (i-1,j),\ (i,j+1),\ (i,j-1).$$

Without further weights, the corresponding precision matrix is (3.31), and the full conditionals of x_i are given in (3.32). In the interior of the lattice, $\pi(x_{ij}|\boldsymbol{x}_{-ij},\kappa)$ is normal with mean

$$\frac{1}{4}\left(x_{i+1,j} + x_{i-1,j} + x_{i,j+1} + x_{i,j-1}\right) \tag{3.35}$$

and precision 4κ. Of course, one could also introduce weights in the formulation as in (3.34).

However, here an extended *anisotropic* model can be used, weighting the horizontal and vertical neighbors differently. More specifically,

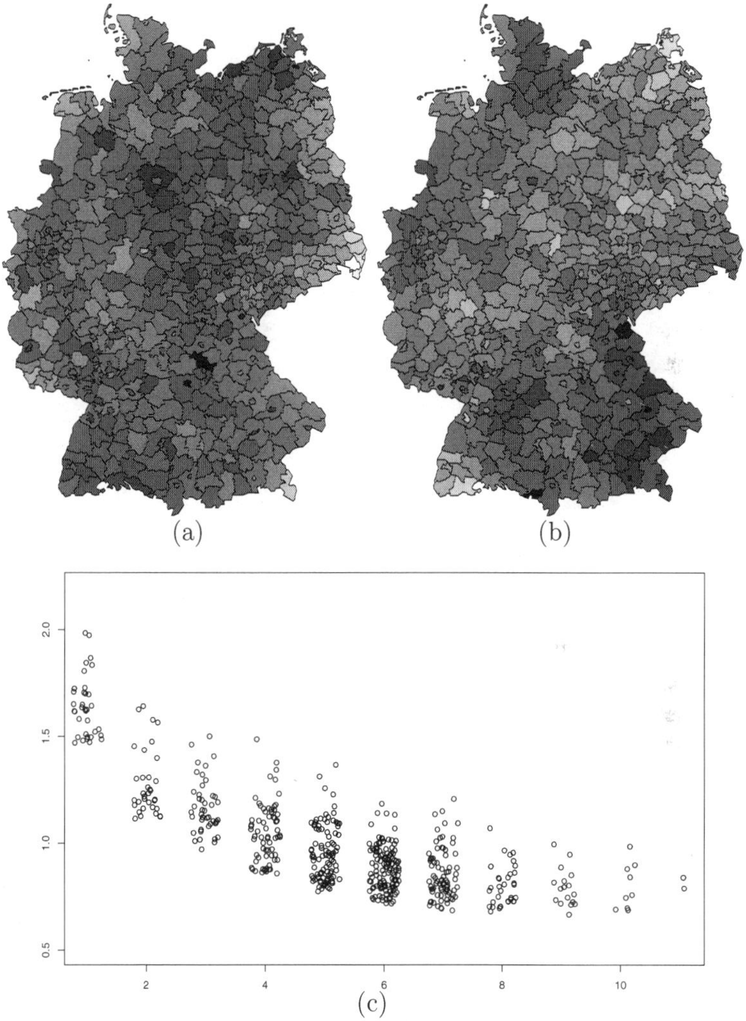

Figure 3.4 *Figures (a) and (b) display two samples from an IGMRF with density (3.30) where two regions sharing a common border are considered as neighbors. Figure (c) displays the variance in relation to the number of neighbors, demonstrating that the variance decreases if the number of neighbors increases.*

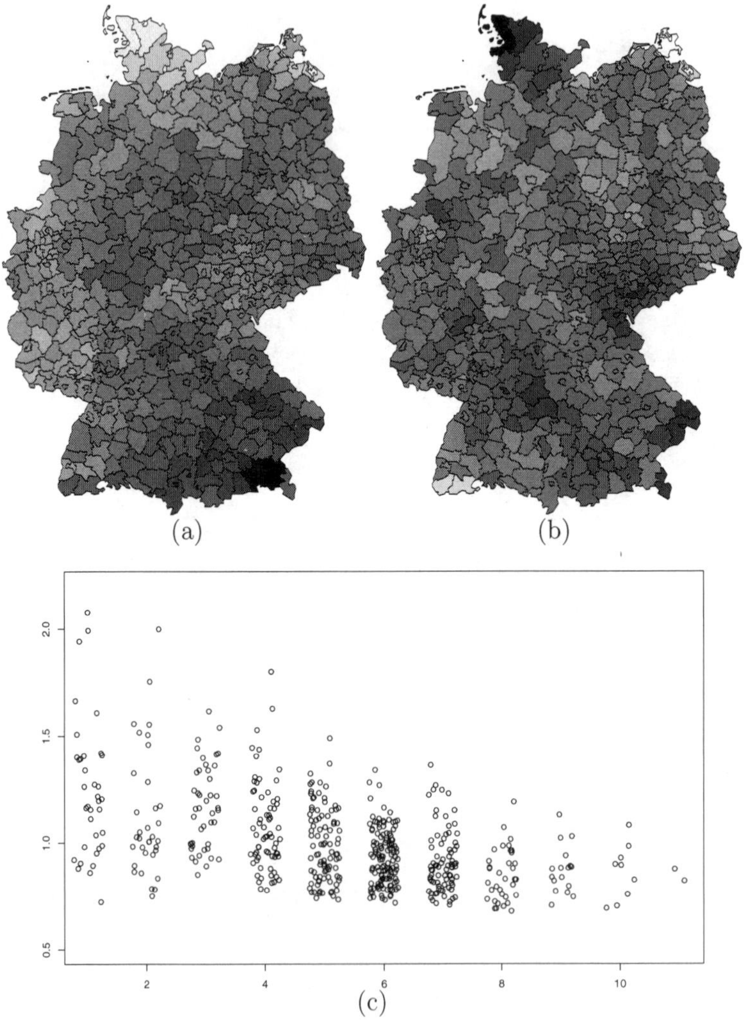

Figure 3.5 *Figures (a) and (b) display two samples of* $\boldsymbol{x}^{\perp}$ *from an IGMRF with density (3.33) where two regions sharing a common border are considered as neighbors and weights are used based in the distance between centroids. Figure (c) displays the variance in relation to the number of neighbors, demonstrating that the variances decreases if the number of neighbors increases.*

suppose the conditional mean is

$$\frac{1}{4}\left(\alpha'(x_{i+1,j} + x_{i-1,j}) + \alpha''(x_{i,j+1} + x_{i,j-1})\right) \tag{3.36}$$

with positive parameters α' and α'', constrained to fulfill $\alpha' + \alpha'' = 2$. The conditional precision is still 4κ. This model can also be obtained by using independent first-order increments in each direction and conditioning on the sums over closed loops being zero (Besag and Higdon, 1999), see Künsch (1999) for a rigorous proof in the infinite lattice case.

In applications, α' (or α'') can be treated as an unknown parameter, so the degree of anisotropy can be estimated from the data. To estimate α' it is necessary to compute $|\boldsymbol{Q}|^*$ of the corresponding precision matrix $\boldsymbol{Q}$ in the density (3.18). Note that $\boldsymbol{Q}$ can be written as a Kronecker product

$$\boldsymbol{Q} = \alpha' \boldsymbol{R}_{n_1} \otimes \boldsymbol{I}_{n_2} + \alpha'' \boldsymbol{I}_{n_1} \otimes \boldsymbol{R}_{n_2},$$

where $\boldsymbol{R}_n$ is the structure matrix (3.22) of the RW1 model of dimension $n \times n$ and $\boldsymbol{I}_m$ is the identity matrix of dimension $m \times m$. Hence, using (3.23) and a result for such sums of Kronecker products, the eigenvalues of $\boldsymbol{Q}$ can be calculated without imposing toroidal boundary conditions,

$$\lambda_{ij} = 2\Big(\alpha'(1 - \cos(\pi(i-1)/n_1)) + \alpha''(1 - \cos(\pi(j-1)/n_2))\Big)$$

for $i = 1, \dots, n_1$ and $j = 1, \dots, n_2$. The generalized determinant $|\boldsymbol{Q}|^*$ can then be easily computed.

Recently Mondal and Besag (2004) have shown that there is a close relationship between first-order IGMRFs on regular (infinite) lattices and the (continuous) *de Wijs process* (Chilès and Delfiner, 1999, Matheron, 1971). In fact, the de Wijs process can be interpreted as the limit of model (3.35) on a sequence of ever finer lattices that are superimposed on the study region. There is also a corresponding result for the asymmetric model (3.36). This is an important result, as it provides a link to an underlying continuous process, similar to the close correspondence between a first-order random walk and the Wiener process. The de Wijs process has good analytic properties and was for this reason widely used by the pioneers of geostatistics. It also has a remarkable property, often called the *de Wijs formula*: Let V be a set in $\mathbb{R}^d$ and $\{v_i\}$ a partition of V, where V and each v_i are geometrically similar. Let v be one of $\{v_i\}$ selected uniformly. Then, the variance between the mean of the process in v and V (the so-called *dispersion variance*) is proportional to the log ratio of the volumes of V and v regardless of the scale.

An alternative limit argument for the first-order IGRMF

The first-order IGMRF is an improper GMRF of rank $n-1$ where $\boldsymbol{Q}\boldsymbol{1} = \boldsymbol{0}$. In Section 3.2 we argue for this construction as the limit when the precision of $\boldsymbol{x}^{\parallel} = \boldsymbol{1}^T\boldsymbol{x}$ tends to zero. In particular, $\boldsymbol{Q}$ can be seen as the limit of the (proper) precision matrix

$$\boldsymbol{Q}(\gamma) = \boldsymbol{Q} + \gamma\boldsymbol{1}\boldsymbol{1}^T, \quad \gamma > 0,$$

as $\gamma \to 0^+$, see (3.19). Consider instead $\boldsymbol{Q}$ as the limit of

$$\widetilde{\boldsymbol{Q}}(\gamma) = \boldsymbol{Q} + \gamma\boldsymbol{I},$$

which is *not* in coherence with the general discussion in Section 3.2. Note that $\boldsymbol{Q}(\gamma)$ is a completely dense matrix while $\widetilde{\boldsymbol{Q}}(\gamma)$ is sparse if $\boldsymbol{Q}$ is.

To see that this is an alternative formulation to obtain IGMRFs in the limit, let $\boldsymbol{x} \sim \mathcal{N}(\boldsymbol{0}, \widetilde{\boldsymbol{Q}}(\gamma)^{-1})$. Then

$$||\text{Prec}(\boldsymbol{x} \mid \boldsymbol{1}^T\boldsymbol{x}) - \boldsymbol{Q}|| = c\gamma. \tag{3.37}$$

where $||\cdot||$ is either the weak norm, in which case $c = \sqrt{(n-1)/n}$ or the strong norm, in which case $c = 1$, see Definition 2.7 for the definition of the weak and strong norm.

For example, suppose we use the precision matrix of a first-order IGMRF and want to make it proper. The arguments in Section 3.2 then suggest to use $\boldsymbol{Q}(\gamma)$ with a small γ, with the consequence that $\boldsymbol{Q}(\gamma)$ is a completely dense matrix. However, (3.37) suggests that a good approximation is to use $\widetilde{\boldsymbol{Q}}(\gamma)$, which *maintains the sparsity* of the precision matrix $\boldsymbol{Q}$. The error in the approximation measured in either the weak or strong norm is $c\gamma$.

To verify (3.37), let $\{(\lambda_i, \boldsymbol{e}_i)\}$ denote the eigenvalue/vector pairs for of $\boldsymbol{Q}$ for $i = 1, \ldots, n$ where $\lambda_1 = 0$ and $\boldsymbol{e}_1 = \boldsymbol{1}$. The corresponding pairs of $\widetilde{\boldsymbol{Q}}(\gamma)$ are

$$(\gamma, \boldsymbol{1}), \ (\lambda_2 + \gamma, \boldsymbol{e}_2), \ldots, (\lambda_n + \gamma, \boldsymbol{e}_n).$$

Note that the eigenvectors remain the same. The eigenvalue/vector pairs of the conditional precision matrix of $\boldsymbol{x}$, conditional on $\boldsymbol{1}^T\boldsymbol{x}$, are given in Section 3.2 as

$$(0, \boldsymbol{1}), \ (\lambda_2 + \gamma, \boldsymbol{e}_2), \ldots, (\lambda_n + \gamma, \boldsymbol{e}_n).$$

and (3.37) follows.

3.4 IGMRFs of higher order

We will now discuss higher-order IGMRFs on the line and on regular lattices. Higher-order IGMRFs have a rank deficiency larger than one and can be defined on the line or in higher dimensions. The main idea is to extend the class of functions for which the improper density is invariant. We first consider (polynomial) IGMRFs of order k, which are invariant to the addition of all polynomials with degree less than or equal to $k-1$. For example, a second-order IGMRF is invariant to the addition of a first-order polynomial, i.e., a line in one dimension and a plane in two dimensions. In Section 3.4.3 we consider examples of nonpolynomial IGMRFs.

3.4.1 IGMRFs of higher order on the line

Let $s_1 < s_2 < \cdots < s_n$ denote the ordered locations of $\boldsymbol{x} = (x_1, \ldots, x_n)$ and define $\boldsymbol{s} = (s_1, \ldots, s_n)^T$. An IGMRF of order k on the line is an improper GMRF of rank $n-k$, where the columns of the design matrix $\boldsymbol{S}_{k-1}$, as defined in (3.4), are a basis for the null space of the precision matrix $\boldsymbol{Q}$.

Definition 3.5 (IGMRF of kth order on the line) *An IGMRF of order k is an improper GMRF of rank $n-k$, where $\boldsymbol{Q}\boldsymbol{S}_{k-1} = \boldsymbol{0}$, with $\boldsymbol{S}_{k-1}$ defined as in (3.4).*

The rank deficiency of $\boldsymbol{Q}$ is k. The condition $\boldsymbol{Q}\boldsymbol{S}_{k-1} = \boldsymbol{0}$ simply means that

$$-\frac{1}{2}(\boldsymbol{x} + \boldsymbol{p}_{k-1})^T \boldsymbol{Q}(\boldsymbol{x} + \boldsymbol{p}_{k-1}) = -\frac{1}{2}\boldsymbol{x}^T \boldsymbol{Q}\boldsymbol{x}$$

for any coefficients $\beta_0, \ldots, \beta_{k-1}$ in (3.3). The density (3.18) is hence invariant to the addition of any polynomial of degree $k-1$, $\boldsymbol{p}_{k-1}$, to $\boldsymbol{x}$.

An alternative view is to decompose $\boldsymbol{x}$ as

$$\begin{aligned} \boldsymbol{x} &= \text{trend}(\boldsymbol{x}) + \text{residuals}(\boldsymbol{x}) \\ &= \boldsymbol{H}_{k-1}\boldsymbol{x} + (\boldsymbol{I} - \boldsymbol{H}_{k-1})\boldsymbol{x}, \end{aligned} \tag{3.38}$$

where the 'trend' is of degree $k-1$. Note that the 'trend' corresponds to $\boldsymbol{x}^{\parallel}$ and the 'residuals' corresponds to $\boldsymbol{x}^{\perp}$ in (3.14). The projection matrix

$$\boldsymbol{H}_{k-1} = \boldsymbol{S}_{k-1}(\boldsymbol{S}_{k-1}^T \boldsymbol{S}_{k-1})^{-1} \boldsymbol{S}_{k-1}^T$$

projects $\boldsymbol{x}$ down to the space spanned by the columns of $\boldsymbol{S}_{k-1}$, and $\boldsymbol{I} - \boldsymbol{H}_{k-1}$ to the space orthogonal to that. The matrix $\boldsymbol{H}_{k-1}$ is commonly named the *hat matrix* and $\boldsymbol{I} - \boldsymbol{H}_{k-1}$ is symmetric, idempotent, satisfying $\boldsymbol{S}_{k-1}^T(\boldsymbol{I} - \boldsymbol{H}_{k-1}) = \boldsymbol{0}$ and has rank $n-k$.

If we reconsider the quadratic term using the decomposition (3.38), we obtain

$$-\frac{1}{2}\boldsymbol{x}^T \boldsymbol{Q}\boldsymbol{x} = -\frac{1}{2}\left((\boldsymbol{I} - \boldsymbol{H}_{k-1})\boldsymbol{x}\right)^T \boldsymbol{Q}\left((\boldsymbol{I} - \boldsymbol{H}_{k-1})\boldsymbol{x}\right),$$

where the simplification is due to $\boldsymbol{Q}\boldsymbol{H}_{k-1} = \boldsymbol{0}$. The interpretation is that the density of a kth order IGMRF only depends on the 'residuals' after removing any polynomial 'trend' of degree $k-1$.

The second-order random walk model for regular locations

Assume now that $s_i = i$ for $i = 1, \ldots, n$, so the distance between consecutive nodes is constant and equal to 1. Following the forward difference approach, we may now use the second-order increments

$$\Delta^2 x_i \sim \mathcal{N}(0, \kappa^{-1})$$

for $i = 1, \ldots, n-2$, to define the joint density of $\boldsymbol{x}$:

$$\begin{aligned} \pi(\boldsymbol{x}) &\propto \kappa^{(n-2)/2} \exp\left(-\frac{\kappa}{2} \sum_{i=1}^{n-2} (x_i - 2x_{i+1} + x_{i+2})^2 \right) \\ &= \kappa^{(n-2)/2} \exp\left(-\frac{1}{2} \boldsymbol{x}^T \boldsymbol{Q} \boldsymbol{x} \right), \end{aligned} \tag{3.39}$$

where the precision matrix is

$$\boldsymbol{Q} = \kappa \begin{pmatrix} 1 & -2 & 1 & & & & & & & \\ -2 & 5 & -4 & 1 & & & & & & \\ 1 & -4 & 6 & -4 & 1 & & & & & \\ & 1 & -4 & 6 & -4 & 1 & & & & \\ & & \ddots & \ddots & \ddots & \ddots & \ddots & & & \\ & & & 1 & -4 & 6 & -4 & 1 & & \\ & & & & 1 & -4 & 6 & -4 & 1 & \\ & & & & & 1 & -4 & 5 & -2 & \\ & & & & & & 1 & -2 & 1 & \end{pmatrix}. \tag{3.40}$$

We can verify directly that $\boldsymbol{Q}\boldsymbol{S}_1 = \boldsymbol{0}$ and that the rank of $\boldsymbol{Q}$ is $n-2$. Hence this is an example of an IGMRF of second-order, invariant to the addition of any line to $\boldsymbol{x}$. This model is known as the second-order random walk model, which we denote by RW2(κ) or simply RW2.

Remark. Although this is the RW2 model defined and used in the literature, in Section 3.5 we will demonstrate that we cannot extend it consistently to the case where the locations are irregular. Similar problems occur if we increase the resolution from n to $2n$ locations, say. Therefore, in Section 3.5 an alternative derivation with the desired continuous time interpretation will be presented, where we are able to correct consistently for irregular locations.

The full conditionals of the second-order random walk are easy to read off from $\boldsymbol{Q}$. The conditional mean and precision is

$$\begin{aligned} \mathrm{E}(x_i \mid \boldsymbol{x}_{-i}, \kappa) &= \frac{4}{6}(x_{i+1} + x_{i-1}) - \frac{1}{6}(x_{i+2} + x_{i-2}), \\ \mathrm{Prec}(x_i \mid \boldsymbol{x}_{-i}, \kappa) &= 6\kappa, \end{aligned}$$

respectively, for $2 < i < n-2$ with obvious modifications in the other cases. Some intuition about the coefficients in the conditional mean can be gained if we consider the second-order polynomial

$$p(j) = \beta_0 + \beta_1 j + \frac{1}{2}\beta_2 j^2$$

and compute the coefficients by a local least-squares fit to the points

$$(i-2, x_{i-2}),\ (i-1, x_{i-1}),\ (i+1, x_{i+1}),\ (i+2, x_{i+2}).$$

Just as the conditional mean of the first-order random walk equals the local first-order polynomial interpolation, here it turns out that $\mathrm{E}(x_i|\boldsymbol{x}_{-i}, \kappa) = p(i)$.

If we fix the second-order random walk at the locations $1, \ldots, i$, future values have the conditional moments

$$\begin{aligned} \mathrm{E}(x_{i+k} \mid x_1, \ldots, x_i, \kappa) &= (1+k)x_i - kx_{i-1}, \\ \mathrm{Prec}(x_{i+k} \mid x_1, \ldots, x_i, \kappa) &= \kappa/(1 + 2^2 + \cdots + k^2), \end{aligned} \tag{3.41}$$

where $2 \le i < i + k \le n$. Hence the conditional mean is the linear extrapolation based on the last two observations x_{i-1} and x_i, with *cubically* increasing variance, since $1+2^2+\cdots+k^2 = k(k+1)(2k+1)/6$.

Based on the discussion in Section 3.3.1, we note that the precision matrix $\boldsymbol{Q}$ consists of terms $-\Delta^4$ apart from the corrections near the boundary, meaning that

$$x_{i-2} - 4x_{i-1} + 6x_i - 4x_{i-1} + x_{i+2}$$

can be interpreted as an estimate of the negative 4th derivative of an underlying continuous function $x(t)$ at location $t = i$ making use of the observed values at $\{i-2, \ldots, i+2\}$.

Figure 3.4.1 displays some samples of $\boldsymbol{x}^{\perp}$ from the RW2 model using $n = 99$, the variance and the correlation between $x^{\perp}_{n/2}$ and $x^{\perp}_i$ for $i = 1, \ldots, n$. The samples are now much smoother than those obtained in Figure 3.2 using the RW1 model. On the other hand, the variability in the variance is somewhat larger, especially near the boundary. The correlation structure is also more prominent, as the two constraints induce a strong negative correlation.

The second-order random walk model for irregular locations

We will now discuss possible modifications to extend the definition of the RW2 model to irregular locations. Those proposals are somewhat *ad hoc*, so later, in Section 3.5, we present a theoretically sounder, but also slightly more involved approach based on an underlying continuous integrated Wiener process. The two models presented here are to be considered as simpler alternatives that 'weight' in some sense the RW2 model, with weights typically proportional to the inverse distances between consecutive locations.

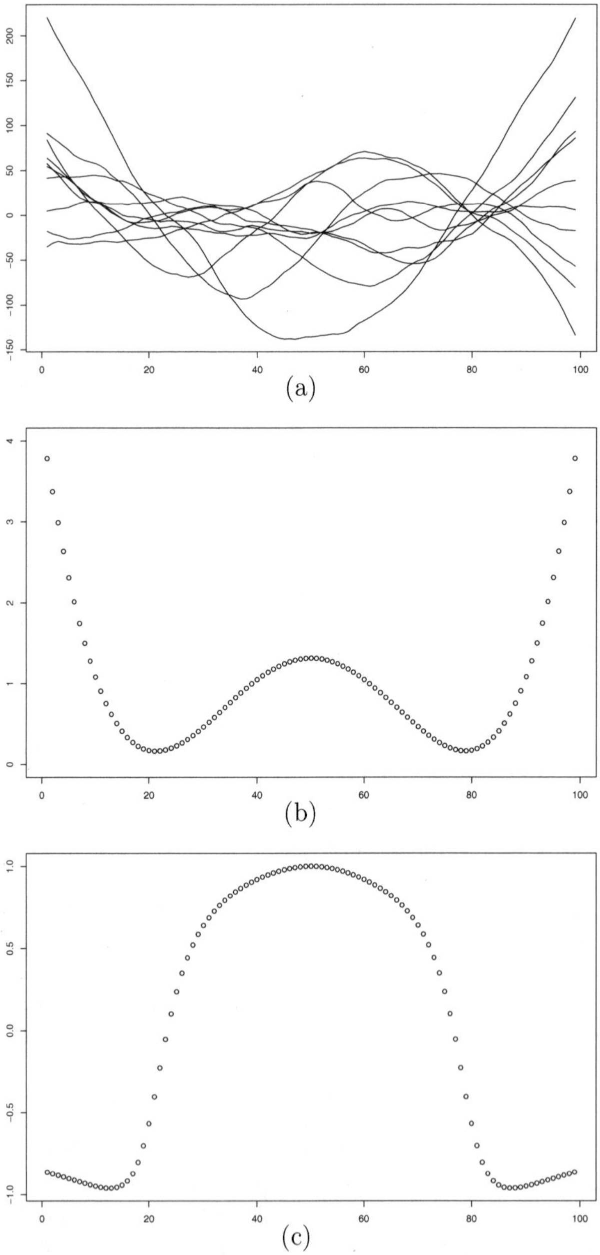

Figure 3.6 *Illustrations of the properties of the RW2 model with* $n = 99$*: (a) displays* 10 *samples of* $\boldsymbol{x}^{\perp}$*, (b) displays* $\mathrm{Var}(x_i^{\perp})$ *for* $i = 1, \ldots, n$ *normalized with the average variance, and (c) displays* $\mathrm{Corr}(x_{n/2}^{\perp}, x_i^{\perp})$ *for* $i = 1, \ldots, n$.

A weighted version of the RW2 model can be obtained in many ways. One simple approach is to rewrite the second-order difference in (3.39) as

$$x_i - 2x_{i+1} + x_{i+2} = (x_{i+2} - x_{i+1}) - (x_{i+1} - x_i)$$

and to replace this with a weighted version

$$w_{i+1}(x_{i+2} - x_{i+1}) - w_i(x_{i+1} - x_i),$$

where $w_i > 0$. This leads to the joint density

$$\pi(\boldsymbol{x} \mid \kappa) \propto \exp\left(-\frac{\kappa}{2}\sum_{i=1}^{n-2} w_{i+1}^2\left(x_{i+2} - (1 + \frac{w_i}{w_{i+1}})x_{i+1} + \frac{w_i}{w_{i+1}}x_i\right)^2\right).$$

assuming a diffuse prior for x_1 and x_2. It is clear that for $i = 3, \ldots, n$,

$$\begin{aligned} \mathrm{E}\left(x_i \mid x_{i-1}, x_{i-2}, \kappa\right) &= \left(1 + \frac{w_{i-2}}{w_{i-1}}\right)x_{i-1} - \frac{w_{i-2}}{w_{i-1}}x_{i-2} \\ \mathrm{Var}\left(x_i \mid x_{i-1}, x_{i-2}, \kappa\right) &= 1/(\kappa w_{i-1}^2). \end{aligned}$$

If w_i is an inverse distance between the locations s_{i+1} and s_i, i.e., $w_i = 1/\delta_i$, the conditional mean is the linear extrapolation of the values x_{i-2} and x_{i-1} to time i. The conditional mean is a consistent generalization of (3.41), but the conditional variance is quadratic in δ_{i-1} rather than cubic.

An alternative approach fixes this problem. Start with the unweighted directed model

$$x_{i+1} \mid x_1, \ldots, x_i, \kappa \sim \mathcal{N}(2x_i - x_{i-1}, \kappa^{-1})$$

from which a generalization of equations (3.41) is derived,

$$\begin{aligned} \mathrm{E}(x_{i+k} \mid x_i, x_{i-s}) &= \left(1 + \frac{k}{s}\right)x_i - \frac{k}{s}x_{i-s} \\ \mathrm{Var}(x_{i+k} \mid x_i, x_{i-s}) &= \frac{k(k+s)(2ks+1)}{6\kappa s}. \end{aligned}$$

Extending the integer 'distances' k and s to real valued distances, this would thus suggests defining a RW2 model for irregular locations as

$$\begin{aligned} \mathrm{E}(x_i \mid x_{i-1}, x_{i-2}) &= \left(1 + \frac{\delta_{i-1}}{\delta_{i-2}}\right)x_{i-1} - \frac{\delta_{i-1}}{\delta_{i-2}}x_{i-2} \\ \mathrm{Var}(x_i \mid x_{i-1}, x_{i-2}) &= \frac{\delta_{i-1}(\delta_{i-1} + \delta_{i-2})(2\delta_{i-1}\delta_{i-2} + 1)}{6\kappa\delta_{i-2}} \end{aligned}$$

for $i = 3, \ldots, n$. Again, the conditional mean is the linear extrapolation of the values x_{i-1} and x_{i-2}, but the conditional variance now has a different form and is cubic in the δ's.

3.4.2 IGMRFs of higher order on regular lattices★

We now consider the construction of polynomial IGMRFs of higher order on regular lattices of dimension $d > 1$. Basically, the idea is that the precision matrix is orthogonal to a polynomial design matrix of a certain degree. We first define this in general and then consider special cases for $d = 2$.

The general construction

An IGMRF of kth order in d dimensions is an improper GMRF of rank $n - m_{k-1,d}$, where the columns of the polynomial design matrix $\boldsymbol{S}_{k-1,d}$ are a basis of the null space of the precision matrix $\boldsymbol{Q}$.

Definition 3.6 (IGMRFs of order k in dimension d) *An IGMRF of order k in dimension d, is an improper GMRF of rank $n - m_{k-1,d}$ where $\boldsymbol{Q}\boldsymbol{S}_{k-1,d} = \boldsymbol{0}$ with $m_{k-1,d}$ and $\boldsymbol{S}_{k-1,d}$ as defined in (3.6) and (3.9).*

A second-order polynomial IGMRF in two dimensions

Let us consider a regular lattice $\mathcal{I}_{\boldsymbol{n}}$ in $d = 2$ dimensions. To construct a second-order IGMRF we choose the increments

$$(x_{i+1,j} + x_{i-1,j} + x_{i,j+1} + x_{i,j-1}) - 4x_{i,j}. \tag{3.42}$$

The motivation for this choice is that (3.42) is

$$\left(\Delta^2_{(1,0)} + \Delta^2_{(0,1)}\right) x_{i-1,j-1},$$

where we generalize the Δ operator in the obvious way to account for direction, so that $\Delta_{(1,0)}$ is the forward difference in direction $(1, 0)$ and similar to $\Delta_{(0,1)}$. Adding the first-order polynomial

$$p_{1,2}(i, j) = \beta_{00} + \beta_{10} i + \beta_{01} j, \tag{3.43}$$

i.e., a simple plane, to $\boldsymbol{x}$ will cancel in (3.42) for any choice of coefficients β_{00}, β_{10}, and β_{01}.

The precision matrix (apart from boundary effects) should have nonzero elements corresponding to

$$-\left(\Delta^2_{(1,0)} + \Delta^2_{(0,1)}\right)^2 = -\left(\Delta^4_{(1,0)} + 2\Delta^2_{(1,0)}\Delta^2_{(0,1)} + \Delta^4_{(0,1)}\right),$$

which is a negative difference approximation to the *biharmonic* differential operator

$$\left(\frac{\partial^2}{\partial x^2} + \frac{\partial^2}{\partial y^2}\right)^2 = \frac{\partial^4}{\partial x^4} + 2\frac{\partial^4}{\partial x^2 \partial y^2} + \frac{\partial^4}{\partial y^4}.$$

The fundamental solution of the biharmonic equation

$$\left(\frac{\partial^4}{\partial x^4} + 2\frac{\partial^4}{\partial x^2 \partial y^2} + \frac{\partial^4}{\partial y^4}\right)\phi(x,y) = 0$$

is the *thin plate spline*, which is the two-dimensional analogue of the cubic spline in one dimension, see, for example, Bookstein (1989), Gu (2002), or Green and Silverman (1994).

Starting from (3.42), we now want to compute the coefficients in $\boldsymbol{Q}$. Although a manual calculation is possible, we want to automate this process to easily compute more refined second-order IGMRFs. To compute the coefficients in the interior only, we wrap out the lattice $\mathcal{I}_{\boldsymbol{n}}$ on a torus $\mathcal{T}_{\boldsymbol{n}}$, and then decompose $\boldsymbol{Q}$ into $\boldsymbol{Q} = \boldsymbol{D}^T\boldsymbol{D}$, where $\boldsymbol{D}$ is a block circulant matrix with base

$$\begin{pmatrix} -4 & 1 & & & 1 \\ 1 & & & & \\ & & & & \\ & & & & \\ 1 & & & & \end{pmatrix}.$$

Therefore also $\boldsymbol{Q}$ will be circulant with base $\boldsymbol{q}$, say, which we can compute using (2.49). The result is

$$\boldsymbol{q} = \begin{pmatrix} 20 & -8 & 1 & & \\ -8 & 2 & & & \\ 1 & & & & \\ & & & & \\ & & & & \end{pmatrix}, \tag{3.44}$$

where only the upper left part of the base is shown. To write out the conditional expectations, it is convenient to use a graphical notation, where, for example, (3.42) looks like

$$\begin{array}{ccc} \circ & \bullet & \circ \\ \bullet & \circ & \bullet \\ \circ & \bullet & \circ \end{array} \; - 4 \begin{array}{ccc} \circ & \circ & \circ \\ \circ & \bullet & \circ \\ \circ & \circ & \circ \end{array}. \tag{3.45}$$

The format is to calculate the sum over all x_{ij}'s in the locations of the '$\bullet$'. The '$\circ$''s are there only to fix the spatial configuration. When this notation is used within a sum, then the sum-index denotes the center node.

Returning to (3.44), then (3.42) gives the following full conditionals in the interior

$$\begin{aligned} \mathrm{E}(x_{ij} \mid \boldsymbol{x}_{-ij}) &= \frac{1}{20}\left(8 \begin{array}{ccccc} \circ&\circ&\circ&\circ&\circ \\ \circ&\circ&\bullet&\circ&\circ \\ \circ&\bullet&\circ&\bullet&\circ \\ \circ&\circ&\bullet&\circ&\circ \\ \circ&\circ&\circ&\circ&\circ \end{array} - 2 \begin{array}{ccccc} \circ&\circ&\circ&\circ&\circ \\ \circ&\bullet&\circ&\bullet&\circ \\ \circ&\circ&\circ&\circ&\circ \\ \circ&\bullet&\circ&\bullet&\circ \\ \circ&\circ&\circ&\circ&\circ \end{array} - 1 \begin{array}{ccccc} \circ&\circ&\bullet&\circ&\circ \\ \circ&\circ&\circ&\circ&\circ \\ \bullet&\circ&\circ&\circ&\bullet \\ \circ&\circ&\circ&\circ&\circ \\ \circ&\circ&\bullet&\circ&\circ \end{array}\right) \\ \mathrm{Prec}(x_{ij} \mid \boldsymbol{x}_{-ij}) &= 20\kappa. \end{aligned}$$

The coefficients of $\boldsymbol{Q}$ that are affected by the boundary are found from expanding the quadratic term

$$-\frac{\kappa}{2}\sum_{i=2}^{n_1-1}\sum_{j=2}^{n_2-1}\left(\begin{smallmatrix}\circ & \bullet & \circ\\ \bullet & \circ & \bullet\\ \circ & \bullet & \circ\end{smallmatrix} - 4\begin{smallmatrix}\circ & \circ & \circ\\ \circ & \bullet & \circ\\ \circ & \circ & \circ\end{smallmatrix}\right)^2, \tag{3.46}$$

but there are different approaches to actually compute them. We now switch notation for a moment and index x_{ij}'s using one index only, x_i. The first approach is to write the quadratic term in (3.46) as

$$\begin{aligned}-\frac{1}{2}\sum_k\left(\sum_i w_{ki}x_i\right)^2 &= -\frac{1}{2}\sum_k\sum_i w_{ki}x_i\sum_j w_{kj}x_j\\ &= -\frac{1}{2}\sum_i\sum_j x_ix_jQ_{ij}\end{aligned} \tag{3.47}$$

for some weights w_{ki}'s and therefore

$$Q_{ij} = \sum_k w_{ki}w_{kj}. \tag{3.48}$$

Some bookkeeping is usually needed as only the nonzero w_{ij}'s and Q_{ij}'s need to be stored. This approach is implemented in `GMRFLib`, see Appendix B.

Example 3.7 *Consider the expression*

$$-\frac{1}{2}\left((x_1 - x_2 + x_3)^2 + (x_2 - x_3)^2\right),$$

which corresponds to (3.47) with $w_{11} = 1$, $w_{12} = -1$, $w_{13} = 1$, $w_{21} = 0$, $w_{22} = 1$, $w_{23} = -1$. *Using (3.48), we obtain, for example,*

$$\begin{aligned}Q_{23} &= w_{12}w_{13} + w_{22}w_{23} = -2, \quad \textit{and}\\ Q_{22} &= w_{12}w_{12} + w_{22}w_{22} = 2.\end{aligned}$$

A less elegant but often quite simple approach to compute the Q_{ij}'s is to note that

$$\left.\frac{\partial^2}{\partial x_i \partial x_j}U(\boldsymbol{x})\right|_{\boldsymbol{x}=\boldsymbol{0}} = \begin{cases}Q_{ii} & i = j\\ Q_{ij} & i > j\end{cases}, \tag{3.49}$$

where

$$U(\boldsymbol{x}) = \frac{1}{2}\sum_k\left(\sum_l w_{kl}x_l\right)^2.$$

Let $\mathbf{1}_i$ be a vector with its ith element equal to one and the rest of the elements zero. Using (3.49) Q_{ij} can be expressed as

$$Q_{ij} = \begin{cases}U(\mathbf{1}_i) - 2U(\mathbf{0}) + U(-\mathbf{1}_i) & i = j\\ U(\mathbf{1}_i + \mathbf{1}_j) + U(\mathbf{0}) - U(\mathbf{1}_j) - U(\mathbf{1}_i) & i \neq j.\end{cases} \tag{3.50}$$

Of course, we do not need to evaluate all terms in $U(\cdot)$, only those that contain x_i and/or x_j will suffice. This approach can also be extended to obtain the canonical parameterization (see Definition 2.2) if $U(\cdot)$ is extended to also include linear terms.

Example 3.8 *Reconsider Example 3.7. By using (3.50) we obtain*

$$\begin{aligned} Q_{23} &= \frac{1}{2}\big[((0-1+1)^2+(1-1)^2)+((0-0+0)^2+(0-0)^2) \\ &\quad -((0-0+1)^2+(0-1)^2)-((0-1+0)^2+(1-0)^2)\big] \\ &= -2 \quad \textit{and} \\ Q_{22} &= \frac{1}{2}\big[((0-1+0)^2+(1-0)^2))-2((0-0+0)^2+(0-0)^2) \\ &\quad +((0-(-1)+0)^2+(-1-0)^2)\big] = 2. \end{aligned}$$

Alternative IGMRFs in two dimensions

Although (3.42) is an obvious first choice, it has some drawbacks. First (3.46) does not contain the four corners $x_{1,1}$, x_{1,n_2}, $x_{n_1,1}$, and x_{n_1,n_2}, so we need to add such terms manually. Furthermore, using only the terms

$$\begin{smallmatrix} \circ & \bullet & \circ \\ \bullet & \bullet & \bullet \\ \circ & \bullet & \circ \end{smallmatrix}$$

to obtain a difference approximation to

$$\frac{\partial^2}{\partial x^2}+\frac{\partial^2}{\partial y^2} \tag{3.51}$$

is not optimal. The discretization error is quite different in the direction 45 degrees to the main directions, hence we could expect a 'directional effect' in our model. The common way out is to use difference approximations where the discretization error is isotropic instead of anisotropic, so it does not depend on the rotation of the coordinate system. The classic (numerical) choice is known under the name *Mehrstellen stencil* and given as

$$-\frac{10}{3}\begin{smallmatrix} \circ & \circ & \circ \\ \circ & \bullet & \circ \\ \circ & \circ & \circ \end{smallmatrix} + \frac{2}{3}\begin{smallmatrix} \circ & \bullet & \circ \\ \bullet & \circ & \bullet \\ \circ & \bullet & \circ \end{smallmatrix} + \frac{1}{6}\begin{smallmatrix} \bullet & \circ & \bullet \\ \circ & \circ & \circ \\ \bullet & \circ & \bullet \end{smallmatrix}. \tag{3.52}$$

Using these differences as increments, we will still obtain nonzero terms in $\boldsymbol{Q}$ that approximate the biharmonic differential operator; however, the approximation is better and isotropic. The corresponding full conditionals are now similar to obtain as for the increments (3.42),

$$\begin{aligned} \mathrm{E}(x_{ij}\mid \boldsymbol{x}_{-ij}) &= \frac{1}{468}\left(144\begin{smallmatrix} \circ&\circ&\circ&\circ&\circ \\ \circ&\circ&\bullet&\circ&\circ \\ \circ&\bullet&\circ&\bullet&\circ \\ \circ&\circ&\bullet&\circ&\circ \\ \circ&\circ&\circ&\circ&\circ \end{smallmatrix} - 18\begin{smallmatrix} \circ&\circ&\bullet&\circ&\circ \\ \circ&\circ&\circ&\circ&\circ \\ \bullet&\circ&\circ&\circ&\bullet \\ \circ&\circ&\circ&\circ&\circ \\ \circ&\circ&\bullet&\circ&\circ \end{smallmatrix}\right. \\ &\quad \left. +8\begin{smallmatrix} \circ&\circ&\circ&\circ&\circ \\ \circ&\bullet&\circ&\bullet&\circ \\ \circ&\circ&\circ&\circ&\circ \\ \circ&\bullet&\circ&\bullet&\circ \\ \circ&\circ&\circ&\circ&\circ \end{smallmatrix} - 8\begin{smallmatrix} \circ&\bullet&\circ&\bullet&\circ \\ \bullet&\circ&\circ&\circ&\bullet \\ \circ&\circ&\circ&\circ&\circ \\ \bullet&\circ&\circ&\circ&\bullet \\ \circ&\bullet&\circ&\bullet&\circ \end{smallmatrix} - 1\begin{smallmatrix} \bullet&\circ&\circ&\circ&\bullet \\ \circ&\circ&\circ&\circ&\circ \\ \circ&\circ&\circ&\circ&\circ \\ \circ&\circ&\circ&\circ&\circ \\ \bullet&\circ&\circ&\circ&\bullet \end{smallmatrix}\right) \end{aligned}$$

$$\mathrm{Prec}(x_{ij} \mid \boldsymbol{x}_{-ij}) \;=\; 13\kappa.$$

As the coefficients in $\boldsymbol{Q}$ will approximate the biharmonic differential operator, it is tempting to *start* with such an approximation, for example, defining

$$\mathrm{E}(x_{ij} \mid \boldsymbol{x}_{-ij}) \;=\; \frac{1}{15}\left(\frac{16}{3}\,\square \;-\; \frac{2}{3}\,\square \right.$$
$$\left. -\,\frac{5}{6}\,\square \;-\; \frac{1}{12}\,\square \right) \qquad (3.53)$$

$$\mathrm{Prec}(x_{ij} \mid \boldsymbol{x}_{-ij}) \;=\; 15\kappa.$$

The big disadvantage is that the corresponding increments *do not* depend on only a few neighbors, but on all $\boldsymbol{x}$ in a larger neighborhood than the neighborhood in (3.53). This can be verified by solving $\boldsymbol{Q} = \boldsymbol{D}^T\boldsymbol{D}$ for $\boldsymbol{D}$ using (2.49).

If we use all neighbors in a 5×5 window around i to approximate (3.51) isotropically, then we may use the increment

$$-\frac{9}{2}\,\square \;+\; \frac{16}{15}\,\square \;+\; \frac{2}{15}\,\square$$
$$-\,\frac{1}{15}\,\square \;-\; \frac{1}{120}\,\square$$

for which $\mathrm{E}(x_{ij} \mid \boldsymbol{x}_{-ij}, \kappa)$ is

$$\frac{1}{358\,420}\left(132\,096\,\square \quad -25\,568\,\square \right.$$
$$+\,2\,048\,\square \quad -66\,\square$$
$$\left. -14\,944\,\square \quad -1\,792\,\square \right)$$

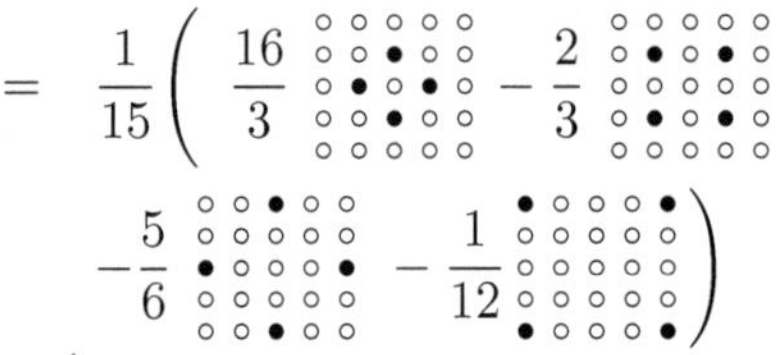

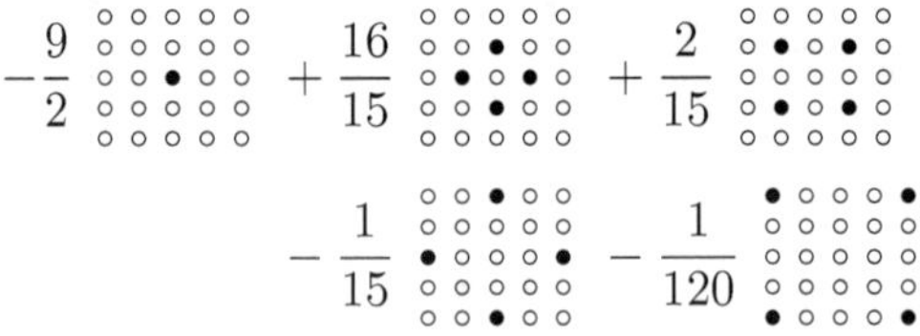

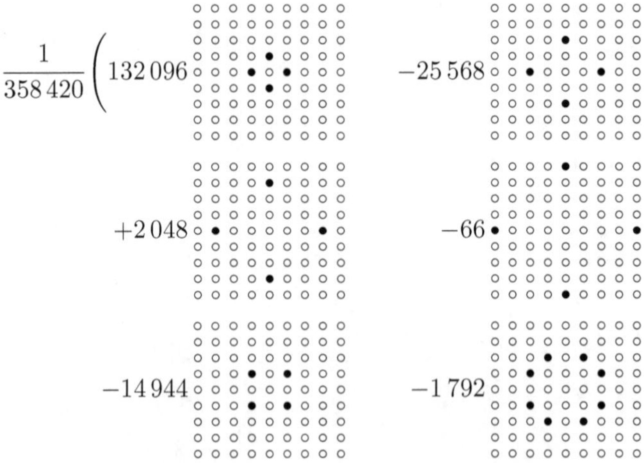

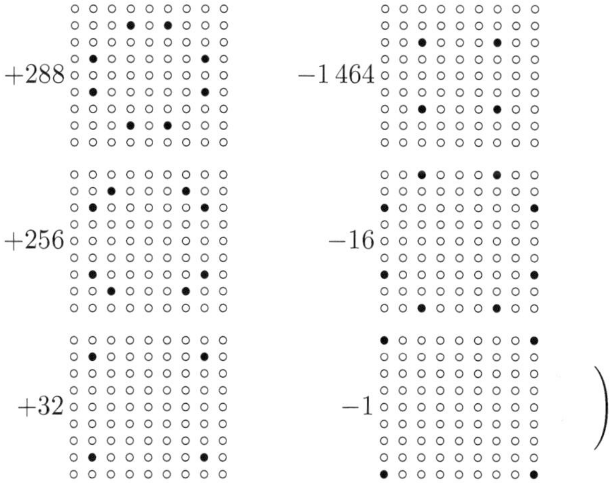

and

$$\mathrm{Prec}(x_{ij} \mid \boldsymbol{x}_{-ij}, \kappa) = \frac{17\,921}{720}\kappa.$$

The expressions for the full conditionals easily get quite complicated using higher-order stencils.

Example 3.9 *Figure 3.7 displays two samples using the two simpler schemes (3.45) and (3.52) on a torus of dimension* 256×256. *We use the torus only for computational reasons. By studying the corresponding correlation matrices, we find that the deviation is largest on the direction* 45 *degrees to the horizontal axis, as expected. This is due to the fact that (3.45) is anisotropic while (3.52) is isotropic. The difference is between* -10^{-4} *and* 10^{-4}, *hence quite small and not of any practical importance. As (3.52) includes four corner terms that (3.45) misses on a regular lattice, we generally recommend using (3.52).*

Remark. If we consider GMRFS defined via (3.45) and (3.52) on the torus rather than the lattice, we obtain (strictly speaking) a first-order and not a second-order IGMRF, because the rank of the precision matrix will be one. Note that we cannot adjust for polynomials of order larger than zero, since those are in general not cyclic.

An alternative model, proposed by Besag and Kooperberg (1995), starts directly with the normal full conditionals defined through

$\mathrm{E}(x_{ij} \mid \boldsymbol{x}_{-ij}) = \frac{1}{4}$ 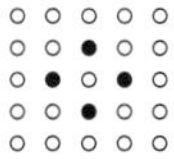 $+\frac{1}{8}$ 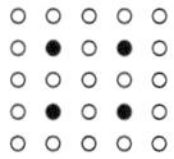 $-\frac{1}{8}$ 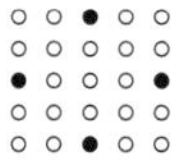 (3.54)

$\mathrm{Prec}(x_{ij} \mid \boldsymbol{x}_{-ij}) = \kappa.$

The motivation for the specific form in (3.54) is that the least-squares locally quadratic fit (3.8) through these twelve points generates these coefficients. Furthermore, the model is invariant to the addition of a

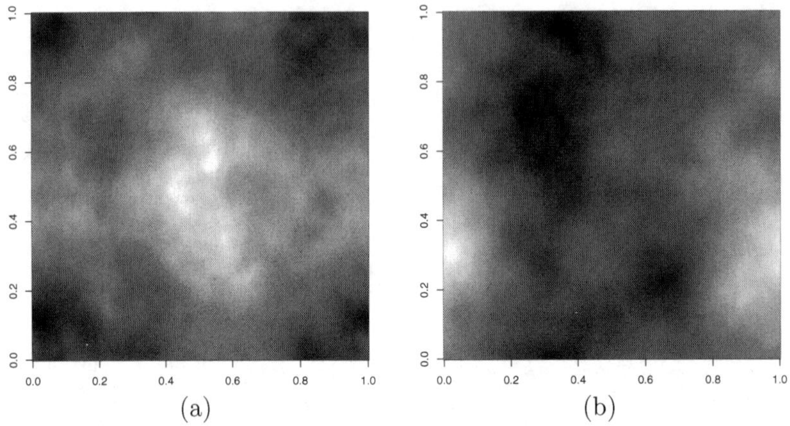

Figure 3.7 *Two samples from an IGMRF on a torus with dimension* 256×256*, where (a) used increments (3.45) and (b) used increments (3.52).*

plane of the form (3.43). However, there is no representation of this IGMRF based on simple increments so the model has the same drawback as (3.53).

3.4.3 Nonpolynomial IGMRFs of higher order

In general, the matrix $\boldsymbol{S}$ that spans the null space of $\boldsymbol{Q}$, i.e., $\boldsymbol{QS} = \boldsymbol{0}$, does not need to be a polynomial design matrix. It can, for example, be constructed so that the IGMRF has more than one unspecified overall level. We will now describe two examples, where such a generalization can be useful. As a side product, the first introduces a general device to construct higher-order IGMRFs using the Kronecker product.

Construction of IGMRFs using the Kronecker product

A useful approach to construct an IGMRF of higher order is to define its structure matrix as the *Kronecker product* of structure matrices of lower-order IGMRFs.

For illustration, consider a regular lattice $\mathcal{I}_{\boldsymbol{n}}$. Now define the structure matrix $\boldsymbol{R}$ of an IGMRF model of $\boldsymbol{x}$ as the *Kronecker product* of two RW1 structure matrices of the form (3.22) of dimension n_1 and n_2, respectively:

$$\boldsymbol{R} = \boldsymbol{R}_1 \otimes \boldsymbol{R}_2.$$

Clearly $\boldsymbol{R}$ has rank $(n_1 - 1)(n_2 - 1)$. It can easily be shown that this specification corresponds to a model with *differences of differences* as increments:

$$\Delta_{(1,0)}\Delta_{(0,1)}x_{ij} \sim \mathcal{N}(0, \kappa^{-1}), \tag{3.55}$$

for $i = 1, \dots, n_1 - 1$, and $j = 1, \dots, n_2 - 1$. Here $\Delta_{(1,0)}$ and $\Delta_{(0,1)}$ are defined as in Section 3.4.2, hence

$$\Delta_{(1,0)}\Delta_{(0,1)}x_{ij} = \begin{smallmatrix} \circ & \bullet \\ \bullet & \circ \end{smallmatrix} - \begin{smallmatrix} \bullet & \circ \\ \circ & \bullet \end{smallmatrix}. \tag{3.56}$$

From (3.56) we see that the IGMRF defined through (3.55) is invariant to the addition of constants to any rows and columns. This is an example of an IGMRF with more than one unspecified level. The density of $\boldsymbol{x}$ is

$$\begin{aligned} \pi(\boldsymbol{x} \mid \kappa) \quad &\propto \quad \kappa^{\frac{(n_1-1)(n_2-1)}{2}} \\ &\times \quad \exp\left(-\frac{\kappa}{2} \sum_{i=1}^{n_1-1} \sum_{j=1}^{n_2-1} (\Delta_{(1,0)}\Delta_{(0,1)}x_{ij})^2 \right) \end{aligned} \tag{3.57}$$

with $\Delta_{(1,0)}\Delta_{(0,1)}x_{ij} = x_{i+1,j+1} - x_{i+1,j} - x_{i,j+1} + x_{ij}$. Note that the conditional mean of x_{ij} in the interior depends on its eight nearest sites and is

$$\frac{1}{2} \begin{smallmatrix} \circ & \bullet & \circ \\ \bullet & \circ & \bullet \\ \circ & \bullet & \circ \end{smallmatrix} - \frac{1}{4} \begin{smallmatrix} \bullet & \circ & \bullet \\ \circ & \circ & \circ \\ \bullet & \circ & \bullet \end{smallmatrix},$$

which equals a least-squares locally quadratic fit through these eight neighbors. The conditional precision is 4κ.

We note that the representation of the precision matrix $\boldsymbol{Q}$ as the Kronecker product $\boldsymbol{Q} = \kappa(\boldsymbol{R}_1 \otimes \boldsymbol{R}_2)$ is also useful for computing $|\boldsymbol{Q}|^*$, because (extending (3.1))

$$|\boldsymbol{R}_1 \otimes \boldsymbol{R}_2|^* = (|\boldsymbol{R}_1|^*)^{n_2-1} \, (|\boldsymbol{R}_2|^*)^{n_1-1},$$

where $n_1 - 1$ and $n_2 - 1$ is the rank of $\boldsymbol{R}_1$ and $\boldsymbol{R}_2$, respectively.

Such a model is useful for smoothing a spatial surface while accommodating arbitrary row and column effects. Alternatively, one may incorporate sum-to-zero constraints on all rows and columns. This model is straightforward to generalize to a torus and then corresponds to the Kronecker product of two circular RW1 models. Figure 3.8 displays two samples from this model on a torus of dimension 256×256 using these constraints.

Under suitable sum-to-zero constraints, the Kronecker product construction is useful as a general device to specify *interaction models*, as proposed in Clayton (1996). For example, to define an IGMRF on a space $\times$ time domain, one might take the Kronecker product of the structure matrix of a RW1 model (3.22) and the structure matrix of a spatial IGMRF of first order, as defined in equation (3.31).

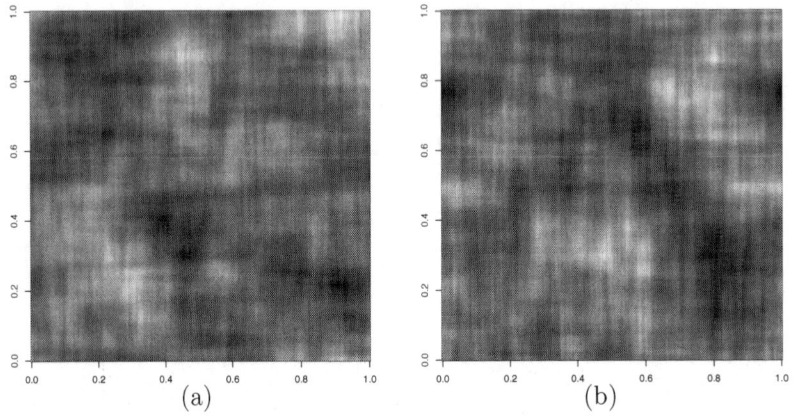

Figure 3.8 *Two samples from model (3.55) defined on a torus of dimension* 256×256 *lattice with sum-to-zero constraints on each row and column.*

An IGMRF model for seasonal variation

Assume the location of the nodes i are all positive integers, i.e., $i = 1, 2, \ldots, n$. Assume seasonal data are given with some periodicity of period length m, say. For example, $m = 12$ for monthly data with a yearly cycle.

A simple model for seasonal variation is obtained by assuming that the sums $x_i + x_{i+1} + \cdots + x_{i+m-1}$ are the increments with precision κ, $i = 1, \ldots, n - m + 1$. The joint density is

$$\pi(\boldsymbol{x} \mid \kappa) \propto \kappa^{\frac{n-m+1}{2}} \exp\left(-\frac{\kappa}{2} \sum_{i=1}^{n-m+1} (x_i + x_{i+1} + \cdots + x_{i+m-1})^2 \right). \tag{3.58}$$

For example, for $m = 4$ the corresponding precision matrix $\boldsymbol{Q}$ has the form

$$\boldsymbol{Q} = \kappa \begin{pmatrix}
1 & 1 & 1 & 1 & & & & & & & \\
1 & 2 & 2 & 2 & 1 & & & & & & \\
1 & 2 & 3 & 3 & 2 & 1 & & & & & \\
1 & 2 & 3 & 4 & 3 & 2 & 1 & & & & \\
 & 1 & 2 & 3 & 4 & 3 & 2 & 1 & & & \\
 & & \ddots & \ddots & \ddots & \ddots & \ddots & \ddots & \ddots & & \\
 & & & 1 & 2 & 3 & 4 & 3 & 2 & 1 & \\
 & & & & 1 & 2 & 3 & 4 & 3 & 2 & 1 \\
 & & & & & 1 & 2 & 3 & 3 & 2 & 1 \\
 & & & & & & 1 & 2 & 2 & 2 & 1 \\
 & & & & & & & 1 & 1 & 1 & 1
\end{pmatrix}. \tag{3.59}$$

The bandwidth of $\boldsymbol{Q}$ is $m-1$ while the rank of $\boldsymbol{Q}$ is $n-m+1$. Thus, the rank deficiency is larger than 1, but this is not a polynomial IGMRF. Instead, $\boldsymbol{QS} = \boldsymbol{0}$ for

$$\boldsymbol{S} = \begin{pmatrix} 1 & 0 & 0 \\ 0 & 1 & 0 \\ 0 & 0 & 1 \\ -1 & -1 & -1 \\ 1 & 0 & 0 \\ 0 & 1 & 0 \\ 0 & 0 & 1 \\ -1 & -1 & -1 \\ 1 & 0 & 0 \\ \vdots & \vdots & \vdots \end{pmatrix}, \tag{3.60}$$

i.e., $\pi(\boldsymbol{x})$ is invariant to the addition of any vector

$$\boldsymbol{c} = (c_1, c_2, c_3, c_4, c_1, c_2, c_3, c_4, c_1, \ldots)^T$$

to $\boldsymbol{x}$, as long as $\sum_{i=1}^{4} c_i = 0$ is fulfilled. This model is quite often used in structural time-series analysis because it is completely nonparametric and flexible; nonparametric, since no parametric form is assumed for the seasonal pattern, flexible, because the seasonal pattern is allowed to change over time. The latter point is evident from (3.58), where the seasonal effects do not sum up to zero, as in a fixed seasonal model, but to normal white noise.

Similar constructions could be made in a spatial context, for example, for a two-dimensional seasonal pattern on regular lattices. As a simple example, one could consider a model where the increments are all sums of the form

$$\begin{matrix} \bullet & \bullet \\ \bullet & \bullet \end{matrix} \quad \text{or} \quad \begin{matrix} \bullet & \bullet & \bullet \\ \bullet & \bullet & \bullet \\ \bullet & \bullet & \bullet \end{matrix}.$$

This can easily be extended to sums over all $m_1 \times m_2$ submatrices. Note that this model can be constructed using the Kronecker product of two seasonal structure matrices of period length m_1 and m_2, respectively.

3.5 Continuous-time random walks*

In this section we introduce a class of random walk models, which satisfy two important properties. First, they are consistent with respect to the choice of the locations and the resolution. Secondly, they have a Markov property that makes the precision matrix sparse so we can do efficient computations. The idea is that we view the unknown parameters $\boldsymbol{x}$ as realizations of an integrated Wiener process in continuous time. We will denote this class of models as *continuous-time* random walks of kth order, which we abbreviate as CRWk.

We have already seen one particular example of a CRWk, the RW1 model for irregular locations with density defined in (3.26). However, it will become clear that the CRW2 is different from the irregular RW2 and this is also the case for higher-order models.

The starting point is to consider a realization of an underlying continuous-time process $\eta(t)$, a $(k-1)$-fold integrated Wiener process where k is a positive integer. We will describe this concept first for general k, and then work out the details for the important case $k = 2$. The cases $k = 3$ and 4 will be briefly sketched at the end.

Definition 3.7 (A $(k-1)$-fold integrated Wiener process) *Let $\eta(t)$ be a $(k-1)$-fold integrated standard Wiener process*

$$\eta(t) = \int_0^t \frac{(t-h)^{k-1}}{(k-1)!} \, dW(h), \tag{3.61}$$

where $W(h)$ is a standard Wiener process. Let $\boldsymbol{x} = (x_1, \ldots, x_n)^T$ be a realization of $\eta(t)$ at the locations $0 < s_1 < s_2 < \ldots < s_n$. Let $\eta(0)$ have a diffuse prior, $\mathcal{N}(0, \tau^{-1})$, where $\tau \to 0$. Then the density $\pi(\boldsymbol{x})$ of $\boldsymbol{x}$ is a standard CRWk model and the density of $\boldsymbol{x}/\sqrt{\kappa}$ is a CRWk model with precision κ.

Due to the $(k-1)$-fold integration, a CRWk model will also be an IGMRF of order k.

Although we have not written up the density of a CRWk model explicitly, the covariance matrix of the Gaussian density of a conditional standard CRWk model is found from

$$\text{Cov}(\eta(t), \eta(s) \mid \eta(0) = 0) = \int_0^s \frac{(s-h)^{k-1}(t-h)^{k-1}}{((k-1)!)^2} \, dh$$

for $0 < s < t$. Furthermore, $E(\eta(t)|\eta(0) = 0) = 0$ for all $t > 0$. However, due to the correlation structure of $\eta(t)$, the conditional mean

$$\text{E}\left(\eta(s_i) \mid \eta(s_1), \ldots, \eta(s_{i-1}),\ \eta(s_{i+1}), \ldots, \eta(s_n)\right)$$

does not simplify and we need to take the values at all other locations into account. In other words, the precision matrix will be a completely dense matrix. However, the conditional densities simplify if we *augment* the process with its derivatives,

$$\boldsymbol{\eta}(t) = (\eta(t), \eta^{(1)}(t), \ldots, \eta^{(k-1)}(t))^T,$$

where $\eta^{(m)}$ is the mth derivative of $\eta(t)$,

$$\eta^{(m)}(t) = \int_0^t \frac{(t-h)^{k-1-m}}{(k-1-m)!} \, dW(h),$$

for $m = 0, \ldots, k-1$. The simplification is due to the following argument. Let $0 < s < t$ and define

$$\begin{aligned} u^{(m)}(s,t) &= \eta^{(m)}(t) - \eta^{(m)}(s) \\ &= \int_s^t \frac{(t-h)^{k-1-m}}{(k-1-m)!} dW(h) \end{aligned}$$

for $m = 0, \ldots, k-1$. Note that $u^{(0)}(s,t) = \eta(t) - \eta(s)$. Then consider the evolution from 0 to t, as first from 0 to s, then from s to t,

$$\begin{aligned} u^{(m)}(0,t) &= \left(\int_0^s + \int_s^t\right) \frac{(t-h)^{k-1-m}}{(k-1-m)!} dW(h) \\ &= u^{(m)}(s,t) + \int_0^s \frac{(t-h)^{k-1-m}}{(k-1-m)!} dW(h) \\ &= u^{(m)}(s,t) \\ &\quad + \int_0^s \frac{((t-s)+(s-h))^{k-1-m}}{(k-1-m)!} dW(h) \\ &= u^{(m)}(s,t) \\ &\quad + \int_0^s \sum_{j=0}^{k-1-m} \binom{k-1-m}{j} \\ &\qquad \frac{(t-s)^j (s-h)^{k-1-m-j}}{(k-1-m)!} dW(h) \\ &= u^{(m)}(s,t) \\ &\quad + \sum_{j=0}^{k-1-m} \frac{(t-s)^j}{j!} \int_0^s \frac{(s-h)^{k-1-m-j}}{(k-1-m-j)!} dW(h) \\ &= u^{(m)}(s,t) + \sum_{j=0}^{k-1-m} \frac{(t-s)^j}{j!} u^{(m+j)}(0,s). \end{aligned} \tag{3.62}$$

If we define

$$\boldsymbol{u}(s,t) = (u^{(0)}(s,t), u^{(1)}(s,t), \ldots, u^{(k-1)}(s,t))^T$$

then (3.62) can be written as

$$\boldsymbol{u}(0,t) = \boldsymbol{T}(s,t)\boldsymbol{u}(0,s) + \boldsymbol{u}(s,t), \tag{3.63}$$

where

$$\boldsymbol{T}(s,t) = \begin{pmatrix} 1 & \frac{t-s}{1!} & \frac{(t-s)^2}{2!} & \frac{(t-s)^3}{3!} & \cdots & \frac{(t-s)^{k-1}}{(k-1)!} \\ & 1 & \frac{t-s}{1!} & \frac{(t-s)^2}{2!} & \cdots & \frac{(t-s)^{k-2}}{(k-2)!} \\ & & 1 & \frac{t-s}{1!} & \cdots & \frac{(t-s)^{k-3}}{(k-3)!} \\ & & & \ddots & & \vdots \\ & & & & 1 & \frac{t-s}{1!} \\ & & & & & 1 \end{pmatrix}.$$

It is perhaps easiest to interpret (3.63) conditionally. For known $\boldsymbol{u}(0,s)$ we can add the normal distributed vector $\boldsymbol{u}(s,t)$ to $\boldsymbol{T}(s,t)\boldsymbol{u}(0,s)$ in order to obtain $\boldsymbol{u}(0,t)$. Since

$$\mathrm{E}(\boldsymbol{u}(s,t) \mid \boldsymbol{u}(0,s)) = \boldsymbol{0},$$

we may write

$$\boldsymbol{u}(0,t) \mid \boldsymbol{u}(0,s) \sim \mathcal{N}(\boldsymbol{T}(s,t)\boldsymbol{u}(0,s),\ \boldsymbol{\Sigma}(s,t)). \tag{3.64}$$

Element ij of $\boldsymbol{\Sigma}(s,t)$, $\Sigma_{ij}(s,t)$, is

$$\begin{aligned}\Sigma_{ij}(s,t) &= \int_s^t \frac{(t-h)^{k-i}}{(k-i)!}\,\frac{(t-h)^{k-j}}{(k-j)!}\,dh \\ &= \frac{(t-s)^{2k+1-i-j}}{(2k+1-i-j)\,(k-i)!\,(k-j)!}.\end{aligned}$$

The practical use of this result is to use the derived model for $\boldsymbol{\eta}(t)$ at the locations of interest under a diffuse prior for the initial conditions $\boldsymbol{\eta}(0)$, and then integrate out all the derivatives of order $1,\ldots,k-1$. However, for simulation-based inference using MCMC methods, we will simulate the full vector $\boldsymbol{\eta}(t)$ at all locations, and simply ignore the derivatives in the analysis.

Note also that the augmented model is computationally fast, as the bandwidth for the corresponding precision matrix is $2k-1$, hence compared to the (inconsistent) RW2 model, the computational cost is about

$$kn \times (2k-1)^2, \qquad \text{versus} \quad n \times 2^2,$$

using the computational complexity of band-matrices, see Section 2.4. For $k=2$, the computational effort for a CRW2 model is about $18/4 = 4.5$ times the costs required for a RW2 model. However, if n is not too large, the practical cost will be very similar.

We will now derive the necessary details for the important case $k=2$, including the precision matrix for irregular locations. We assume for simplicity that $\kappa = 1$. Let $\boldsymbol{t} = (s_1,\ldots,s_n)$ be the locations of $\boldsymbol{x} =$

$(x_1, \ldots, x_n)^T$ and recall (3.25). Then (3.64) gives

$$\begin{pmatrix} \eta(s_{i+1}) \\ \eta^{(1)}(s_{i+1}) \end{pmatrix} = \begin{pmatrix} 1 & \delta_i \\ 0 & 1 \end{pmatrix} \begin{pmatrix} \eta(s_i) \\ \eta^{(1)}(s_i) \end{pmatrix} + \boldsymbol{u}(\delta_i),$$

where

$$\boldsymbol{u}(\delta_i) \sim \mathcal{N}\left(\begin{pmatrix} 0 \\ 0 \end{pmatrix}, \begin{pmatrix} \delta_i^3/3 & \delta_i^2/2 \\ \delta_i^2/2 & \delta_i \end{pmatrix} \right).$$

So,

$$\log \pi(\boldsymbol{\eta}(s_{i+1}) \mid \boldsymbol{\eta}(s_i)) = -\frac{1}{2}\boldsymbol{u}(\delta_i)^T \begin{pmatrix} 12/\delta_i^3 & -6/\delta_i^2 \\ -6/\delta_i^2 & 4/\delta_i \end{pmatrix} \boldsymbol{u}(\delta_i),$$

where

$$\boldsymbol{u}(\delta_i) = \begin{pmatrix} \eta(s_{i+1}) \\ \eta^{(1)}(s_{i+1}) \end{pmatrix} - \begin{pmatrix} 1 & \delta_i \\ 0 & 1 \end{pmatrix} \begin{pmatrix} \eta(s_i) \\ \eta^{(1)}(s_i) \end{pmatrix}.$$

Define the matrices

$$\begin{aligned} \boldsymbol{A}_i &= \begin{pmatrix} 12/\delta_i^3 & 6/\delta_i^2 \\ 6/\delta_i^2 & 4/\delta_i \end{pmatrix} \\ \boldsymbol{B}_i &= \begin{pmatrix} -12/\delta_i^3 & 6/\delta_i^2 \\ -6/\delta_i^2 & 2/\delta_i \end{pmatrix} \\ \boldsymbol{C}_i &= \begin{pmatrix} 12/\delta_i^3 & -6/\delta_i^2 \\ -6/\delta_i^2 & 4/\delta_i \end{pmatrix} \end{aligned}$$

for $i = 1, \ldots, n-1$, then some straightforward algebra shows that

$$(\eta(s_1), \eta^{(1)}(s_1), \eta(s_2), \eta^{(1)}(s_2), \ldots, \eta(s_n), \eta^{(1)}(s_n))^T$$

has precision matrix

$$\begin{pmatrix} \boldsymbol{A}_1 & \boldsymbol{B}_1 & & & & \\ \boldsymbol{B}_1^T & \boldsymbol{A}_2 + \boldsymbol{C}_1 & \boldsymbol{B}_2 & & & \\ & \boldsymbol{B}_2^T & \boldsymbol{A}_3 + \boldsymbol{C}_2 & \boldsymbol{B}_3 & & \\ & & \ddots & \ddots & \ddots & \\ & & & & & \boldsymbol{B}_{n-1} \\ & & & & \boldsymbol{B}_{n-1}^T & \boldsymbol{C}_{n-1} \end{pmatrix}. \quad (3.65)$$

Here we have used diffuse initial conditions for $(\eta(s_1), \eta^{(1)}(s_1))^T$. The precision matrix for a CRW2 model with precision κ is found by scaling (3.65) by κ. Note that κ for the CRW2 model is the precision for the first-order increments, while for the RW2 model κ is the precision for the second-order increments.

The null space of $\boldsymbol{Q}$ is spanned by the two vectors

$$(s_1, 1, s_2, 1, s_3, 1, \ldots, s_n, 1)^T \quad \text{and} \quad (1, 0, 1, 0, 1, 0, \ldots, 1, 0)^T,$$

which can be verified directly. The density is invariant to the addition of any constant to the locations, and to the addition of an arbitrary constant to the derivatives with an obvious correction for the locations.

When the locations are equidistant and $\delta_i = 1$ for all i, then the matrices $\boldsymbol{A}_i$, $\boldsymbol{B}_i$, and $\boldsymbol{C}_i$ do not depend on i, and are equal to

$$\boldsymbol{A}_i = \begin{pmatrix} 12 & 6 \\ 6 & 4 \end{pmatrix} \quad \boldsymbol{B}_i = \begin{pmatrix} -12 & 6 \\ -6 & 2 \end{pmatrix} \quad \boldsymbol{C}_i = \begin{pmatrix} 12 & -6 \\ -6 & 4 \end{pmatrix}.$$

Figure 3.5 displays 10 samples of $\boldsymbol{x}^{\perp}$ (ignoring the derivative) from the CRW2 model for $\kappa = 1$ and $n = 99$, the marginal variances and the correlations between $x^{\perp}_{n/2}$ and $x^{\perp}_i$ for $i = 1, \ldots, n$. Note that we fixed $s_{50} = 50$ and require that $s_i + s_{100-i} = 100$ holds for all locations $i = 1, \ldots, n$, as in Figure 3.3.1. The samples show a very similar behavior as those obtained in Figure 3.4.1 using the (equally spaced) RW2 model. The (theoretical) variances and correlations $\mathrm{Corr}(x^{\perp}_{n/2}, x^{\perp}_i)$ are also very similar.

When we go to higher-order models, then the precision matrix for

$$(\eta(s_1), \ldots, \eta^{(k-1)}(s_1), \ldots, \eta(s_n), \ldots, \eta^{(k-1)}(s_n))^T$$

has the same structure as (3.65), but the matrices $\boldsymbol{A}_i$, $\boldsymbol{B}_i$, and $\boldsymbol{C}_i$ will differ. For completeness, we will give the result for $k = 3$:

$$\boldsymbol{A}_i = \begin{pmatrix} 720/\delta_i^5 & 360/\delta_i^4 & 60/\delta_i^3 \\ 360/\delta_i^4 & 192/\delta_i^3 & 36/\delta_i^2 \\ 60/\delta_i^3 & 36/\delta_i^2 & 9/\delta_i \end{pmatrix},$$

$$\boldsymbol{B}_i = \begin{pmatrix} -720/\delta_i^5 & 360/\delta_i^4 & -60/\delta_i^3 \\ -360/\delta_i^4 & 168/\delta_i^3 & -24/\delta_i^2 \\ -60/\delta_i^3 & 24/\delta_i^2 & -3/\delta_i \end{pmatrix}$$

$$\boldsymbol{C}_i = \begin{pmatrix} 720/\delta_i^5 & -360/\delta_i^4 & 60/\delta_i^3 \\ -360/\delta_i^4 & 192/\delta_i^3 & -36/\delta_i^2 \\ 60/\delta_i^3 & -36/\delta_i^2 & 9/\delta_i \end{pmatrix},$$

and for $k = 4$,

$$\boldsymbol{A}_i = \begin{pmatrix} 100800/\delta_i^7 & 50400/\delta_i^6 & 10080/\delta_i^5 & 840/\delta_i^4 \\ 50400/\delta_i^6 & 25920/\delta_i^5 & 5400/\delta_i^4 & 480/\delta_i^4 \\ 10080/\delta_i^5 & 5400/\delta_i^4 & 1200/\delta_i^3 & 120/\delta_i^2 \\ 840/\delta_i^4 & 480/\delta_i^3 & 120/\delta_i^2 & 16/\delta_i \end{pmatrix}$$

$$\boldsymbol{B}_i = \begin{pmatrix} -100800/\delta_i^7 & 50400/\delta_i^6 & -10080/\delta_i^5 & -840/\delta_i^4 \\ -50400/\delta_i^6 & 24480/\delta_i^5 & -4680/\delta_i^4 & -360/\delta_i^3 \\ -10080/\delta_i^5 & 4680/\delta_i^4 & -840/\delta_i^3 & -60/\delta_i^2 \\ -840/\delta_i^4 & 360/\delta_i^3 & -60/\delta_i^2 & 4/\delta_i \end{pmatrix}$$

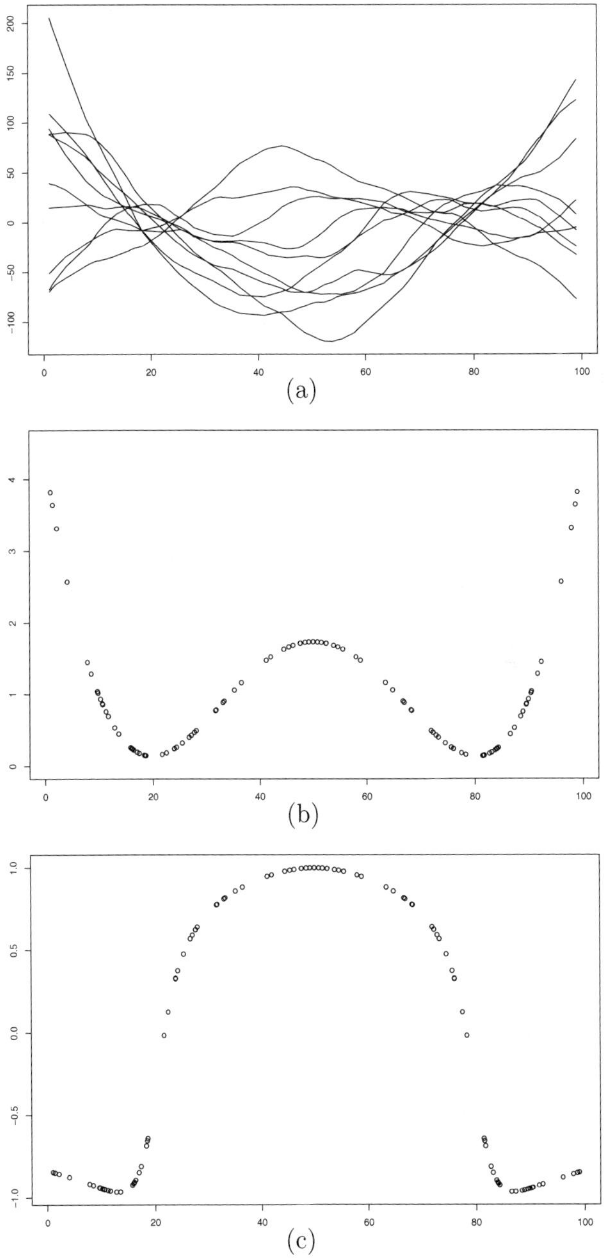

Figure 3.9 *Illustrations of the properties of the CRW2 model with* $n = 99$ *and irregular locations: (a) displays* 10 *samples of* $\boldsymbol{x}^{\perp}$ *(ignoring the derivative), (b) displays* $\mathrm{Var}(x_i^{\perp})$ *for* $i = 1, \ldots, n$ *normalized with the average variance, and (c) displays* $\mathrm{Corr}(x_{n/2}^{\perp}, x_i^{\perp})$ *for* $i = 1, \ldots, n$.

$$\boldsymbol{C}_i = \begin{pmatrix} 100800/\delta_i^7 & -50400/\delta_i^6 & 10080/\delta_i^5 & -840/\delta_i^4 \\ -50400/\delta_i^6 & 25920/\delta_i^5 & -5400/\delta_i^4 & 480/\delta_i^3 \\ 10080/\delta_i^5 & -5400/\delta_i^4 & 1200/\delta_i^3 & -120/\delta_i^2 \\ -840/\delta_i^4 & 480/\delta_i^3 & -120/\delta_i^2 & 16/\delta_i \end{pmatrix}.$$

It is straightforward to generate these and further results for higher order using a proper tool for symbolic computation.

3.6 Bibliographic notes

The properties of Kronecker products are extracted from Searle (1982) and Harville (1997).

RW1, RW2, and seasonal models of the form (3.58) are frequently used in time series analysis, see Harvey (1989), West and Harrison (1997) and Kitagawa and Gersch (1996).

Continuous intrinsic models in geostatistics have been pioneered by Matheron (1973) and are now part of the standard literature, see, for example, Cressie (1993), Chilés and Delfiner (1999) and Lantuéjoul (2002). Künsch (1987) generalizes stationary autoregressions on a two-dimensional infinite lattice to intrinsic autoregressions using related ideas in the geostatistics literature and derives a spectral approximation to the log likelihood of intrinsic autoregressions. This approximation is similar to the Whittle (1954) approximation (2.32) and is discussed further by Kent and Mohammadzadeh (1999). Besag and Kooperberg (1995) argue for the use of intrinsic autoregressions, derive some theoretical results, and discuss corrections of undesirable second-order characteristics for small arrays and nonlattice applications. For maximum likelihood estimation in intrinsic models, see also Mardia (1990). The first-order IGMRF on irregular lattices (3.30) was made popular by Besag et al. (1991).

Our presentation of IGMRFs is based on the construction on finite graphs leaving the interesting infinite lattice case, which is similar to Section 2.6.5, undiscussed. Although intrinsic models are traditionally invariant to polynomial trends, this is not always the case and is our motivation for Section 3.4.3.

The construction of IGMRFs has similarities to the construction of splines (Gu, 2002, Wahba, 1990). This has motivated Section 3.2 and Section 3.4.2. As an example about making IGMRFs proper which is justified by (3.37), see Fernández and Green (2002). For analytical results about the eigenstructure of some structure matrices, also for higher dimension, see Gorsich et al. (2002). Interaction models based on the Kronecker product are discussed in Clayton (1996), see Knorr-Held (2000a) and Schmid and Held (2004) for further extensions.

An excellent source for the numerical stencils used in Section 3.4.2 is

Patra and Karttunen (2004).

Lindgren and Rue (2004) discuss the construction of IGMRF's on triangulated spheres for applications in environmental statistics.

The construction of the CRW2 in Section 3.5 is from Wecker and Ansley (1983) who consider fitting polynomial splines to a nonequally spaced time series using algorithms derived from the Kalman filter, see also Jones (1981) and Kohn and Ansley (1987). Wecker and Ansley (1983) base their work on the results of Wahba (1978) about the connection between the posterior expectation of a diffuse integrated Wiener process and polynomial splines. See, for example, Shepp (1966) for background on the $(k-1)$-fold integrated Wiener process.

CHAPTER 4

Case studies in hierarchical modeling

One of the main areas where GMRF models are used in statistics are hierarchical models. Here, GMRFs serve as a convenient formulation to model stochastic dependence between parameters, and thus implicitly, dependence between observed data. The dependence can be of various kinds, such as temporal, spatial or even spatiotemporal.

A *hierarchical GMRF model* is characterized through several *stages* of observables and parameters. A typical scenario is as follows. In the *first stage* we will formulate a distributional assumption for the observables, dependent on latent parameters. If we have observed a time series of binary observations $\boldsymbol{y}$, we may assume a Bernoulli model with unknown probability p_i for y_i, $i = 1, \dots, n$: $y_i \sim \mathcal{B}(p_i)$. Given the parameters of the observation model, we assume the observations to be *conditionally independent*. In the *second stage* we assign a prior model for the unknown parameters, here p_i. This is where GMRFs enter. For example, we could choose an autoregressive model for the logit-transformed probabilities $x_i = \text{logit}(p_i)$. Finally, a prior distribution is assigned to unknown parameters (or hyperparameters) of the GMRF, such as the precision parameter κ of the GMRF $\boldsymbol{x}$. This is the *third stage* of a hierarchical model. There may be further stages if necessary.

In a regression context, our simple example would thus correspond to a generalized linear model where the *intercept* is varying over some domain according to a GMRF with unknown hyperparameters. More generally, so-called *generalized additive models* can be fitted using GMRFs. We will give examples of such models later. GMRFs are also useful in extended regression models where *covariate effects* are allowed to vary over some domain. Such models have been termed *varying coefficient models* (Hastie and Tibshirani, 1990) and the domain over which the effect is allowed to vary is called the *effect modifier*. Again, GMRF models can be used to analyze this class of models, see Fahrmeir and Lang (2001a) for a generalized additive model based on GMRFs involving varying coefficients.

For statistical inference we will mainly use Markov chain Monte Carlo (MCMC) techniques, e.g., Robert and Casella (1999) or Gilks et al.

(1996). This will involve simulation from (possibly large) GMRFs, and we will make extensive use of the methods discussed in Chapter 2. In Section 4.1 we will give a brief summary of MCMC methods.

The hierarchical approach to model dependent data has been dominant in recent years. However, traditional Markov random field models have been used in a nonhierarchical setting as a direct model for the observed data, not as a model for unobserved parameters, compare, for example, Besag (1974) or Künsch (1987). Such direct modeling approaches have shown not to be flexible enough for applied data analysis. For example, Markov random field models for Poisson observations, so-called auto-Poisson models, can only model negative dependence between neighboring sites (Besag, 1974). In contrast, a hierarchical model with Poisson observations and a latent GMRF on the (log) rates is able to capture positive dependence between observations.

There is a vast literature on various applications of GMRFs in hierarchical models. We therefore do not attempt to cover the whole area, but only give examples of applications in different settings. First we will look at normal responses. Inference in this class of models is fairly straightforward. In the temporal domain, this model class corresponds to so-called *state-space-models* (Harvey, 1989, West and Harrison, 1997) and we will outline analogies in our inferential methods and those used in traditional state-space models, such as the Kalman filter and smoother and the so-called *forward-filtering-backward-sampling* algorithms for inference via MCMC.

The second class of models is characterized through the fact that the sampling algorithms for statistical inference are similar to the normal case, once we introduce so-called *auxiliary variables*. This is typically achieved within a so-called *scale mixtures of normals* model formulation. There are two main areas in this class: The first is to account for nonnormal (but still continuous) distributions and typically uses Student-t_ν distributions for the observational error distribution or for the independent increments defining GMRFs. The second area are models for binary and multicategorical responses. In particular, we will discuss how probit and logit regression models can be implemented using an auxiliary variable approach. Finally, we will discuss models where such auxiliary variable approaches are not available. Those include, for example, Poisson regression models or certain regression models for survival data. We will discuss how GMRF approximations can be used to facilitate MCMC via the Metropolis-Hastings algorithm.

4.1 MCMC for hierarchical GMRF models

For further understanding and to introduce our notation, it is necessary to give a brief introduction to Markov chain Monte Carlo (MCMC) methods before discussing strategies for block updating in hierarchical GMRF models.

4.1.1 A brief introduction to MCMC

Suppose θ is an unknown scalar parameter, and we are interested in the posterior distribution $\pi(\theta|y)$ after observing some data y. We suppress the dependence on y in this section and simply write $\pi(\theta)$ for the posterior (target) distribution. The celebrated Metropolis-Hastings algorithm, which forms the basis of most MCMC algorithms, can be used to generate a Markov chain $\theta^{(1)}, \theta^{(2)}, \ldots, \theta^{(k)}, \ldots$ that converges (under mild regularity conditions) to $\pi(\theta)$:

1. Start with some arbitrary starting value $\theta^{(0)}$ where $\pi(\theta^{(0)}) > 0$. Set $k = 1$.
2. Generate a *proposal* θ^* from some *proposal kernel* $q(\theta^*|\theta^{(k-1)})$ that in general depends on the current value $\theta^{(k-1)}$ of the simulated Markov chain. Set $\theta^{(k)} = \theta^*$ with probability
$$\alpha = \min\left\{1, \frac{\pi(\theta^*)}{\pi(\theta^{(k-1)})}\frac{q(\theta^{(k-1)}|\theta^*)}{q(\theta^*|\theta^{(k-1)})}\right\};$$
otherwise set $\theta^{(k)} = \theta^{(k-1)}$.
3. Set $k = k + 1$ and go back to 2.

Step 2 is often called the *acceptance step*, because the proposed value θ^* is accepted with probability α as the new value of the Markov chain.

Depending on the specific choice of the proposal kernel $q(\theta^*|\theta)$, very different algorithms result. There are two important subclasses: if $q(\theta^*|\theta)$ does not depend on the current value of θ, i.e., $q(\theta^*|\theta) = q(\theta^*)$, the proposal is called an *independence proposal*. Another important class can be obtained if $q(\theta^*|\theta) = q(\theta|\theta^*)$, in which case the acceptance probability α simplifies to the ratio of the target density, evaluated at the proposed new and the old value. These includes so-called *random-walk proposals*, where $q(\theta^*|\theta)$ is symmetric around the current value θ. Typical examples of random-walk proposals add a mean zero uniform or normal (or any other symmetric) distribution to the current value of θ.

A trivial case occurs if the proposal distribution equals the target distribution, i.e., $q(\theta^*|\theta) = \pi(\theta^*)$, the acceptance probability α then always equals one. So direct independent sampling from $\pi(\theta)$ is a special case of the Metropolis-Hastings algorithm. However, if $\pi(\theta)$ is nonstandard, it may not be straightforward to sample from $\pi(\theta)$ directly.

The beauty of the Metropolis-Hastings algorithm is that we can use (under suitable regularity conditions) *any* distribution as the proposal and the algorithm will still converge to the target distribution. However, the rate of convergence toward $\pi(\theta)$ and the degree of dependence between successive samples of the Markov chain (its *mixing* properties) will depend on the chosen proposal.

MCMC algorithms are often controlled through the acceptance probability α or its expected value $\mathrm{E}(\alpha)$, assuming the chain is in equilibrium. For a random-walk proposal, a too large value of $\mathrm{E}(\alpha)$ implies that the proposal distribution is too narrow around the current value, so effectively the Markov chain will only make small steps. On the other hand, if $\mathrm{E}(\alpha)$ is very small, nearly all proposals are rejected and the algorithm will stay too long in certain values of the target distribution. Some intermediate values of $\mathrm{E}(\alpha)$ in the interval 25 to 40% often work well in practice, see also Roberts et al. (1997) for some analytical results. Therefore, the spread of random-walk proposals is chosen so that $E(\alpha)$ is in this interval. For an independence proposal, the situation is different as a high $\mathrm{E}(\alpha)$ indicates that the proposal distribution $q(\theta^*)$ approximates the target $\pi(\theta)$ quite well.

The simple algorithm described above forms the basis of nearly all MCMC methods. However, usually we are interested in a multivariate (posterior) distribution $\pi(\boldsymbol{\theta})$ of a random vector $\boldsymbol{\theta}$ of high dimension, not in a single parameter θ. Some modifications are necessary to apply the above algorithm to the multivariate setting.

Historically, most MCMC algorithms have been based on updating each scalar component θ_i, $i = 1, \ldots, p$ of $\boldsymbol{\theta}$, conditional on the values of the other parameter $\boldsymbol{\theta}_{-i}$, using the Metropolis-Hastings algorithm. Essentially, we apply the Metropolis-Hastings algorithm in turn to every component θ_i of $\boldsymbol{\theta}$ with arbitrary proposal kernels $q_i(\theta_i^*|\theta_i, \boldsymbol{\theta}_{-i})$. As long as we update each component of $\boldsymbol{\theta}$, this algorithm will converge to the target distribution $\pi(\boldsymbol{\theta})$.

Of particular prominence is the so-called Gibbs sampler algorithm, where each component θ_i is updated with a random variable from its *full conditional* $\pi(\theta_i|\boldsymbol{\theta}_{-i})$. Note that this is a special case of the component-wise Metropolis-Hastings algorithm, since α simply equals unity in this case. Of course, any other proposal kernel can be used, where all terms are now conditional on the current values of $\boldsymbol{\theta}_{-i}$, to update θ_i. For more details on these algorithms see, for example, Tierney (1994) or Besag et al. (1995).

However, it was immediately realized that such *single-site* updating can be disadvantageous if parameters are highly dependent in the posterior distribution $\pi(\boldsymbol{\theta})$. The problem is that the Markov chain may move around very slowly in its target (posterior) distribution. A general approach to circumvent this problem is to update parameters in larger blocks, $\boldsymbol{\theta}_j$, say, where the bold face indicates that $\boldsymbol{\theta}_j$ is a vector of components of $\boldsymbol{\theta}$. The choice of blocks are often controlled by what is possible to do in practice. Ideally, we should choose a small number of blocks with large dependence within the blocks but with less dependence between blocks. The extreme case is to update all parameters $\boldsymbol{\theta}$ in one block.

Blocking is particularly easy if $\boldsymbol{\theta}$ is a GMRF. Then we can apply one of the algorithms discussed in Section 2.3 to simulate $\boldsymbol{\theta}$ in one step. However, in Bayesian hierarchical models, more parameters are involved typically. For example, $\boldsymbol{\theta} = (\kappa, \boldsymbol{x})$ where an unknown scalar precision parameter κ may be of interest additional to the GMRF $\boldsymbol{x}$. It is tempting to form two blocks in these cases, where we update from the full conditionals of each block; sample $\boldsymbol{x}$ from $\pi(\boldsymbol{x}|\kappa)$ and subsequently sample κ from $\pi(\kappa|\boldsymbol{x})$. However, there is often strong dependence between κ and $\boldsymbol{x}$ in the posterior, and then a joint Metropolis-Hastings update of κ and $\boldsymbol{x}$ is preferable. In the following examples we update the GMRF $\boldsymbol{x}$ (or parts of it) and its associated hyperparameters in one block, which is not as difficult as it seems. Why this modification is important and why it works is discussed next.

4.1.2 Blocking strategies

To illustrate and discuss our strategy for block updating in hierarchical GMRF models, we will start discussing a simple (normal) example where explicit analytical results are possible to obtain. This will illustrate why a joint update of $\boldsymbol{x}$ and its hyperparameters is important. At the end, we discuss the more general case.

A simple example

Before we compare analytical results about the rate of convergence for various sampling schemes, we need to define it. Let $\boldsymbol{\theta}^{(1)}, \boldsymbol{\theta}^{(2)}, \ldots$ denote a Markov chain with target distribution $\pi(\boldsymbol{\theta})$ and initial value $\boldsymbol{\theta}^{(0)} \sim \pi(\boldsymbol{\theta})$. The rate of convergence of the Markov chain can be characterized by studying how quickly $\mathrm{E}(h(\boldsymbol{\theta}^{(t)})|\boldsymbol{\theta}^{(0)})$ approaches the stationary value $\mathrm{E}(h(\boldsymbol{\theta}))$ for all square π-integrable functions $h(\cdot)$. Let ρ be the minimum number such that for all $h(\cdot)$ and for all $r > \rho$

$$\lim_{k\to\infty} \mathrm{E}\left[\left(\mathrm{E}\left(h(\boldsymbol{\theta}^{(k)}) \mid \boldsymbol{\theta}^{(0)}\right) - \mathrm{E}\left(h(\boldsymbol{\theta})\right)\right)^2 r^{-2k}\right] = 0. \tag{4.1}$$

We say that ρ is the *rate of convergence*. For normal target distribution it is sufficient to consider linear functions $h(\cdot)$ only (see for example Roberts and Sahu (1997)).

Assume now that $\boldsymbol{x}$ is a first-order autoregressive process

$$x_t - \mu = \gamma(x_{t-1} - \mu) + \nu_t, \quad t = 2, \ldots, n, \tag{4.2}$$

where $|\gamma| < 1$, $\{\nu_t\}$ are iid normals with zero mean and variance σ^2, and $x_1 \sim \mathcal{N}(\mu, \frac{\sigma^2}{1-\gamma^2})$, which is the stationary distribution of x_i. Let γ, σ^2, and μ be fixed parameters. We can of course sample from this model directly, but here we want to apply an MCMC algorithm to generate samples from $\pi(\boldsymbol{x})$.

At each iteration a single-site Gibbs sampler will sample x_t from the full conditional $\pi(x_t|\boldsymbol{x}_{-t})$ for $t = 1, \ldots, n$,

$$x_t \mid \boldsymbol{x}_{-t} \sim \begin{cases} \mathcal{N}(\mu + \gamma(x_2 - \mu),\ \sigma^2) & t = 1, \\ \mathcal{N}(\mu + \frac{\gamma}{1+\gamma^2}(x_{t-1} + x_{t+1} - 2\mu),\ \frac{\sigma^2}{1+\gamma^2}) & t = 2, \ldots, n-1, \\ \mathcal{N}(\mu + \gamma(x_{n-1} - \mu),\ \sigma^2) & t = n. \end{cases}$$

For large n, the rate of convergence of this algorithm is (Pitt and Shephard, 1999, Theorem 1)

$$\rho = 4\frac{\gamma^2}{(1+\gamma^2)^2}. \tag{4.3}$$

For $|\gamma|$ close to one the rate of convergence can be slow: If $\gamma = 1 - \delta$ for small $\delta > 0$, then $\rho = 1 - \delta^2 + \mathcal{O}(\delta^3)$. The reason is that strong dependency within $\boldsymbol{x}$ allow for larger moves in the joint posterior. To circumvent this problem, we may update $\boldsymbol{x}$ in one block. This is possible as $\boldsymbol{x}$ is a GMRF. The block algorithm converges immediately and provides iid samples from the joint density.

We now relax the assumptions of fixed hyperparameters. Consider a hierarchical formulation where the mean of x_t, μ, is unknown and assigned with a standard normal prior,

$$\mu \sim \mathcal{N}(0, 1) \quad \text{and} \quad \boldsymbol{x} \mid \mu \sim \mathcal{N}(\mu\mathbf{1}, \boldsymbol{Q}^{-1}),$$

where $\boldsymbol{Q}$ is the precision matrix of the GMRF $\boldsymbol{x}|\mu$. The joint density of $(\mu, \boldsymbol{x})$ is normal. We have two natural blocks, μ and $\boldsymbol{x}$.

Since both full conditionals $\pi(\mu|\boldsymbol{x})$ and $\pi(\boldsymbol{x}|\mu)$ are normal, it is tempting to form a two-block Gibbs sampler and to update μ and $\boldsymbol{x}$ with samples from their full conditionals,

$$\begin{aligned} \mu^{(k)}|\boldsymbol{x}^{(k)} &\sim \mathcal{N}\left(\frac{\mathbf{1}^T\boldsymbol{Q}\boldsymbol{x}^{(k-1)}}{1 + \mathbf{1}^T\boldsymbol{Q}\mathbf{1}},\ \left(1 + \mathbf{1}^T\boldsymbol{Q}\mathbf{1}\right)^{-1}\right) \\ \boldsymbol{x}^{(k)}|\mu^{(k)} &\sim \mathcal{N}(\mu^{(k)}\mathbf{1},\ \boldsymbol{Q}^{-1}). \end{aligned} \tag{4.4}$$

The presence of the hyperparameter μ will slow down the convergence compared to the case when μ is fixed. Due to the nice structure of (4.4) we can characterize explicitly the marginal chain of $\{\mu^{(k)}\}$.

Theorem 4.1 *The marginal chain $\mu^{(1)}, \mu^{(2)}, \ldots$ from the two-block Gibbs sampler defined in (4.4) and started in equilibrium, is a first-order autoregressive process*

$$\mu^{(k)} = \phi\mu^{(k-1)} + \epsilon_k,$$

where

$$\phi = \frac{\mathbf{1}^T \boldsymbol{Q}\mathbf{1}}{1 + \mathbf{1}^T \boldsymbol{Q}\mathbf{1}}$$

and $\epsilon_k \overset{iid}{\sim} \mathcal{N}(0, 1-\phi^2)$.

Proof. It follows directly that the marginal chain $\mu^{(1)}, \mu^{(2)}, \ldots$ is a first-order autoregressive process; the marginal distribution of μ is normal with zero mean and unit variance, the chain has the Markov property

$$\pi(\mu^{(k)} \mid \mu^{(1)}, \ldots, \mu^{(k-1)}) = \pi(\mu^{(k)} \mid \mu^{(k-1)})$$

and the density of $(\mu^{(1)}, \mu^{(2)}, \ldots, \mu^{(k)})$ is normal for $k = 1, 2, \ldots$. The coefficient ϕ is found by computing the covariance at lag 1,

$$\begin{aligned}
\mathrm{Cov}(\mu^{(k)}, \mu^{(k+1)}) &= \mathrm{E}\left(\mu^{(k)}\mu^{(k+1)}\right) \\
&= \mathrm{E}\left(\mu^{(k)}\mathrm{E}\left(\mu^{(k+1)} \mid \mu^{(k)}\right)\right) \\
&= \mathrm{E}\left(\mu^{(k)}\mathrm{E}\left(\mathrm{E}\left(\mu^{(k+1)} \mid \boldsymbol{x}^{(k)}\right) \mid \mu^{(k)}\right)\right) \\
&= \mathrm{E}\left(\mu^{(k)}\mathrm{E}\left(\frac{\mathbf{1}^T \boldsymbol{Q}\boldsymbol{x}^{(k)}}{1 + \mathbf{1}^T \boldsymbol{Q}\mathbf{1}} \mid \mu^{(k)}\right)\right) \\
&= \frac{\mathbf{1}^T \boldsymbol{Q}\mathbf{1}}{1 + \mathbf{1}^T \boldsymbol{Q}\mathbf{1}}\mathrm{Var}(\mu^{(k)}),
\end{aligned}$$

which is known to be ϕ times the variance $\mathrm{Var}(\mu^{(k)})$ for a first-order autoregressive process. The variance of ϵ_k is determined such that the variance of $\mu^{(k)}$ is 1. □

It can be shown that for a two-block Gibbs sampler the marginal chains and the joint chain have the same rate of convergence (Liu et al., 1994, Thm. 3.2). Applying (4.1) to the marginal chain $\mu^{(1)}, \mu^{(2)}, \ldots$, we see that the rate of convergence is $\rho = |\phi|$.

It is illustrative to discuss the behavior of the two-block Gibbs sampler for large n. For the autoregressive model (4.2), $Q_{ii} = (1+\gamma^2)/\sigma^2$ except for $i = 1$ and n where it is $1/\sigma^2$, and $Q_{i,i+1} = -\gamma/\sigma^2$. This implies that $\mathbf{1}^T \boldsymbol{Q}\mathbf{1}$ is asymptotically equal to $n(1-\gamma)^2/\sigma^2$, so that

$$\phi = \frac{n(1-\gamma)^2/\sigma^2}{1 + n(1-\gamma)^2/\sigma^2} = 1 - \frac{\mathrm{Var}(x_t)}{n}\frac{1-\gamma^2}{(1-\gamma)^2} + \mathcal{O}(1/n^2).$$

When n is large, ϕ is close to 1 and the chain will both mix and converge slowly even though we use a two-block Gibbs sampler. The minimum number of iterations, k^*, needed before the correlation between $\mu^{(k)}$ and $\mu^{(k+l)}$ is smaller than $\zeta = 0.05$, say, is

$$k^*/|\log(\zeta)| = -1/\log(\phi) = \frac{1}{2} + n\frac{1}{\mathrm{Var}(x_t)}\frac{(1-\gamma)^2}{1-\gamma^2} + \mathcal{O}(1/n).$$

Since $(1-\gamma)^2/(1-\gamma^2)$ is strictly decreasing in the interval $(-1, 1)$, we conclude the following:

- For constant $\mathrm{Var}(x_t)$, k^* increases for decreasing γ.
- For constant γ, k^* increases for decreasing $\mathrm{Var}(x_t)$.

One might be tempted to believe that k^* should increase for increasing γ^2 due to (4.3). However, since we update $\boldsymbol{x}$ in one block this is not the case: The variance of $\mu|\boldsymbol{x}$ in (4.4) increases for increasing γ and increasing $\mathrm{Var}(x_t)$, and this weakens the dependence between μ and $\boldsymbol{x}$.

Figure 4.1(a) shows a simulation with length 1 000 of the marginal chain for μ using $n = 100$, $\sigma^2 = 1/10$, $\phi = 0.9$, and the plot of the pairs $(\mu^{(k)}, \mathbf{1}^T\boldsymbol{Q}\boldsymbol{x}^{(k)})$ in Figure 4.1(b). The reason for the slow mixing (and convergence) of the μ chain is the strong dependence between $\mu^{(k)}$ and $\mathbf{1}^T\boldsymbol{Q}\boldsymbol{x}^{(k)}$, the sufficient statistics of $\mu^{(k)}$ in the full conditional (4.4). The two-block Gibbs sampler only moves either horizontally (update μ) or vertically (update $\boldsymbol{x}$). Note that this is just the same as in the standard example sampling from a two-dimensional normal distribution using Gibbs sampling (see for example Gilks et al. (1996, Chapter 1)).

The discussion so far has only revealed the seemingly obvious, that blocking improves mainly within the block. If there is strong dependence *between* blocks, the MCMC algorithm may still suffer from slow convergence. Our solution is to update $(\mu, \boldsymbol{x})$ jointly. Since μ is univariate, we can use a simple scheme for updating μ as long as we delay the accept/reject step until $\boldsymbol{x}$ also is updated. The joint proposal is generated as follows:

$$\begin{aligned} \mu^* &\sim q(\mu^* \mid \mu^{(k-1)}) \\ \boldsymbol{x}^*|\mu^* &\sim \mathcal{N}(\mu^*\mathbf{1}, \boldsymbol{Q}^{-1}) \end{aligned} \tag{4.5}$$

and then we accept/reject $(\mu^*, \boldsymbol{x}^*)$ jointly. Here, $q(\mu^*|\mu^{(k-1)})$ can be a simple random-walk proposal or some other suitable proposal distribution. To see how a joint update of $(\mu, \boldsymbol{x})$ can be helpful, consider Figure 4.1(b). A proposal from $\mu^{(k-1)}$ to μ^* may take μ^* out of the diagonal, while sampling $\boldsymbol{x}^*$ from $\pi(\boldsymbol{x}^*|\mu^*)$ will take it back into the diagonal again. Hence the mixing (and convergence) can be very much improved.

Assuming $q(\cdot|\cdot)$ in (4.5) is a symmetric proposal, the acceptance

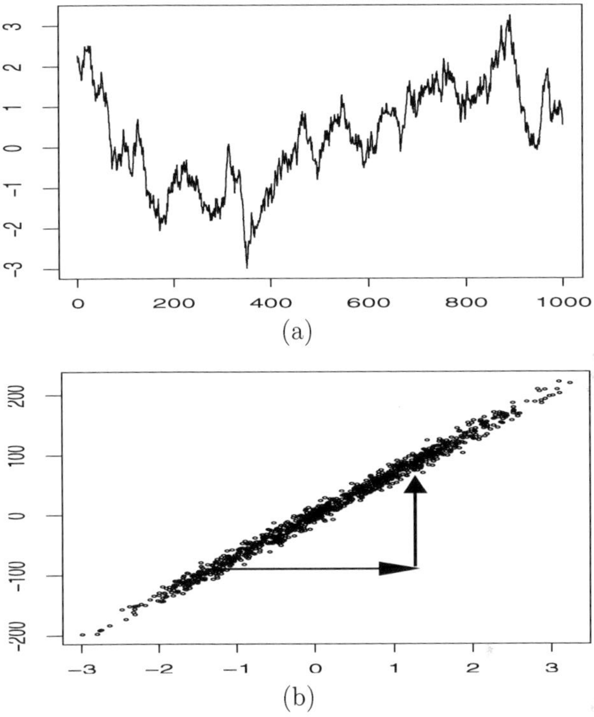

Figure 4.1 *Figure (a) shows the marginal chain for μ over 1000 iterations of the marginal chain for μ using $n = 100$, $\sigma^2 = 1/10$ and $\phi = 0.9$. The algorithm updates successively μ and $\boldsymbol{x}$ from their full conditionals. Figure (b) displays the pairs $(\mu^{(k)}, \mathbf{1}^T\boldsymbol{Q}\boldsymbol{x}^{(k)})$, with $\mu^{(k)}$ on the horizontal axis. The slow mixing (and convergence) of μ is due to the strong dependence with $\mathbf{1}^T\boldsymbol{Q}\boldsymbol{x}^{(k)}$ as only horizontal and vertical moves are allowed. The arrows illustrate how a joint update can improve the mixing (and convergence).*

probability for $(\mu^*, \boldsymbol{x}^*)$ becomes

$$\alpha = \min\left\{1,\ \exp(-\frac{1}{2}((\mu^*)^2 - (\mu^{(k-1)})^2))\right\}. \tag{4.6}$$

Note that only the *marginal density* of μ is needed in (4.6): Since we sample $\boldsymbol{x}$ from its full conditional, we effectively integrate $\boldsymbol{x}$ out of the joint density $\pi(\mu, \boldsymbol{x})$. The minor modification to delay the accept/reject step until $\boldsymbol{x}$ is updated as well can give a large improvement.

A further extension of (4.2) is to condition on observed normal data $\boldsymbol{y}$. The distribution of interest is then $\pi(\boldsymbol{x}, \mu|\boldsymbol{y})$. Assume that

$$\boldsymbol{y} \mid \boldsymbol{x}, \mu \sim \mathcal{N}(\boldsymbol{x},\ \boldsymbol{H}^{-1}),$$

where $\boldsymbol{H}$ is a known precision matrix. Consider the two-block Gibbs sampler which first updates μ from $\pi(\mu|\boldsymbol{x})$ and then $\boldsymbol{x}$ from $\pi(\boldsymbol{x}|\mu, \boldsymbol{y})$. It is straightforward to extend Theorem 4.1 to this case. The marginal chain $\{\mu^{(k)}\}$ is still a first-order autoregressive process, but the autoregressive parameter ϕ is now

$$\phi = \frac{\mathbf{1}^T \boldsymbol{Q}(\boldsymbol{Q}+\boldsymbol{H})^{-1}\boldsymbol{Q}\mathbf{1}}{1+\mathbf{1}^T\boldsymbol{Q}\mathbf{1}}. \tag{4.7}$$

Assume for simplicity that $\boldsymbol{H}$ is a diagonal matrix with κ on the diagonal, meaning that $y_i|\boldsymbol{x}, \mu \sim \mathcal{N}(x_i, 1/\kappa)$. We can (with little effort) compute (4.7) using the asymptotically negligible circulant approximation to $\boldsymbol{Q}$, see Section 2.6.4, to obtain the limiting value of ϕ as $n \to \infty$,

$$\phi = \frac{1}{1+\kappa\sigma^2/(1-\gamma)^2}.$$

In this case $\phi < 1$ for all n when $\kappa > 0$. The two-block Gibbs sampler does no longer converge arbitrarily slow as n increases. However, in practice the convergence is often still slow and a joint update will be of great advantage.

Block algorithms for hierarchical GMRF models

Let us now consider a more general setup that contains all the forthcoming case studies: Some hyperparameters $\boldsymbol{\theta}$ control the GMRF $\boldsymbol{x}$ of size n and some of the nodes of $\boldsymbol{x}$ are observed by data $\boldsymbol{y}$. The posterior is then

$$\pi(\boldsymbol{x}, \boldsymbol{\theta} \mid \boldsymbol{y}) \propto \pi(\boldsymbol{\theta})\ \pi(\boldsymbol{x} \mid \boldsymbol{\theta})\ \pi(\boldsymbol{y} \mid \boldsymbol{x}, \boldsymbol{\theta}).$$

Assume for a moment that we are able to sample from $\pi(\boldsymbol{x}|\boldsymbol{\theta}, \boldsymbol{y})$, i.e., the full conditional of $\boldsymbol{x}$ is a GMRF. We will later discuss options when this full conditional is not a GMRF. The following algorithm is now a direct generalization of (4.5) and updates $(\boldsymbol{\theta}, \boldsymbol{x})$ in one block:

$$\begin{aligned} \boldsymbol{\theta}^* &\sim q(\boldsymbol{\theta}^* \mid \boldsymbol{\theta}^{(k-1)}) \\ \boldsymbol{x}^* &\sim \pi(\boldsymbol{x} \mid \boldsymbol{\theta}^*, \boldsymbol{y}). \end{aligned} \tag{4.8}$$

The proposal $(\boldsymbol{\theta}^*, \boldsymbol{x}^*)$ is then accepted/rejected jointly. We denote this as the *one-block* algorithm.

If we consider only the $\boldsymbol{\theta}$ chain, then we are in fact sampling from the posterior marginal $\pi(\boldsymbol{\theta}|\boldsymbol{y})$ using the proposal (4.8). This is evident from the acceptance probability for the joint proposal, which is

$$\alpha = \min\left\{1, \frac{\pi(\boldsymbol{\theta}^* \mid \boldsymbol{y})}{\pi(\boldsymbol{\theta}^{(k-1)} \mid \boldsymbol{y})}\right\}.$$

The dimension of $\boldsymbol{\theta}$ is typically low and often between 1 and 5, say. Hence the proposed algorithm should not experience any serious mixing problems. By sampling $\boldsymbol{x}$ from $\pi(\boldsymbol{x}|\boldsymbol{\theta}^*, \boldsymbol{y})$, we are in fact integrating $\boldsymbol{x}$ out of $\pi(\boldsymbol{x}, \boldsymbol{\theta}|\boldsymbol{y})$. The computational cost per iteration depends on n (the dimension of $\boldsymbol{x}$) and is (usually) dominated by the cost of sampling from $\pi(\boldsymbol{x}|\boldsymbol{\theta}^*, \boldsymbol{y})$. The fast algorithms in Section 2 for GMRFs are therefore very useful.

However, the one-block algorithm is not always feasible for the following reasons:

1. The full conditional of $\boldsymbol{x}$ can be a GMRF with a precision matrix that is not sparse. This will prohibit a fast factorization, hence a joint update is feasible but not computationally efficient.
2. The data can be nonnormal so the full conditional of $\boldsymbol{x}$ is not a GMRF and sampling $\boldsymbol{x}^*$ using (4.8) is not possible (in general).

The first problem can often be approached using *subblocks* of $(\boldsymbol{\theta}, \boldsymbol{x})$, following an idea of Knorr-Held and Rue (2002). Assume a natural splitting exists for both $\boldsymbol{\theta}$ and $\boldsymbol{x}$ into

$$(\boldsymbol{\theta}_a, \boldsymbol{x}_a),\ (\boldsymbol{\theta}_b, \boldsymbol{x}_b) \quad \text{and} \quad (\boldsymbol{\theta}_c, \boldsymbol{x}_c), \tag{4.9}$$

say. The sets a, b, and c do not need to be disjoint. One class of examples where such an approach is fruitful is (geo-)additive models where a, b, and c represent three different covariate effects with their respective hyperparameters. We will discuss such an example in Section 4.2.2. In this class of models the full conditional of $\boldsymbol{x}_a$ has a sparse precision matrix and similarly for $\boldsymbol{x}_b$ and $\boldsymbol{x}_c$. The subblock approach is then to update each subblock in (4.9), using

$$\begin{aligned} \boldsymbol{\theta}_a^* &\sim q(\boldsymbol{\theta}_a^* \mid \boldsymbol{\theta}) \\ \boldsymbol{x}_a^* &\sim \pi(\boldsymbol{x}_a \mid \boldsymbol{x}_{-a}, \boldsymbol{\theta}_a^*, \boldsymbol{\theta}_{-a}, \boldsymbol{y}) \end{aligned} \tag{4.10}$$

(dropping the superscript $(k-1)$ from here on) and then accepting/rejecting $(\boldsymbol{\theta}_a^*, \boldsymbol{x}_a^*)$ jointly. Subblocks b and c are updated similarly. We denote this as the *subblock* algorithm.

When the observed data is nonnormal, the full conditional of $\boldsymbol{x}$ will not be a GMRF. However, we may use *auxiliary variables* $\boldsymbol{z}$ such that the conditional distribution of $\boldsymbol{x}$ is still a GMRF. Typical examples include logit and probit regression models for binary and multicategorical data, and Student-t_ν distributed observations. We will discuss the auxiliary variables approach in Section 4.3.

A more general idea is to construct a *GMRF approximation* to $\pi(\boldsymbol{x}|\boldsymbol{\theta}, \boldsymbol{y})$ using a second-order Taylor expansion. Such an approximation can be surprisingly accurate in many cases and can be interpreted as integrating $\boldsymbol{x}$ *approximately* out of $\pi(\boldsymbol{x}, \boldsymbol{\theta}|\boldsymbol{y})$. Using a GMRF approximation will generalize the one-block and subblock algorithms. The most prominent example is Poisson regression, which we discuss in Section 4.4.

The subblock approach can obviously also be used in connection with auxiliary variables and the GMRF approximation, when the full conditional has a nonsparse precision matrix and its distribution is not a GMRF.

4.2 Normal response models

We now look at hierarchical GMRF models, where the response variable is assumed to be normally distributed with conditional mean given as a linear function of underlying unknown parameters $\boldsymbol{x}$. These unknown parameters are assumed to follow a (possibly intrinsic) GMRF *a priori*, typically with additional unknown hyperparameters $\boldsymbol{\kappa}$. It is clear that the conditional posterior $\pi(\boldsymbol{x}|\boldsymbol{y}, \boldsymbol{\kappa})$ is still a GMRF so direct sampling from this distribution is particularly easy using the algorithms described in Section 2.3.

We will now describe two case studies. The first is concerned with the analysis of a time series, using latent trend and seasonal components and additional covariate information. It is not important that the latent GMRF is defined on a *temporal* domain rather than a spatial domain, but in this special case there are close connections to algorithms based on the Kalman filter. We will describe the analogies briefly at the end of this example. The second example describes the spatial analysis of rent prices in Munich using a geoadditive model with nonparametric and fixed covariate effects.

4.2.1 Example: Drivers data

In this example we consider a regular time series giving the monthly totals of car drivers in Great Britain killed or seriously injured January 1969 to December 1984 (Harvey, 1989). This time series has length $n = 192$ and exhibits a strong seasonal pattern. One of our objectives is to predict the pattern in the next $m = 12$ months.

We first assume that the square root counts y_i, $i = 1, \ldots, n$, are conditionally independent normal variables,

$$y_i \sim \mathcal{N}(s_i + t_i,\ 1/\kappa_{\boldsymbol{y}}),$$

where the mean is a sum of a smooth trend t_i and a seasonal effect s_i. We assume $\boldsymbol{s} = (s_1, \ldots, s_{n+m})$ follows the seasonal model (3.58) with season length 12 and precision $\kappa_{\boldsymbol{s}}$ and $\boldsymbol{t} = (t_1, \ldots, t_{n+m})$ follows the RW2 model (3.39) with precision $\kappa_{\boldsymbol{t}}$. The trend $\boldsymbol{t}$ and the seasonal effect $\boldsymbol{s}$ are assumed to be independent. Note that no observations y_i are available for $i = n+1, \ldots, n+m$, but we can still include the corresponding parameters of $\boldsymbol{s}$ and $\boldsymbol{t}$ for prediction.

Let $\boldsymbol{\kappa}$ denote the three precisions $\kappa_{\boldsymbol{y}}$, $\kappa_{\boldsymbol{s}}$, and $\kappa_{\boldsymbol{t}}$, which is the vector of hyperparameters in this model. The task is to do inference for $(\boldsymbol{\kappa}, \boldsymbol{s}, \boldsymbol{t})$ using the one-block algorithm (4.8). For illustration, we will do this explicitly by first deriving the joint density $\pi(\boldsymbol{s}, \boldsymbol{t}, \boldsymbol{y}|\boldsymbol{\kappa})$ and then condition on $\boldsymbol{y}$. The joint density is

$$\pi(\boldsymbol{s}, \boldsymbol{t}, \boldsymbol{y} \mid \boldsymbol{\kappa}) = \pi(\boldsymbol{y} \mid \boldsymbol{s}, \boldsymbol{t}, \kappa_{\boldsymbol{y}})\, \pi(\boldsymbol{s} \mid \kappa_{\boldsymbol{s}})\, \pi(\boldsymbol{t} \mid \kappa_{\boldsymbol{t}}), \tag{4.11}$$

a GMRF with precision matrix $\boldsymbol{Q}$, say. We then use (3.47) and (3.48) to derive the desired precision matrix of the conditional distribution $\pi(\boldsymbol{s}, \boldsymbol{t}|\boldsymbol{y}, \boldsymbol{\kappa})$. Due to Theorem 2.5, this precision matrix is simply a principal matrix of $\boldsymbol{Q}$. Note that (4.11) is improper but the conditional distribution of interest is proper.

The details are as follows. First partition the precision matrix $\boldsymbol{Q}$ as

$$\boldsymbol{Q} = \begin{pmatrix} \boldsymbol{Q}_{ss} & \boldsymbol{Q}_{st} & \boldsymbol{Q}_{sy} \\ \boldsymbol{Q}_{ts} & \boldsymbol{Q}_{tt} & \boldsymbol{Q}_{ty} \\ \boldsymbol{Q}_{ys} & \boldsymbol{Q}_{yt} & \boldsymbol{Q}_{yy} \end{pmatrix},$$

where $\boldsymbol{Q}_{ss}$ and $\boldsymbol{Q}_{tt}$ are of dimension $(n+m) \times (n+m)$ while $\boldsymbol{Q}_{yy}$ is of dimension $n \times n$. The dimensions of the other entries follow immediately.

Since $\boldsymbol{s}$ and $\boldsymbol{t}$ are *a priori* independent, let us start with the density $\pi(\boldsymbol{y}|\boldsymbol{s}, \boldsymbol{t}, \boldsymbol{\kappa})$, which might add some dependence between $\boldsymbol{s}$ and $\boldsymbol{t}$. The conditional density of the data $\boldsymbol{y}$ is

$$\pi(\boldsymbol{y} \mid \boldsymbol{s}, \boldsymbol{t}, \kappa_{\boldsymbol{y}}) \propto \exp\left(-\frac{\kappa_{\boldsymbol{y}}}{2} \sum_{i=1}^{n} (y_i - s_i - t_i)^2 \right). \tag{4.12}$$

We immediately see that (4.12) induces dependence between y_i and s_i, y_i and t_i, and s_i and t_i. Specifically, $\boldsymbol{Q}_{yy}$ is a diagonal matrix with entries $\kappa_{\boldsymbol{y}}$, $\boldsymbol{Q}_{st}$ is diagonal where the first n entries are $\kappa_{\boldsymbol{y}}$ and the other m entries are zero, while $\boldsymbol{Q}_{sy}$ and $\boldsymbol{Q}_{ty}$, both of dimension $(n+m) \times n$, have nonzero entries $-\kappa_{\boldsymbol{y}}$ only at elements with the same row and column index. The terms $\boldsymbol{Q}_{ss}$ and $\boldsymbol{Q}_{tt}$ are the sum of two terms, one part from the prior and an additional term on the diagonal due to (4.12). Finally, $\boldsymbol{Q}_{ss}$ is the analog of (3.59) with seasonal length 12 rather than 4 and precision $\kappa_{\boldsymbol{s}}$ plus a diagonal matrix with $\kappa_{\boldsymbol{y}}$ on the diagonal, while $\boldsymbol{Q}_{tt}$ equals (3.40) with precision $\kappa_{\boldsymbol{t}}$ plus a diagonal matrix with $\kappa_{\boldsymbol{y}}$ on the diagonal.

The density $\pi(\boldsymbol{s}, \boldsymbol{t}|\boldsymbol{y}, \boldsymbol{\kappa})$ can now be found using Theorem 2.5 or Lemma 2.1. It is easiest to represent in its canonical parameterization:

$$\begin{pmatrix} \boldsymbol{s} \\ \boldsymbol{t} \end{pmatrix} \mid \boldsymbol{y}, \boldsymbol{\kappa} \sim \mathcal{N}_C\left(-\begin{pmatrix} \boldsymbol{Q}_{sy} \\ \boldsymbol{Q}_{ty} \end{pmatrix} \boldsymbol{y},\ \begin{pmatrix} \boldsymbol{Q}_{ss} & \boldsymbol{Q}_{st} \\ \boldsymbol{Q}_{ts} & \boldsymbol{Q}_{tt} \end{pmatrix} \right).$$

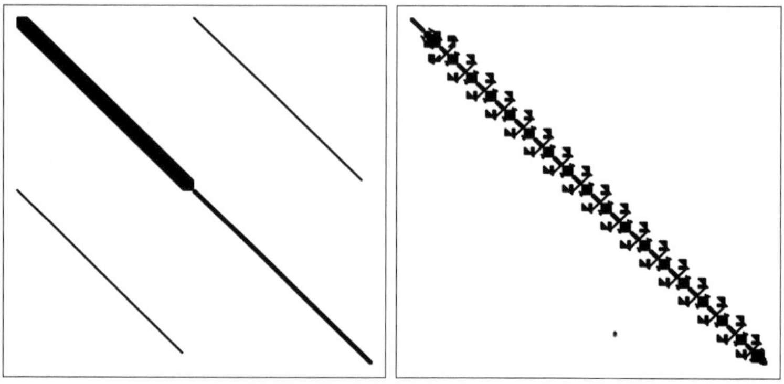

(a) (b)

Figure 4.2 *(a) The precision matrix $\boldsymbol{Q}$ of $\boldsymbol{s}, \boldsymbol{t} | \boldsymbol{y}, \boldsymbol{\kappa}$ in the original ordering, and (b) after appropriate reordering to obtain a band matrix with small bandwidth. Only the nonzero terms are shown and those are indicated by a dot.*

The nonzero structure of this precision matrix is displayed in Figure 4.2 before and after suitable reordering to reduce the bandwidth. Note that before reordering, the submatrices of the seasonal model $\boldsymbol{Q}_{ss}$ and the RW2 model $\boldsymbol{Q}_{tt}$ are clearly visible.

Additional to $\boldsymbol{s}$ and $\boldsymbol{t}$ we also want to perform Bayesian inference on the unknown precision parameters $\boldsymbol{\kappa}$. Under a Poisson model for observed counts, the square root counts are approximately normal with constant variance 1/4, but, in order to allow for overdispersion, we assume a $\mathcal{G}(4,4)$ prior for $\kappa_{\boldsymbol{y}}$. For $\kappa_{\boldsymbol{t}}$ we use a $\mathcal{G}(1, 0.0005)$ prior and for $\kappa_{\boldsymbol{s}}$ a $\mathcal{G}(1, 0.1)$ prior. All of these priors are assumed to be independent. Of course, other choices could be made as well.

We now propose a new configuration $(\boldsymbol{s}, \boldsymbol{t}, \boldsymbol{\kappa})$ using the one-block algorithm (4.8). Specifically, we do

$$\begin{aligned}
\kappa_{\boldsymbol{s}}^* &\sim q(\kappa_{\boldsymbol{s}}^* \mid \kappa_{\boldsymbol{s}}) \\
\kappa_{\boldsymbol{t}}^* &\sim q(\kappa_{\boldsymbol{t}}^* \mid \kappa_{\boldsymbol{t}}) \\
\kappa_{\boldsymbol{y}}^* &\sim q(\kappa_{\boldsymbol{y}}^* \mid \kappa_{\boldsymbol{y}}) \\
\begin{pmatrix} \boldsymbol{s}^* \\ \boldsymbol{t}^* \end{pmatrix} &\sim \pi(\boldsymbol{s}, \boldsymbol{t} \mid \boldsymbol{\kappa}^*, \boldsymbol{y}).
\end{aligned}$$

To update the precisions, we will make a simple choice and follow a suggestion from Knorr-Held and Rue (2002). Let $\kappa_{\boldsymbol{s}}^* = f\kappa_{\boldsymbol{s}}$ where the scaling factor f have density

$$\pi(f) \propto 1 + 1/f, \quad \text{for} \quad f \in [1/F, F] \tag{4.13}$$

and zero otherwise. Here, $F > 1$ is a tuning parameter. Other choices

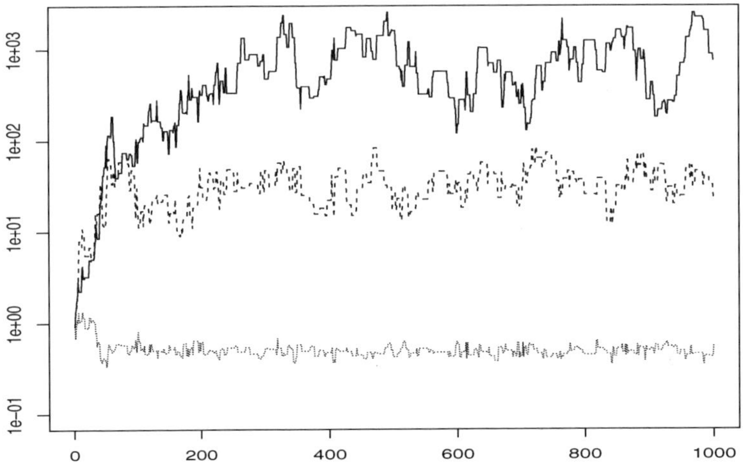

Figure 4.3 *Trace plot showing the log of the three precisions* κ_t *(top, solid line),* κ_s *(middle, dashed line) and* κ_y *(bottom, dotted line) for the first* 1000 *iterations. The acceptance rate was about* 30%.

for the proposal distribution may be more appropriate but we avoid a discussion here to ease the presentation. The choice (4.13) is reasonable, as the log is often a variance stabilizing transformation for precisions. It is also convenient, as

$$\frac{q(\kappa_s^* \mid \kappa_s)}{q(\kappa_s \mid \kappa_s^*)} = 1,$$

since the density of κ_s^* is hence proportional to $(\kappa_s^* + \kappa_s)/(\kappa_s^* \kappa_s)$ on $\kappa_s[1/F, F]$. We use the same proposal for κ_t and κ_y as well.

The joint proposal $(\boldsymbol{\kappa}^*, \boldsymbol{s}^*, \boldsymbol{t}^*)$ is accepted/rejected jointly with probability,

$$\begin{aligned}
\alpha &= \min\left\{1, \frac{\pi(\boldsymbol{s}^*, \boldsymbol{t}^*, \boldsymbol{\kappa}^* \mid \boldsymbol{y})}{\pi(\boldsymbol{s}, \boldsymbol{t}, \boldsymbol{\kappa} \mid \boldsymbol{y})} \times \frac{\pi(\boldsymbol{s}, \boldsymbol{t} \mid \boldsymbol{\kappa}, \boldsymbol{y})}{\pi(\boldsymbol{s}^*, \boldsymbol{t}^* \mid \boldsymbol{\kappa}^*, \boldsymbol{y})}\right\} \\
&= \min\left\{1, \frac{\pi(\boldsymbol{\kappa}^* \mid \boldsymbol{y})}{\pi(\boldsymbol{\kappa} \mid \boldsymbol{y})}\right\}.
\end{aligned}$$

Only the posterior marginals for $\boldsymbol{\kappa}$ are involved since we are integrating out $(\boldsymbol{s}, \boldsymbol{t})$. It is important not to forget to include correct normalization constants (using the generalized determinant) of the IGMRF priors for $\boldsymbol{s}$ and $\boldsymbol{t}$ when computing the acceptance probability.

We have chosen the tuning parameters such that the acceptance rate was approximately 30% using the same factor F for all three precisions. The mixing of all precision parameters was quite good, see Figure 4.3. More refined methods could be used, for example, incorporating the

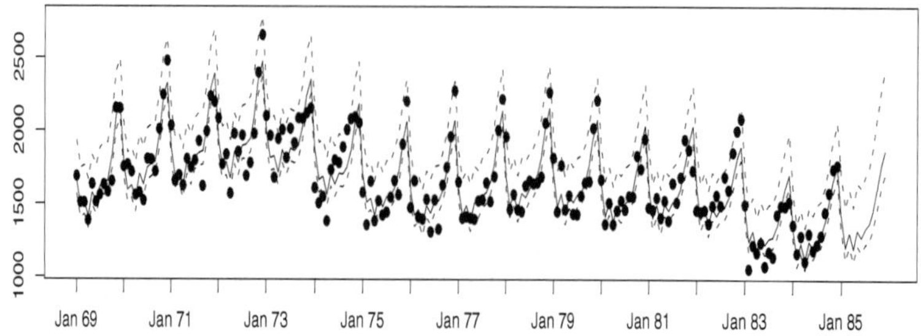

Figure 4.4 *Observed and predicted counts (posterior median within* 2.5 *and* 97.5% *quantiles) for the drivers data without the covariate*

posterior dependence between the precision parameters in the proposal distribution for $\boldsymbol{\kappa}^*$, but that is not necessary in this example. Note that the algorithm is effectively a simple Metropolis-Hastings algorithm for a three-dimensional density. We neither expect or experience any problems regarding convergence nor mixing of this algorithm.

Figure 4.4 displays the data and quantiles of the *predictive* distribution of $\boldsymbol{y}$ (posterior median within 2.5 and 97.5% quantiles), back on the original (squared) scale. The predictive distribution has been obtained by simply adding zero-mean normal noise with precision $\kappa_{\boldsymbol{y}}^{(j)}$ to each sample of $\boldsymbol{s}^{(j)} + \boldsymbol{t}^{(j)}$; here j denotes the jth sample obtained from the MCMC algorithm. Note that the figure also displays the predictive distribution for 1985 where no data are available. There is evidence for overdispersion with a posterior median of the observational precision $\kappa_{\boldsymbol{y}}$ equal to 0.49 rather than 4 that would have been expected under a Poisson observation model.

We now extend the above model to include a regression parameter for the effect of compulsory wearing of seat belts due to a law introduced on January 31, 1983. The mean response of y_i is now

$$E(y_i \mid \boldsymbol{s}, \boldsymbol{t}, \beta) = \begin{cases} s_i + t_i & i = 1, \ldots, 169 \\ s_i + t_i + \beta & i = 170, \ldots, 204. \end{cases}$$

The additional parameter β, for which we assume a normal prior with zero precision can easily be added to $\boldsymbol{s}$ and $\boldsymbol{t}$ to obtain a larger GMRF. It is straightforward to sample the GMRF $(\boldsymbol{s}, \boldsymbol{t}, \beta)|(\boldsymbol{y}, \boldsymbol{\kappa})$ in one block so we essentially use the same algorithm as without the covariate.

The posterior median of β is -5.0 (95% CI: $[-6.8, -3.2]$), very close to the observed difference in square root counts of the corresponding two periods shortly before and after January 31, 1983, which is -5.07. The inclusion of this covariate has some effect on the estimated precision parameters, which change from 0.49 to 0.54 for $\kappa_{\boldsymbol{y}}$, from 495 to 1283 for $\kappa_{\boldsymbol{x}}$, and from 28.8 to 27.6 for $\kappa_{\boldsymbol{s}}$ (all posterior medians). The sharp

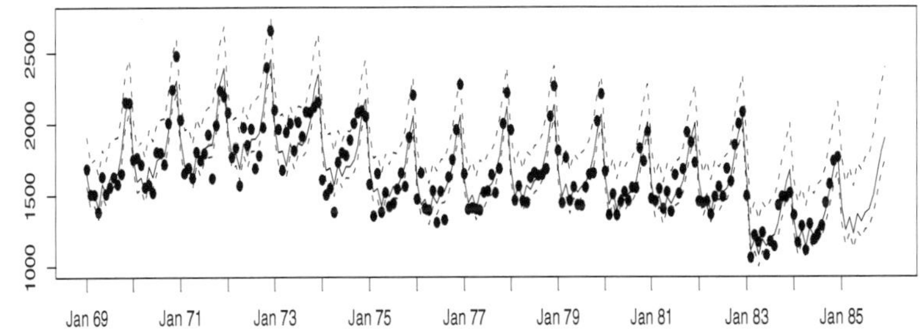

Figure 4.5 *Observed and predicted counts for the drivers data with the seat belt covariate.*

increase in the random-walk precision indicates that the previous model was undersmoothing the overall trend, because it ignores the information about the seat belt law, where a sudden nonsmooth drop in incidence could be expected. Note also that the overdispersion has now slightly decreased.

Observed and predicted counts can be seen in Figure 4.5 and the slightly better fit of this model before and after January 1983 is visible.

The connection to methods based on the Kalman filter

An alternative method for direct simulation from the conditional posterior $\pi(\boldsymbol{s}, \boldsymbol{t}, \beta | \boldsymbol{y}, \boldsymbol{\kappa})$ is to use the *forward-filtering-backward-sampling* (FFBS) algorithm by Carter and Kohn (1994) and Frühwirth-Schnatter (1994). This requires forcing our model into a state-space form, which is possible, but requires a high-dimensional state space and a degenerate stochastic model with deterministic relationships between the parameters. These tricks are necessary in order to apply the Kalman filter recursions and apply the FFBS algorithm. Knorr-Held and Rue (2002, Appendix A) have shown that, for nondegenerate Markov models, the GMRF approach using band-Cholesky factorization is equivalent to the FFBS algorithm. The same relation also holds if we run the forward-filtering-backward-sampling on the nondeterministic part of the state-space equations, suggested by Frühwirth-Schnatter (1994) and de Jong and Shephard (1995). However, the GMRF approach using sparse-matrix methods is superior over the Kalman-filter as it will run faster. Sparse-matrix methods also offer great simplification conceptually, extend trivially to general graphs, and can easily deal with conditioning, hard and soft constraints. Moreover, the same computer code can be used for GMRF models in time and in space (or even in space-time) on arbitrary graphs.

4.2.2 Example: Munich rental guide

As a second example for a normal response model, we consider a Bayesian semiparametric regression model with an additional spatial effect. For more introduction to this subject, see Hastie and Tibshirani (2000), Fahrmeir and Lang (2001a) and Fahrmeir and Tutz (2001, section 8.5). This class of models has recently been coined *geoadditive* by Kammann and Wand (2003).

Here we build a model similar to Fahrmeir and Tutz (2001, Example 8.7) for the 2003 Munich rental data. The response variable y_i is the rent (per square meter in Euros) for a flat and the covariates are the spatial location, floor space, year of construction, and various indicator variables such as an indicator for a flat with no central heating, no bathroom, large balcony facing south or west, etc. A regression analysis of such data provides a rental guide that is published by most larger cities. According to German law, the owners can base an increase in the amount they charge on an 'average rent' of a comparable flat. This information is provided in an official rental guide. The dataset we will consider consists of $n = 2\,035$ observations.

Important covariates of each observation include the size of the flat, z^S, which ranges between 17 and 185 square meters, and the year of construction z^C, with values between 1918 and 2001. We adopt a nonparametric modeling approach for the effect of these two covariates. Let $\boldsymbol{s}_S = (s_1^S, \ldots, s_{n_S}^S)$ denote the ordered distinct covariate values of z^S and define similarly $\boldsymbol{s}_C = (s_1^C, \ldots, s_{n_C}^C)$. We now define the corresponding parameter values $\boldsymbol{x}_S$ and $\boldsymbol{x}_C$ and assume that both $\boldsymbol{x}_S$ and $\boldsymbol{x}_C$ follow the CRW2 model (3.61) at the locations $\boldsymbol{s}_S$ and $\boldsymbol{s}_C$, respectively. We estimate a distinct parameter value for each value of the covariates $\boldsymbol{x}_S$ and $\boldsymbol{x}_C$, respectively.

There is also information in which district of Munich each flat is located. In total there are 380 districts in Munich, and we assume an unweighted IGMRF model (3.30) of order one for the spatial effect of a corresponding parameter vector $\boldsymbol{x}_L$ defined at each district. Only 312 of the districts actually contain data in our sample, so for the other districts the estimates are based on extrapolation.

Finally, we also include a number of additional binary indicators as fixed effects. We subsume all these covariates and an entry for the intercept in a vector $\boldsymbol{z}_i$ with parameters $\boldsymbol{\beta}$. We assume $\boldsymbol{\beta}$ to have a diffuse prior. For reasons of identifiability we place sum-to-zero restrictions on both CRW2 models and the spatial IGMRF model.

We now assume that the response variables y_i, $i = 1, \ldots, n$, are normally distributed:

$$y_i \sim \mathcal{N}\left(\mu + x^S(i) + x^C(i) + x^L(i) + \boldsymbol{z}_i^T \boldsymbol{\beta},\ 1/\kappa_{\boldsymbol{y}}\right)$$

with precision $\kappa_{\boldsymbol{y}}$. The notation $x^S(i)$ (and similarly $x^C(i)$ and $x^L(i)$) denote the entry in $\boldsymbol{x}_S$, corresponding to the value of the covariate s^S for observation i.

We used independent gamma $\mathcal{G}(1.0, 0.01)$ priors for all precision parameters in the model $\boldsymbol{\kappa} = (\kappa_{\boldsymbol{y}}, \kappa_S, \kappa_C, \kappa_L)$. The posterior distribution is

$$\begin{aligned} \pi(\boldsymbol{x}^S, \boldsymbol{x}^C, \boldsymbol{x}^L, \mu, \boldsymbol{\beta}, \boldsymbol{\kappa} \mid \boldsymbol{y}) &\propto \pi(\boldsymbol{x}^S \mid \kappa_S)\, \pi(\boldsymbol{x}^C \mid \kappa_C)\, \pi(\boldsymbol{x}^L \mid \kappa_L) \pi(\mu) \\ &\times \pi(\boldsymbol{\beta})\, \pi(\boldsymbol{\kappa})\, \pi(\boldsymbol{y} \mid \boldsymbol{x}^S, \boldsymbol{x}^C, \boldsymbol{x}^L, \boldsymbol{\beta}, \kappa_{\boldsymbol{y}}, \mu). \end{aligned}$$

Similar as in Section 4.2.1, the components $\boldsymbol{x}^S$, $\boldsymbol{x}^C$, $\boldsymbol{x}^L$, and $\boldsymbol{\beta}$ are *a priori* independent. However, conditional on the data they become dependent, but their full conditionals are still GMRFs. To study the dependence structure, let us consider in detail the likelihood term that is proportional to

$$\exp\left(-\frac{\kappa_{\boldsymbol{y}}}{2} \sum_i \left(y_i - \left(\mu + x^S(i) + x^C(i) + x^L(i) + \boldsymbol{z}_i^T \boldsymbol{\beta} \right) \right)^2 \right).$$

The dependence structure introduced is now different from the one in Section 4.2.1. Each combination of covariate values $x^S(i)$, $x^C(i)$, $x^L(i)$, and $\boldsymbol{z}_i$, introduces dependence within this combination and makes the specific term in the precision matrix nonzero. For these data there are 1980 different combinations of the first three covariates and 2011 different combinations if we also include $\boldsymbol{z}_i$.

The precision matrix of $(\boldsymbol{x}^S, \boldsymbol{x}^C, \boldsymbol{x}^L, \mu, \boldsymbol{\beta})$ will therefore be nonsparse, so a joint update will not be computationally efficient ruling out the one-block algorithm. We therefore switch to the subblock algorithm. There is a natural grouping of the variables of interest and we use four subblocks,

$$(\boldsymbol{x}^S, \kappa_S), \quad (\boldsymbol{x}^C, \kappa_C), \quad (\boldsymbol{x}^L, \kappa_L), \quad \text{and} \quad (\boldsymbol{\beta}, \mu, \kappa_{\boldsymbol{y}}).$$

We expect the dependence within each block to be stronger than between the blocks. The last block consists of all fixed effects plus the intercept jointly and the precision $\kappa_{\boldsymbol{y}}$ of the response variable. This can be advantageous, if there is high posterior dependency between the intercept and the fixed covariate effects.

The subblock algorithm now updates each block at a time, using

$$\begin{aligned} \kappa_L^* &\sim q(\kappa_L^* \mid \kappa_L) \\ \boldsymbol{x}^{L,*} &\sim \pi(\boldsymbol{x}^{L,*} \mid \text{the rest}) \end{aligned}$$

and then accepts/rejects $(\kappa_L^*, \boldsymbol{x}^{L,*})$ jointly. The other subblocks are updated similarly. A nice feature of the subblock algorithm is that the full conditional of $\boldsymbol{x}^L$ (and so with $\boldsymbol{x}^S$ and $\boldsymbol{x}^C$) has the same Markov properties as the prior. This will be clear when we now derive

$\pi(\boldsymbol{x}^{L,*}|\text{the rest})$. Introduce 'fake' data $\tilde{\boldsymbol{y}}$

$$\tilde{y}_i = y_i - \left(\mu + x^S(i) + x^C(i) + \boldsymbol{z}_i^T\boldsymbol{\beta}\right),$$

then the full conditional of $\boldsymbol{x}^L$ is

$$\begin{aligned}\pi(\boldsymbol{x}^L \mid \text{the rest}) &\propto \exp(-\frac{\kappa_L}{2}\sum_{i\sim j}(x_i^L - x_j^L)^2) \\ &\times \exp(-\frac{\kappa_{\boldsymbol{y}}}{2}\sum_k \left(\tilde{y}_k - x^L(k)\right)^2).\end{aligned}$$

The data $\tilde{\boldsymbol{y}}$ do not introduce extra dependence between the x_i^L's, as $\tilde{y}_i$ acts as a noisy observation of x_i^L. Denote by n_i the number of neighbors to location i and let $L(i)$ be

$$L(i) = \{k \,:\, x^L(k) = x_i^L\},$$

where its size is $|L(i)|$. The full conditional of $\boldsymbol{x}^L$ is a GMRF with canonical parameters $(\boldsymbol{b}, \boldsymbol{Q})$, where

$$b_i = \kappa_{\boldsymbol{y}} \sum_{k\in L_i} \tilde{y}_k \quad \text{and}$$

$$Q_{ij} = \begin{cases}\kappa_L n_i + \kappa_{\boldsymbol{y}}|L(i)| & \text{if } i = j\\ -\kappa_L & \text{if } i \sim j\\ 0 & \text{otherwise.}\end{cases}$$

The additional sum-to-zero constraint is dealt with as described in Section 2.3.3. Note that this is not equivalent to just recentering $\boldsymbol{x}^{L,*}$.

The value of $\boldsymbol{\kappa}$ for the first 1000 iterations of the subblock algorithm are shown in Figure 4.6. The scaling factor F in (4.13) was tuned to give an acceptance rate between 30% and 40%.

Figure 4.7 displays the estimated nonparametric effects of the floor space ($\boldsymbol{x}^S$) and the year of construction ($\boldsymbol{x}^C$). There is a clear nonlinear effect of $\boldsymbol{x}_S$ with a positive effect for smaller flats. The effect of the variable $\boldsymbol{x}_C$ is approximately linear for flats built after 1940, but older flats, especially those built at the beginning of the 20th century are more expensive than those built around 1940. Note that detailed information about this covariate seems to be available only from 1966 on, whereas before 1966 the variable seems to rather crudely categorized, see Figure 4.7(b). Figure 4.8 displays the posterior median spatial effect, which is similar to earlier analysis. The large estimated spatial effect of the district in the east is due to very few observations in this district, plus the fact that the district has only one neighbor, so the amount of spatial smoothing will be small.This could easily be fixed by adding more neighbors to this district, see the general discussion in Example 3.6.

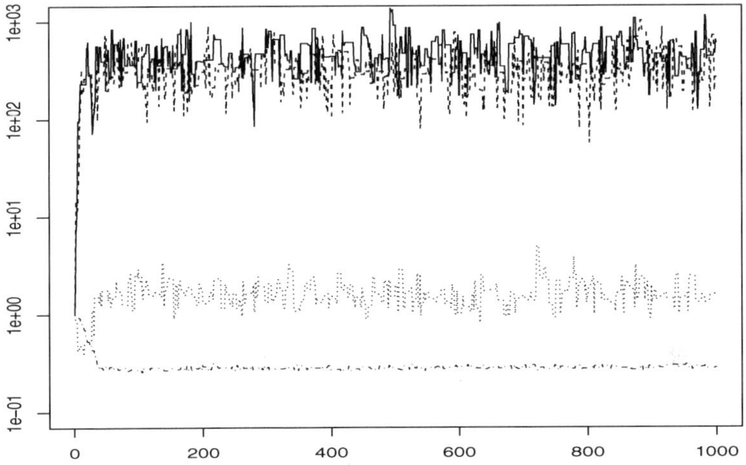

Figure 4.6 *The value of* κ *for the* 1000 *first iterations of the subblock algorithm, where* κ^S *is the solid line (top),* κ^C *is the dashed line,* κ^L *is the dotted line and* κ_y *is the dashed-dotted line (bottom).*

Covariate	post. median	95% credible interval
Intercept μ	8.81	$(8, 60, 9.03)$
Good location	0.48	$(0.26, 0.70)$
Excellent location	1.63	$(0.99, 2.23)$
No hot water	−2.02	$(-2.56, -1.47)$
No central heating	−1.34	$(-1.68, -0.96)$
No tiles in bathroom	−0.54	$(-0.74, -0.30)$
Special bathroom interior	0.55	$(0.23, 0.87)$
Special kitchen interior	1.19	$(0.86, 1.52)$

Table 4.1 *Posterior median and 95% credible interval for fixed effects*

The estimates of the fixed effects $\boldsymbol{\beta}$ and the intercept μ are given in Table 4.1.

4.3 Auxiliary variable models

Auxiliary variables can in some cases be introduced into the model to retrieve GMRF full conditionals that are otherwise lost by nonnormality. We first discuss the construction of scale mixtures of normals and then discuss how auxiliary variables can be useful for replacing normal with Student-t distributions. We give more emphasis to the binomial probit and logit model for categorical data, which we also illustrate with two

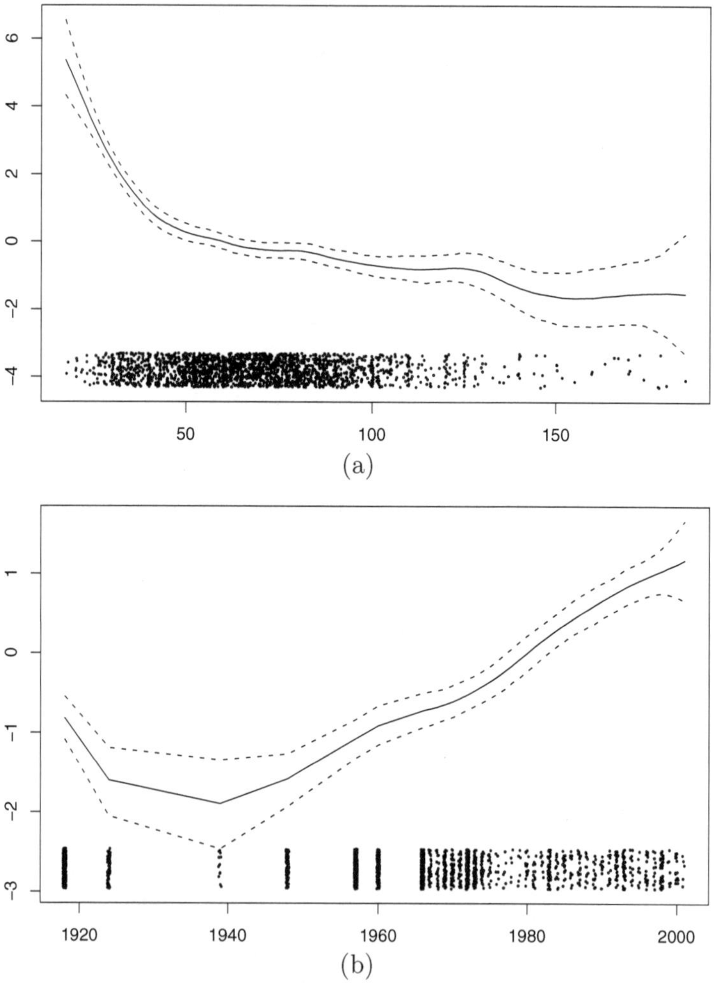

Figure 4.7 *Nonparametric effects for floor space (a) and year of construction (b). The figures show the posterior median within* 2.5 *and* 97.5% *quantiles. The distribution of the observed data is indicated with jittered dots.*

case studies.

4.3.1 Scale mixtures of normals

Scale mixtures of normals play an important role in hierarchical modeling. Suppose $x|\lambda \sim \mathcal{N}(0, \lambda^{-1})$ where $\lambda > 0$ is a precision parameter with some prespecified distribution. Then x is a called *normal scale*

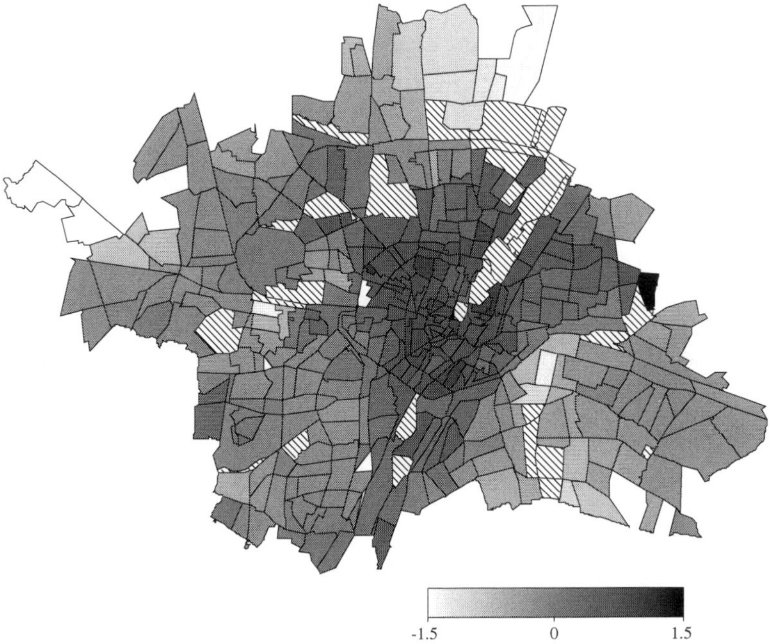

Figure 4.8 *Estimated posterior median effect for the location variable. The shaded areas are districts with no houses, such as parks or fields.*

mixture. Although $\pi(x|\lambda)$ is normal, its marginal density $f(x)$ is not normal unless λ is a constant. However, f is necessarily both symmetric and unimodal. Kelker (1971) and Andrews and Mallows (1974) show the following result establishing necessary and sufficient conditions for $x \sim f(x)$ to have a normal scale mixture representation.

Theorem 4.2 *If x has density $f(x)$ symmetric around 0, then there exist independent random variables z and v, with z standard normal such that $x = z/v$ iff the derivatives of $f(x)$ satisfy*

$$\left(-\frac{d}{dy}\right)^k f(\sqrt{y}) \geq 0$$

for $y > 0$ and for $k = 1, 2, \ldots$.

Many important symmetric random variates are scale mixtures of normals, in particular the Student-t_ν distribution with ν degrees of freedom, which includes the Cauchy distribution as a special case for $\nu = 1$, the Laplace distribution, and the logistic distribution. Table 4.2 gives the corresponding mixing distribution for the precision parameter λ that generates these distributions as scale mixtures of normals.

Distribution of x	Mixing distribution of λ
Student-t_ν	$\mathcal{G}(\nu/2, \nu/2)$
Logistic	$1/(2K)^2$ where K is Kolmogorov-Smirnov distributed
Laplace	$1/(2E)$ where E is exponential distributed

Table 4.2 *Important scale mixtures of normals*

The Kolmogorov-Smirnov distribution and the logistic distribution are defined in Appendix A and are abbreviated as $\mathcal{KS}$ and $\mathcal{L}$, respectively.

The representation of the logistic distribution as a scale mixture of normals will become important for the logistic regression model for binary response data.

4.3.2 Hierarchical-t formulations

Using a scale mixture approach for GMRF models facilitates the analysis of certain models for continuous responses. The main idea is to include in the MCMC algorithm the mixing variable λ, or a vector of such variables $\boldsymbol{\lambda}$, as so-called *auxiliary variables.* The auxiliary variables are not of interest but can be helpful to ensure GMRF full conditionals.

As a simple example consider the RW1 model discussed in Section 3.3.1. Suppose we wish to replace the assumption of normally distributed increments by a Student-t_ν distribution to allow for larger jumps in the sequence $\boldsymbol{x}$. This can be done using $n-1$ independent $\mathcal{G}(\nu/2, \nu/2)$ scale mixture variables λ_i:

$$\Delta x_i \mid \lambda_i \overset{\text{iid}}{\sim} \mathcal{N}(0, (\kappa\lambda_i)^{-1}), \quad i = 1, \ldots, n-1.$$

With observed data $y_i \sim \mathcal{N}(x_i, \kappa_{\boldsymbol{y}}^{-1})$ for $i = 1, \ldots, n$, the posterior density for $(\boldsymbol{x}, \boldsymbol{\lambda})$ is

$$\pi(\boldsymbol{x}, \boldsymbol{\lambda} \mid \boldsymbol{y}) \propto \pi(\boldsymbol{x} \mid \boldsymbol{\lambda})\, \pi(\boldsymbol{\lambda})\, \pi(\boldsymbol{y} \mid \boldsymbol{x}).$$

Note that $\boldsymbol{x}|(\boldsymbol{y}, \boldsymbol{\lambda})$ is now a GMRF while $\lambda_1, \ldots, \lambda_{n-1}|(\boldsymbol{x}, \boldsymbol{y})$ are conditionally independent gamma distributed with parameters $(\nu+1)/2$ and $(\nu + \kappa(\Delta x_i)^2)/2$.

The replacement of a normal distribution with a Student-t_ν distribution is quite popular and is sometimes called the *hierarchical-t formulation.* The same approach can also be used to replace the observational normal distribution with a Student-t_ν distribution. In this case the posterior of $(\boldsymbol{x}, \boldsymbol{\lambda})$ is

$$\pi(\boldsymbol{x}, \boldsymbol{\lambda} \mid \boldsymbol{y}) \propto \pi(\boldsymbol{x})\, \pi(\boldsymbol{y} \mid \boldsymbol{x}, \boldsymbol{\lambda})\, \pi(\boldsymbol{\lambda}).$$

The full conditional of $\boldsymbol{x}$ is a GMRF while $\lambda_1, \ldots, \lambda_n | (\boldsymbol{x}, \boldsymbol{y})$ are conditionally independent.

4.3.3 Binary regression models

Consider a Bernoulli observational model for binary responses with latent parameters that follow a GMRF $\boldsymbol{x}$, which in turn usually depends on further hyperparameters $\boldsymbol{\theta}$. Denote by $\mathcal{B}(p)$ a Bernoulli distribution with probability p for 1 and $1-p$ for 0. The most prominent regression models in this framework are logit and probit models, where

$$y_i \sim \mathcal{B}(g^{-1}(\boldsymbol{z}_i^T \boldsymbol{x})) \tag{4.14}$$

for $i = 1, \ldots, m$. Here $\boldsymbol{z}_i$ is the vector of covariates, assumed to be fixed, and $g(p)$ is a link function:

$$g(p) = \begin{cases} \log(p/(1-p)) & \text{logit link} \\ \Phi(p) & \text{probit link} \end{cases}$$

where $\Phi(\cdot)$ denotes the standard normal distribution function.

These models have an equivalent representation using auxiliary variables $\boldsymbol{w} = (w_1, \ldots, w_m)^T$, where

$$\begin{aligned} \epsilon_i &\overset{\text{iid}}{\sim} G(\epsilon_i) \\ w_i &= \boldsymbol{z}_i^T \boldsymbol{x} + \epsilon_i \\ y_i &= \begin{cases} 1 & \text{if } w_i > 0 \\ 0 & \text{otherwise.} \end{cases} \end{aligned}$$

Here, $G(\cdot)$ is the distribution function of the standard normal distribution in the probit case and of the standard logistic distribution (see Appendix A) in the logit case.

Note that y_i is deterministic conditional on the sign of the stochastic auxiliary variable w_i. The equivalence can be seen immediately from

$$\text{Prob}(y_i = 1) = \text{Prob}(w_i > 0) = \text{Prob}(\boldsymbol{z}_i^T \boldsymbol{x} + \epsilon_i > 0) = G(\boldsymbol{z}_i^T \boldsymbol{x}),$$

using that the density of ϵ_i is symmetric about zero. This auxiliary variable approach was proposed by Albert and Chib (1993) for the probit link, and Chen and Dey (1998) and Holmes and Held (2003) for the logit link.

The motivation for introducing auxiliary variables is to ease the construction of MCMC algorithms. We will discuss this issue in detail in the following, first for the simpler probit link and then for the logit link.

MCMC for probit regression using auxiliary variables

Let $\boldsymbol{x}|\boldsymbol{\theta}$ be a zero mean GMRF of size n and assume first that $\boldsymbol{z}_i^T\boldsymbol{x} = x_i$ and $m = n$. Using the probit link, the posterior distribution is

$$\pi(\boldsymbol{x}, \boldsymbol{w}, \boldsymbol{\theta} \mid \boldsymbol{y}) \propto \pi(\boldsymbol{\theta})\, \pi(\boldsymbol{x} \mid \boldsymbol{\theta})\, \pi(\boldsymbol{w} \mid \boldsymbol{x})\, \pi(\boldsymbol{y} \mid \boldsymbol{w}). \tag{4.15}$$

The full conditional of $\boldsymbol{x}$ is then

$$\pi(\boldsymbol{x} \mid \boldsymbol{\theta}, \boldsymbol{w}) \propto \exp\left(-\frac{1}{2}\boldsymbol{x}^T\boldsymbol{Q}(\boldsymbol{\theta})\boldsymbol{x} - \frac{1}{2}\sum_i (x_i - w_i)^2\right),$$

which is a GMRF:

$$\boldsymbol{x} \mid \boldsymbol{\theta}, \boldsymbol{w} \sim \mathcal{N}_C(\boldsymbol{w},\ \boldsymbol{Q}(\boldsymbol{\theta}) + \boldsymbol{I}). \tag{4.16}$$

The full conditional of $\boldsymbol{w}$ factorizes as

$$\pi(\boldsymbol{w} \mid \boldsymbol{x}, \boldsymbol{y}) = \prod_i \pi(w_i \mid x_i, y_i), \tag{4.17}$$

where $w_i|(\boldsymbol{x}, \boldsymbol{y})$ is a standard normal with mean x_i, but truncated to be positive if $y_i = 1$ or to be negative if $y_i = 0$. A natural approach to sample from (4.15) is to use two subblocks $(\boldsymbol{\theta}, \boldsymbol{x})$ and $\boldsymbol{w}$. The block $(\boldsymbol{\theta}, \boldsymbol{x})$ is sampled using (4.10) and $\boldsymbol{w}$ is updated using the factorization (4.17) and algorithms for truncated normals, see, for example, Robert (1995).

Consider now the general setup (4.14) using a probit link. The full conditional of $\boldsymbol{x}$ is then

$$\boldsymbol{x} \mid \boldsymbol{\theta}, \boldsymbol{w} \sim \mathcal{N}_C\left(\boldsymbol{Z}^T\boldsymbol{w},\ \boldsymbol{Q}(\boldsymbol{\theta}) + \boldsymbol{Z}^T\boldsymbol{Z}\right), \tag{4.18}$$

where the $m \times n$ matrix $\boldsymbol{Z}$ is

$$\boldsymbol{Z} = \begin{pmatrix} \boldsymbol{z}_1^T \\ \boldsymbol{z}_2^T \\ \vdots \\ \boldsymbol{z}_m^T \end{pmatrix}.$$

The full conditional of $\boldsymbol{w}$ is now

$$\pi(\boldsymbol{w} \mid \boldsymbol{x}, \boldsymbol{y}) = \prod_i \pi(w_i \mid \boldsymbol{x}, y_i)$$

where $w_i|(\boldsymbol{x}, y_i)$ is standard normal with mean $\boldsymbol{z}_i^T\boldsymbol{x}$, but truncated to be positive if $y_i = 1$ or to be negative if $y_i = 0$. Sparseness of the precision matrix for $\boldsymbol{x}|(\boldsymbol{\theta}, \boldsymbol{w})$ now depends also on the sparseness of $\boldsymbol{Z}^T\boldsymbol{Z}$. Typically, but not always, if n is large then $\boldsymbol{Z}^T\boldsymbol{Z}$ is sparse, while if n is small then $\boldsymbol{Z}^T\boldsymbol{Z}$ is dense. Hence, sampling $\boldsymbol{x}$ from its full conditional is often computationally feasible.

If m is not too large, we can also integrate out $\boldsymbol{x}$ to obtain

$$\boldsymbol{w} \mid \boldsymbol{y} \sim \mathcal{N}(\boldsymbol{0},\ \boldsymbol{I} + \boldsymbol{Z}\boldsymbol{Q}(\boldsymbol{\theta})^{-1}\boldsymbol{Z}^T)\ \mathbf{1}[\boldsymbol{w}, \boldsymbol{y}]. \tag{4.19}$$

Here $\mathbf{1}[\boldsymbol{w}, \boldsymbol{y}]$ is a shorthand for the truncation induced by $w_i > 0$ if $y_i = 1$ and $w_i < 0$ if $y_i = 0$ for each i. Sampling from a truncated normal in high dimension is hard, therefore an alternative to sample $\boldsymbol{w}$ from $\pi(\boldsymbol{w}|\boldsymbol{x}, \boldsymbol{y})$ is to sample each component w_i from $\pi(w_i|\boldsymbol{w}_{-i}, \boldsymbol{y})$ using (4.19). See Holmes and Held (2003) for further details and a comparison.

MCMC for logistic regression using auxiliary variables

For the logit link we need to introduce one additional vector of length m of auxiliary variables to transform the logistic distributed ϵ_i into a scale mixture of normals. Using the result in Table 4.2 we obtain the following representation

$$\begin{aligned} \psi_i &\overset{\text{iid}}{\sim} \mathcal{KS} \\ \lambda_i &= 1/(2\psi_i)^2 \\ w_i &\overset{\text{iid}}{\sim} \mathcal{N}(\boldsymbol{z}_i^T \boldsymbol{x},\ 1/\lambda_i) \\ y_i &= \begin{cases} 1 & \text{if } w_i > 0 \\ 0 & \text{if } w_i < 0. \end{cases} \end{aligned}$$

The variable ψ_i is introduced for clarity only as it is a deterministic function of λ_i. The posterior of interest is now

$$\pi(\boldsymbol{x}, \boldsymbol{w}, \boldsymbol{\lambda}, \boldsymbol{\theta} \mid \boldsymbol{y}) \propto \pi(\boldsymbol{\theta})\ \pi(\boldsymbol{x} \mid \boldsymbol{\theta})\ \pi(\boldsymbol{\lambda})\ \pi(\boldsymbol{w} \mid \boldsymbol{x}, \boldsymbol{\lambda})\ \pi(\boldsymbol{y} \mid \boldsymbol{w}).$$

The full conditional of $\boldsymbol{x}$ is

$$\boldsymbol{x} \mid \boldsymbol{\theta}, \boldsymbol{w} \sim \mathcal{N}_C\left(\boldsymbol{Z}^T \boldsymbol{\Lambda} \boldsymbol{w},\ \boldsymbol{Q}(\boldsymbol{\theta}) + \boldsymbol{Z}^T \boldsymbol{\Lambda} \boldsymbol{Z}\right) \tag{4.20}$$

where $\boldsymbol{\Lambda} = \text{diag}(\boldsymbol{\lambda})$. The minor adjustments are due to the different precisions of the w_i's rather than all precisions equal to unity as in the probit case.

The full conditional of $\boldsymbol{w}$ factorizes as

$$\pi(\boldsymbol{w} \mid \boldsymbol{x}, \boldsymbol{\lambda}, \boldsymbol{y}) = \prod_i \pi(w_i \mid \boldsymbol{x}, \lambda_i, y_i),$$

where $w_i|(\boldsymbol{x}, \lambda_i, y_i)$ is normal with mean $\boldsymbol{z}_i^T \boldsymbol{x}$ and precision λ_i, but truncated to be positive if $y_i = 1$ and negative if $y_i = 0$. The full conditional of $\boldsymbol{\lambda}$ factorizes similarly:

$$\pi(\boldsymbol{\lambda} \mid \boldsymbol{x}, \boldsymbol{w}) = \prod_i \pi(\lambda_i \mid \boldsymbol{x}, w_i). \tag{4.21}$$

It is a nonstandard task to sample from (4.21) as each term on the rhs involves the Kolmogorov-Smirnov distribution for which the distribution function is only known as an infinite series, see Appendix A. Holmes and Held (2003) describe an efficient and exact approach based on the series method (Devroye, 1986) that avoids the density evaluation. The alternative algorithm proposed in Chen and Dey (1998) approximates the Kolmogorov-Smirnov density by a finite evaluation of this series.

The discussion so far suggests using the subblock approach and constructing an MCMC algorithm updating each of the following three subblocks conditionally on the rest: $(\boldsymbol{\theta}, \boldsymbol{x})$, $\boldsymbol{w}$ and $\boldsymbol{\lambda}$. However, further progress can be made if we merge $\boldsymbol{w}$ and $\boldsymbol{\lambda}$ into one block using

$$\pi(\boldsymbol{w}, \boldsymbol{\lambda} \mid \boldsymbol{x}, \boldsymbol{y}) = \pi(\boldsymbol{w} \mid \boldsymbol{x}, \boldsymbol{y})\, \pi(\boldsymbol{\lambda} \mid \boldsymbol{w}, \boldsymbol{x}), \tag{4.22}$$

where we have integrated $\boldsymbol{\lambda}$ analytically out of $\pi(\boldsymbol{w}|\boldsymbol{x}, \boldsymbol{y}, \boldsymbol{\lambda})$ to obtain $\pi(\boldsymbol{w}|\boldsymbol{x}, \boldsymbol{y})$. It now follows that

$$\pi(\boldsymbol{w} \mid \boldsymbol{x}, \boldsymbol{y}) = \prod_i \pi(w_i \mid \boldsymbol{x}, y_i),$$

where $w_i|(\boldsymbol{x}, y_i)$ is $\mathcal{L}(\boldsymbol{z}_i^T\boldsymbol{x}, 1)$ distributed, but truncated to be positive if $y_i = 1$ and negative if $y_i = 0$. It is easy to sample from this distribution using inversion.

To summarize, we suggest using the subblock algorithm with two subblocks $(\boldsymbol{\theta}, \boldsymbol{x})$ and $(\boldsymbol{w}, \boldsymbol{\lambda})$. The first block is updated using (4.18) together with a simple proposal for $\boldsymbol{\theta}$, while the second block is updated using (4.22), sampling first $\boldsymbol{w}$ then $\boldsymbol{\lambda}$.

4.3.4 Example: Tokyo rainfall data

For illustration, we consider a simple but much analyzed binomial time series, taken from Kitagawa (1987). Each day during the years 1983 and 1984, it was recorded whether there was more than 1 mm rainfall in Tokyo. It is of interest to estimate the underlying probability p_i of rainfall at calendar day $i = 1, \ldots, 366$, which is assumed to be gradually changing with time. Note that for $t = 60$, which corresponds to February 29, only one binary observation is available, while for all other calender days there are two. In total, we have $m = 366 + 365 = 731$ binary observations and an underlying GMRF $\boldsymbol{x}$ of dimension $n = 366$.

In contrast to earlier modeling approaches (Fahrmeir and Tutz, 2001, Kitagawa, 1987), we assume a *circular* RW2 model for $\boldsymbol{x} = g(\boldsymbol{p})$ with precision κ. This explicitly connects the end and the beginning of the time series, because smooth changes between the last week in December and the first week in January are also to be expected. Such a model cannot be directly analyzed with a state-space modeling approach. The precision matrix of $\boldsymbol{x}$ is now a circular precision matrix $\boldsymbol{Q} = \kappa\boldsymbol{R}$ with circulant structure matrix $\boldsymbol{R}$ with base $(6, -4, 1, 0, \ldots, 0, 1, -4)^T$. Comparing $\boldsymbol{Q}$ to the precision matrix (3.40) of the ordinary RW2 model, we see that only the entries in the first two and last two rows and columns are different. Note that the rank of $\boldsymbol{Q}$ is $n-1$ and larger than the rank of the noncircular RW2 model. The precision κ is assigned a $\mathcal{G}(1.0, 0.0001)$ prior.

If we assume a probit link, then the observational model is

$$\begin{aligned} y_{i1}, y_{i2} &\sim \mathcal{B}(g^{-1}(x_i)), \quad i \neq 60 \\ y_{60,1} &\sim \mathcal{B}(g^{-1}(x_{60})). \end{aligned}$$

We have two observations for each calendar day except for February 29. Therefore we use a double index for the observed data. The data only contain information about the sum $y_{i1} + y_{i2}$, so we assign (completely arbitrary) $y_{i1} = 1$ and $y_{i2} = 0$ if the sum is 1.

We now introduce auxiliary variables w_{ij} for each binary response variable y_{ij}. Let m_i denote the number of observations of x_i, which is 2 except for $i = 60$, and let $\boldsymbol{m} = (m_1, \ldots, m_n)^T$. Further, let

$$w_{i\bullet} = \sum_{j=1}^{m_i} w_{ij}$$

and $\boldsymbol{w}_{\bullet} = (w_{1\bullet}, \ldots, w_{n\bullet})^T$. The full conditional of $\boldsymbol{x}$ can now be derived either by extending (4.16) or from (4.18), as

$$\boldsymbol{x} \mid \text{the rest} \sim \mathcal{N}_C\left(\boldsymbol{w}_{\bullet},\ \boldsymbol{Q} + \text{diag}(\boldsymbol{m})\right). \tag{4.23}$$

The full conditional of $\boldsymbol{w}$ is (4.17) where w_{ij} is standard normal with mean x_i, but truncated to be positive if $y_{ij} = 1$ and negative if $y_{ij} = 0$. The MCMC algorithm uses two subblocks $(\kappa, \boldsymbol{x})$ and $\boldsymbol{w}$, and successively updates each block conditional on the rest.

For comparison, we have also implemented the auxiliary approach for the logistic regression model, using a $\mathcal{G}(1.0, 0.000289)$ prior for κ to accommodate the different variance of the logistic distribution compared to the standard normal distribution. Note that the ratio of the expectations of the two priors (logit versus probit) is $0.000289/0.0001 = 2.89$, approximately equal to $\pi^2/3 \cdot (15/16)^2$, the common factor to translate logit into probit results.

We need to introduce another layer of auxiliary variables $\boldsymbol{\lambda}$ to transform the logistic distribution into a scale mixture of normals, see Section 4.3.3. The full conditional of $\boldsymbol{x}$ now follows from (4.20), which is of the form (4.23) using $\widetilde{\boldsymbol{w}}_{\bullet}$ and $\widetilde{\boldsymbol{m}}$ instead, where

$$\widetilde{w}_{i\bullet} = \sum_{j=1}^{m_i} \lambda_{ij} w_{ij}, \quad \text{and} \quad \widetilde{m}_i = \sum_{j=1}^{m_i} \lambda_{ij}.$$

Also in this case we use the subblock algorithm using two blocks $(\kappa, \boldsymbol{x})$ and $(\boldsymbol{w}, \boldsymbol{\lambda})$. The second block is sampled using (4.22) in the correct order, first w_{ij} (for all ij) from the $\mathcal{L}(x_i, 1)$ distribution, truncated to

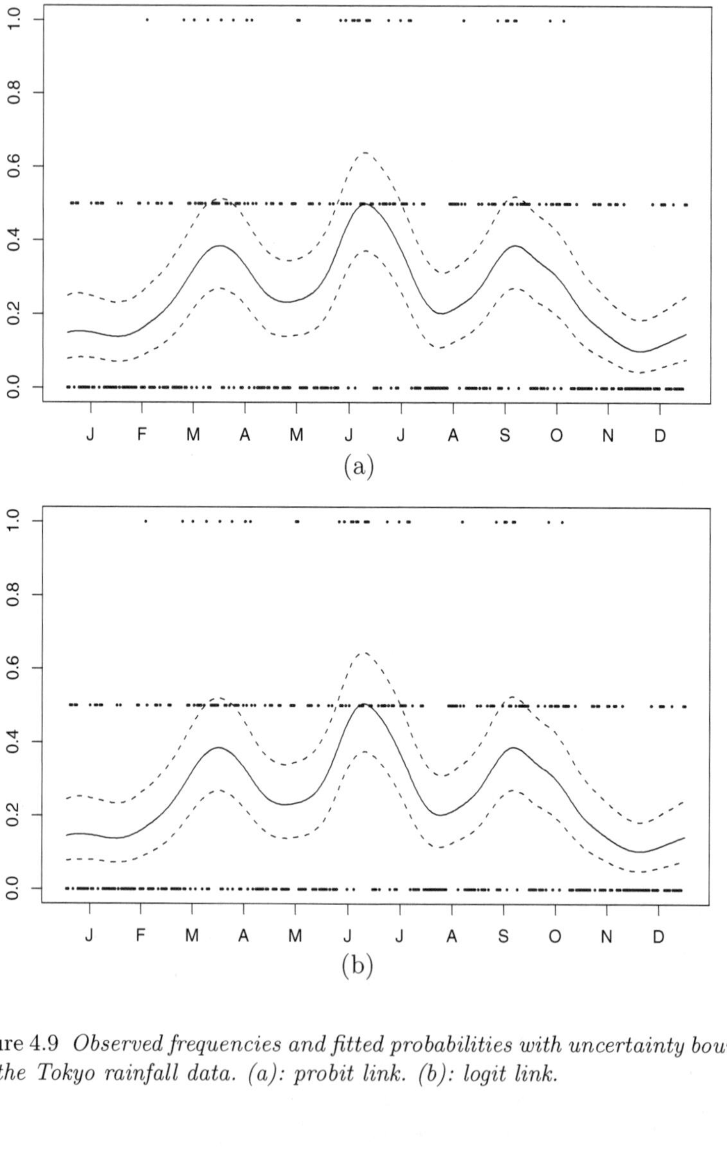

(a)

(b)

Figure 4.9 *Observed frequencies and fitted probabilities with uncertainty bounds for the Tokyo rainfall data. (a): probit link. (b): logit link.*

be positive if $y_{ij} = 1$ and negative if $y_{ij} = 0$, and then λ_{ij} (for all ij) from (4.21).

Figure 4.9(a) displays the binomial frequencies, scaled to the interval $[0, 1]$, and the estimated underlying probabilities p_i obtained from the probit regression approach, while (b) gives the corresponding results obtained with the logistic link function. There is virtually no difference between the results using the two link functions. Note that the credible

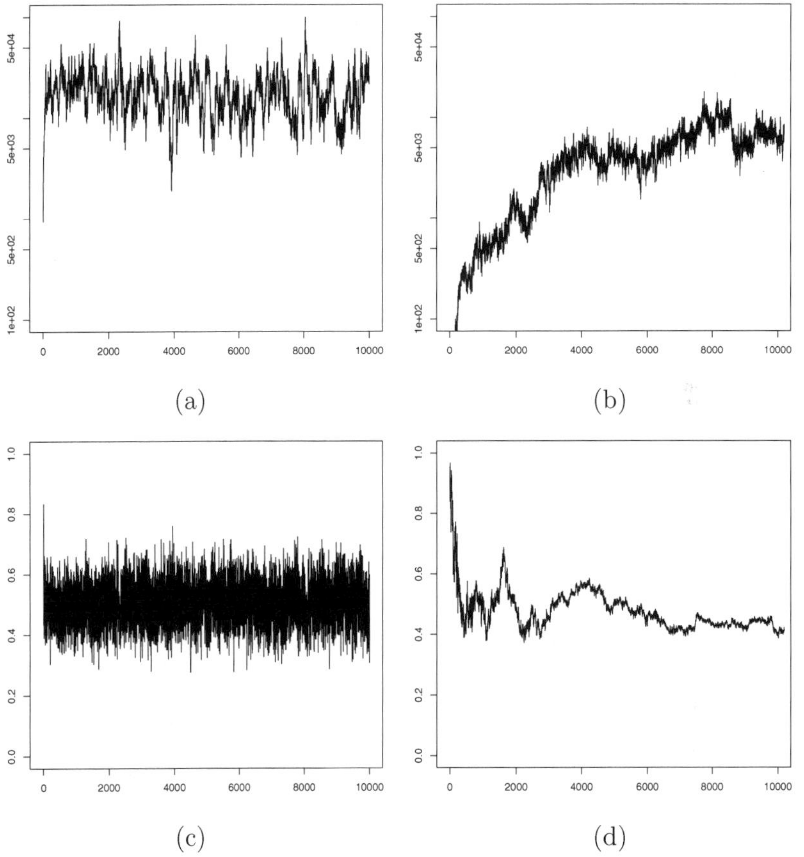

Figure 4.10 *Trace plots using the subblock algorithm and the single-site Gibbs sampler; (a) and (c) show the traces of* $\log \kappa$ *and* $g^{-1}(x_{180})$ *using the subblock algorithm, while (b) and (d) show the traces of* $\log \kappa$ *and* $g^{-1}(x_{180})$ *using the single-site Gibbs sampler.*

intervals do not get wider at the beginning and the end of the time series, due to the circular RW2 model.

We take this opportunity to compare the subblock algorithm with a naïve single-site Gibbs sampler which is know to converge slowly for this problem (Knorr-Held, 1999). It is important to remember that this is not a hard problem nor is the dimension high. We present the results for the logit link. Figure 4.10 shows the trace of $\log \kappa$ using the subblock algorithm in (a) and the single-site Gibbs sampler in (b), and the trace of $g^{-1}(x_{180})$ in (c) and (d). Both algorithms were run for 10000 iterations.

The results clearly demonstrate that the single-site Gibbs sampler has severe problems. The trace of $\log \kappa$ has not yet reached its level even after 10 000 iterations while the subblock algorithm seems to converge after just a few iterations. The computational cost per iteration is comparable. However, trace plots similar to those in Figure 4.10(b) are not at all uncommon using single-site schemes for moderately complex problems, especially after hierarchical models are becoming increasingly popular. The discussion in Section 4.1.2 is relevant here as well.

The sum of repeated independent Bernoulli observations is binomial distributed if the probability for success is constant, hence this example also demonstrates how to use the auxiliary approach for binomial regression.

We end this example with a comment on an alternative MCMC updating algorithms in the probit model. We could add a further auxiliary variable $w_{60,2}$ to make up for the 'missing' second observation for February 29. The precision matrix for the full conditional of $\boldsymbol{x}$ will now be circulant; hence we could have used the fast discrete Fourier transform and Algorithm 2.10 to simulate from it. However, this would not have been possible in the logit model as there the auxiliary variables $\boldsymbol{w}$ have different precisions.

4.3.5 Example: Mapping cancer incidence

As a second example for an auxiliary variable approach, we consider a problem in mapping cancer incidence where the stage of the disease at time of diagnosis is known. For an introduction to the topic see Knorr-Held et al. (2002). Data were available on all incidence cases of cervical cancer in the former East German Republic (GDR) from 1979, stratified by district and age group. Each of the $n = 6\,690$ cases of cervical cancer has been classified into either a premalignant (3755 cases) or a malignant (2935 cases) stage. It is of interest to estimate the spatial variation of the incidence ratio of premalignant to malignant cases in the 216 districts, after adjusting for age effects. Age was categorized into 15 age groups.

Let $y_i = 1$ denote a premalignant case and $y_i = 0$ a malignant case. We assume a logistic binary regression model $y_i \sim \mathcal{B}(p_i)$, $i = 1, \ldots, n$ with

$$\text{logit}(p_i) = \alpha + \beta_{j(i)} + \gamma_{k(i)},$$

where $j(i)$ and $k(i)$ denote age group and district of case i, respectively. The age group effects $\boldsymbol{\beta}$ are assumed to follow a RW2 model while for the spatial effect $\boldsymbol{\gamma}$ we assume that it is the sum of an IGMRF model (3.30) plus additional unstructured variation:

$$\gamma_k = u_k + v_k.$$

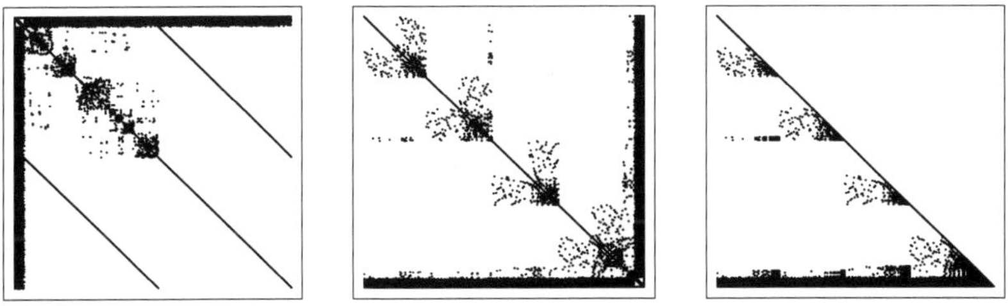

(a) (b) (c)

Figure 4.11 *(a) The precision matrix $\boldsymbol{Q}$ in the original ordering, and (b) the precision matrix after appropriate reordering to reduce to number of nonzero terms in the Cholesky triangle shown in (c). Only the nonzero terms are shown and those are indicated by a dot.*

Here, $\boldsymbol{u}$ follows the IGMRF model (3.30) with precision $\kappa_{\boldsymbol{u}}$ and $\boldsymbol{v}$ is normal with zero mean and diagonal precision matrix with entries $\kappa_{\boldsymbol{v}}$. For the corresponding precision parameters we assume a $\mathcal{G}(1.0, 0.01)$ for both $\kappa_{\boldsymbol{u}}$ and $\kappa_{\boldsymbol{v}}$ and a $\mathcal{G}(1.0, 0.0005)$ for $\kappa_{\boldsymbol{\beta}}$. A diffuse prior is assumed for the overall mean α and sum-to-zero constraints are placed both on $\boldsymbol{\beta}$ and $\boldsymbol{u}$. Note that this is not necessary for the unstructured effect $\boldsymbol{v}$, which has a proper prior with mean zero *a priori*.

Let $\boldsymbol{\kappa} = (\kappa_{\boldsymbol{\beta}}, \kappa_{\boldsymbol{u}}, \kappa_{\boldsymbol{v}})^T$ denote the vector of all precision parameters in the model. Similar to the previous example, we used auxiliary variables $w_1, \dots, w_n$ and $\lambda_1, \dots, \lambda_n$ to facilitate the implementation of the logistic regression model. Figure 4.11 displays the precision matrix of $(\alpha, \boldsymbol{\beta}, \boldsymbol{u}, \boldsymbol{v})|(\boldsymbol{\kappa}, \boldsymbol{w}, \boldsymbol{\lambda})$ before (a) and after (b) appropriate reordering to reduce the number of nonzero terms in the Cholesky triangle (c). Note how the reordering algorithm puts the α and $\boldsymbol{\beta}$ variables at the end, so the remaining variables become conditionally independent after factorization as discussed in Section 2.4.3.

In our MCMC algorithm we group all variables into two subblocks and update all variables in one subblock conditional on the rest. The first subblock consists of $(\alpha, \boldsymbol{\beta}, \boldsymbol{u}, \boldsymbol{v}, \boldsymbol{\kappa})$, while the auxiliary variables $(\boldsymbol{w}, \boldsymbol{\lambda})$ form the other block.

Figure 4.12 displays the estimated age-group effect, which is close to a linear effect on the log odds scale with a slightly increasing slope for increasing age. Figure 4.13 displays the estimates of the spatially structured component $\exp(\boldsymbol{u})$ and the total spatial effect $\exp(\boldsymbol{\gamma}) = \exp(\boldsymbol{u} + \boldsymbol{v})$. It can clearly be seen that the total spatial variation is dominated by the spatially structured component. However, the unstructured component plays a nonnegligible role, as the total pattern is slightly rougher (range 0.34 to 5.2) than the pattern of the spatially structured component alone (range 0.43 to 4.4).

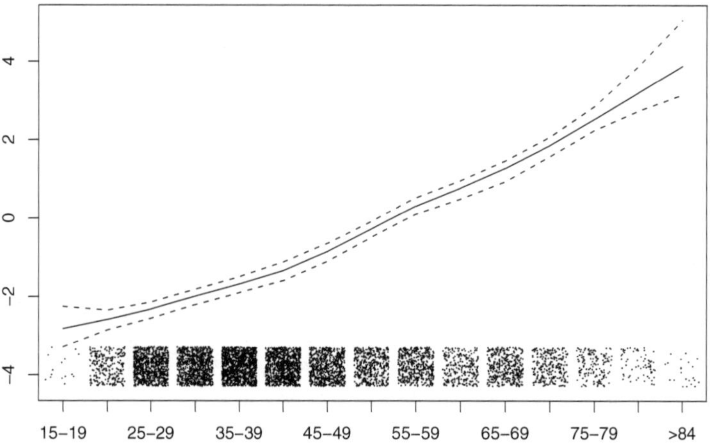

Figure 4.12 *Nonparametric effect of age group. Posterior median of the log odds within 2.5 and 97.5% quantiles. The distribution of the observed covariate is indicated with jittered dots.*

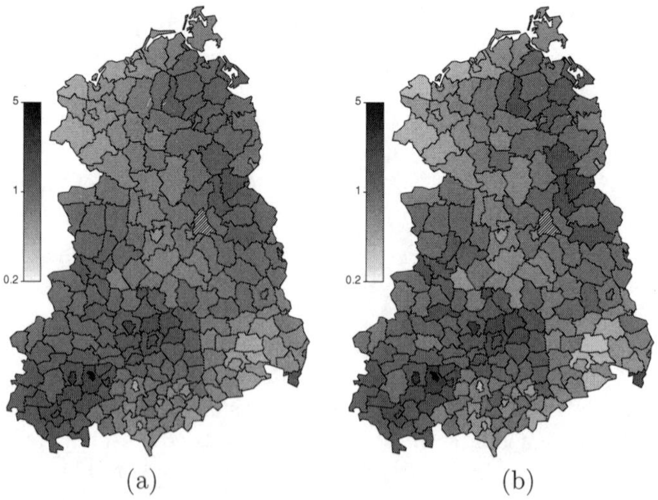

Figure 4.13 *Estimated odds ratio (posterior median) for (a) the spatially structured component u_i and (b) the sum of the spatially structured and unstructured variable $u_i + v_i$. The shaded region is West Berlin.*

The estimates are similar to the results obtained by Knorr-Held et al. (2002), who analyzed the corresponding data from 1975. Incidentally, the intercept α, which is not of central interest, has a posterior median of 0.79 with 95% credible interval $[0.62, 1.00]$.

4.4 Nonnormal response models

We will now look at hierarchical GMRF models where the likelihood is nonnormal. We have seen in the previous section that the binomial model with probit and logit link can be reformulated using auxiliary variables so that the full conditional of the latent GMRF is still a GMRF. However, in other situations such an augmentation of the parameter space is not possible. In these cases, it is useful to approximate the log likelihood using a second-order Taylor expansion and to use this GMRF approximation as a Metropolis-Hastings proposal in an MCMC algorithm. A natural choice is to expand around the current state of the Markov chain. However, the corresponding approximation can be improved by repeating this process and expanding around the mean of this approximation to generate an improved approximation. Such a strategy will ultimately converge (under some regularity conditions), so the mean of the approximation equals the mode of the full conditional. The method has thus much in common with algorithms to calculate the maximum likelihood estimator in generalized linear models, such as Fisher scoring or iteratively reweighted least squares.

4.4.1 The GMRF approximation

The univariate case

The approach taken is best described by a simple example. Suppose there is only one observation y from a Poisson distribution with mean λ. Suppose further that we place a normal prior on $\eta = \log\lambda$, $\eta \sim \mathcal{N}(\mu, \kappa^{-1})$, so the posterior distribution $\pi(\eta|y)$ is

$$\begin{aligned} \pi(\eta \mid y) &\propto \pi(\eta)\,\pi(y \mid \eta) \qquad (4.24)\\ &= \exp\left(-\frac{\kappa}{2}(\eta-\mu)^2 + y\eta - \exp(\eta)\right) = \exp(f(\eta)), \end{aligned}$$

say. In order to approximate $f(\eta)$, a common approach is to construct a quadratic Taylor expansion of the (unnormalized) log-posterior $f(\eta)$ around a suitable value η_0,

$$\begin{aligned} f(\eta) &\approx f(\eta_0) + f'(\eta_0)(\eta-\eta_0) + \frac{1}{2}f''(\eta_0)(\eta-\eta_0)^2 \\ &= a + b\eta - \frac{1}{2}c\eta^2. \end{aligned}$$

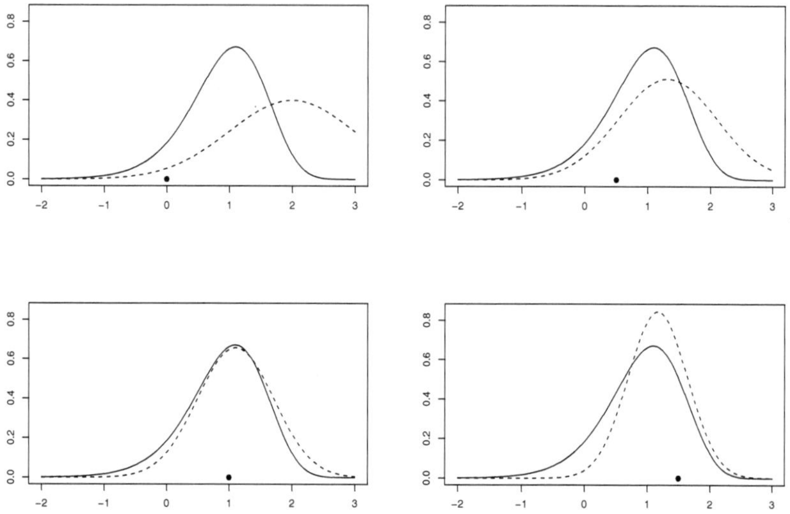

Figure 4.14 *Normal approximation (dashed line) of the posterior density (4.24) (solid line) for* $y = 3$, $\mu = 0$ *and* $\kappa = 0.001$ *based on a quadratic Taylor expansion around* η_0 *for* $\eta_0 = 0, 0.5, 1, 1.5$. *The value of* η_0 *is indicated with a small dot in each plot.*

Here, $b = f'(\eta_0) - f''(\eta_0)\eta_0$ and $c = -f''(\eta_0)$. The value of a is not relevant for the following.

We can now approximate $\pi(\eta|y)$ by $\widetilde{\pi}(\eta|y)$ where

$$\widetilde{\pi}(\eta \mid y) \propto \exp\left(-\frac{1}{2}c\eta^2 + b\eta\right), \tag{4.25}$$

which is in the form of the canonical parametrization $\mathcal{N}_C(b, c)$. Hence $\widetilde{\pi}(\eta|y)$ is a normal distribution with mean $\mu_1(\eta_0) = b/c$ and precision $\kappa_1(\eta_0) = c$. We explicitly emphasize the dependence of μ_1 and κ_1 on η_0 and that all that enters in this construction is the value of $f(\eta)$ and its first and second derivative at η_0. Figure 4.14 illustrates this approximation for $y = 3$, $\mu = 0$, $\kappa = 0.001$, and $\eta_0 = 0, 0.5, 1$, and 1.5. One can clearly see that the approximation is better, the closer η_0 is to the mode of $\pi(\eta|y)$.

The idea is now to use this normal distribution as a proposal distribution in a Metropolis-Hastings step. For η_0 we may simply take the current value of the simulated Markov chain. More specifically, we use a proposal distribution $q(\eta^*|\eta_0)$, which is normal with mean $\mu_1(\eta_0)$ and precision $\kappa_1(\eta_0)$. Following the Metropolis-Hastings algorithm, this proposal is then accepted with probability

$$\alpha = \min\left\{1, \frac{\pi(\eta^*|y)}{\pi(\eta_0|y)} \frac{q(\eta_0|\eta^*)}{q(\eta^*|\eta_0)}\right\}. \tag{4.26}$$

Note that this involves not only the evaluation of $q(\eta^*|\eta_0)$, which is available as a by-product of the construction of the proposal, but also the evaluation of $q(\eta_0|\eta^*)$. Thus, we also have to construct a *second* quadratic approximation around the *proposed* value η^* and evaluate the density of the corresponding normal distribution at the current point η_0 in order to evaluate $q(\eta_0|\eta^*)$ and α.

It is instructive to consider this algorithm in the case where $y|\eta$ is not Poisson, but normal with unknown mean η and known precision λ,

$$\pi(\eta \mid y) \propto \exp\left(-\frac{1}{2}(\lambda + \kappa)\eta^2 + (\lambda y + \kappa\mu)\eta\right).$$

Thus, $\pi(\eta|y)$ is already in the quadratic form of (4.25) and hence normal with mean $(\lambda y + \kappa\mu)/(\lambda + \kappa)$ and precision $(\lambda + \kappa)$. A quadratic approximation to $\log \pi(\eta|y)$ at any point η_0 will just reproduce the same quadratic function independent of η_0. Using this distribution as a proposal distribution leads to $\alpha = 1$ in (4.26). Thus, if $\log \pi(y|\eta)$ is already a quadratic function the algorithm outlined above leads to the mean/mode of $\pi(\eta|y)$ in one step, independent of η_0.

In the more general setting it is well known that iterated applications of such quadratic approximations converge (under regularity conditions) to the mode of $\pi(\eta|y)$. More specifically, we set $\eta_1 = \mu_1$ and repeat the quadratic approximation process to calculate $\mu_2(\eta_1)$ and $\kappa_2(\eta_1)$, then set $\eta_2 = \mu_2(\eta_1)$ and repeat this process until convergence. This algorithm is in fact just the well-known Newton-Raphson algorithm; a slightly modified version is known in statistics as Fisher scoring or iteratively reweighted least squares.

It is seen from Figure 4.14 that the normal approximation to the density $\pi(\eta|y)$ improves, the closer η_0 is to the mode of $\pi(\eta|y)$. This suggests that one may apply iterative quadratic approximations to the posterior density until convergence in order to obtain a normal proposal with mean equal to the posterior mode. Such a proposal should have relatively large acceptance rates and will also be independent of η_0.

On the other hand, it is desirable to avoid iterative algorithms within iterative algorithms such as the Metropolis-Hastings algorithm, as this is likely to slow down the speed of the Metropolis-Hastings algorithm. Nevertheless, it can be advantageous to apply the quadratic approximation not only once, but twice or perhaps even more, in order to improve the approximation of the proposal density to the posterior density. Note that we have also to repeat constructing these iterated approximations around the proposed value η^* in order to evaluate $q(\eta_0|\eta^*)$.

Generalization to the multivariate case

The idea described in the previous section can easily be generalized to a multivariate setting. Suppose for simplicity that there are n conditionally independent observations $y_1, \ldots, y_n$ from a nonnormal distribution where y_i is an indirect observation of x_i. Here $\boldsymbol{x}$ is a GMRF with precision matrix $\boldsymbol{Q}$ and mean $\boldsymbol{\mu}$ possibly depending on further hyperparameters. The full conditional $\pi(\boldsymbol{x}|\boldsymbol{y})$ is then

$$\pi(\boldsymbol{x} \mid \boldsymbol{y}) \propto \exp\left(-\frac{1}{2}(\boldsymbol{x}-\boldsymbol{\mu})^T\boldsymbol{Q}(\boldsymbol{x}-\boldsymbol{\mu}) + \sum_{i=1}^{n} \log \pi(y_i \mid x_i)\right).$$

We now use a second-order Taylor expansion of $\sum_{i=1}^n \log \pi(y_i|x_i)$ around $\boldsymbol{\mu}_0$, say, to construct a suitable GMRF proposal density $\widetilde{\pi}(\boldsymbol{x}|\boldsymbol{y})$. To be specific,

$$\begin{aligned}\widetilde{\pi}(\boldsymbol{x} \mid \boldsymbol{y}) &\propto \exp\left(-\frac{1}{2}\boldsymbol{x}^T\boldsymbol{Q}\boldsymbol{x} + \boldsymbol{\mu}^T\boldsymbol{Q}\boldsymbol{x} + \sum_i (a_i + b_i x_i - \frac{1}{2}c_i x_i^2)\right) \\ &\propto \exp\left(-\frac{1}{2}\boldsymbol{x}^T(\boldsymbol{Q} + \mathrm{diag}(\boldsymbol{c}))\boldsymbol{x} + (\boldsymbol{Q}\boldsymbol{\mu} + \boldsymbol{b})^T\boldsymbol{x}\right), \qquad (4.27)\end{aligned}$$

where c_i might be set to zero if it is negative. The canonical parameterization is

$$\mathcal{N}_C(\boldsymbol{Q}\boldsymbol{\mu} + \boldsymbol{b},\ \boldsymbol{Q} + \mathrm{diag}(\boldsymbol{c}))$$

with mean $\boldsymbol{\mu}_1$, say. The approximation depends on $\boldsymbol{\mu}_0$ as both $\boldsymbol{b}$ and $\boldsymbol{c}$ depend on $\boldsymbol{\mu}_0$. Similar to the univariate case, we can repeat this process and expand around $\boldsymbol{\mu}_1$ to improve the approximation. The improvement is due to $\boldsymbol{\mu}_1$ being closer to the mode of $\pi(\boldsymbol{x}|\boldsymbol{y})$ than $\boldsymbol{x}_0$. This is (most) often the case as $\boldsymbol{\mu}_1$ is one step of the multivariate Newton-Raphson method to locate the mode of $\pi(\boldsymbol{x}|\boldsymbol{y})$. After m iterations when $\boldsymbol{\mu}_m$ equals the mode of $\pi(\boldsymbol{x}|\boldsymbol{y})$, we denote the approximation $\widetilde{\pi}(\boldsymbol{x}|\boldsymbol{y})$ as *the GMRF approximation*.

An important feature of (4.27) is that the GMRF approximation inherits the Markov property of the prior on $\boldsymbol{x}$, which is very useful for MCMC simulation. A closer look at (4.27) reveals that this is because y_i depends on x_i only. If covariates $\boldsymbol{z}_i$ are also included, then y_i will typically depend on $\boldsymbol{z}_i^T\boldsymbol{x}$. In this case the Markov properties may or may not be inherited, depending on the $\boldsymbol{z}_i$'s. Since Taylor expansion is somewhat cumbersome in high dimensions, we will always try to parameterize to ensure that y_i depends only on x_i. We will illustrate this in the following examples.

There are also other strategies for locating the mode of $\log \pi(\boldsymbol{x}|\boldsymbol{y})$ than the Newton-Raphson method. Algorithms based on line search in a specific direction are often based on the gradient of $\log \pi(\boldsymbol{x}|\boldsymbol{y})$ or that part of the gradient orthogonal to previous search directions. Such approaches are particularly feasible as the gradient is

$$\boldsymbol{\nabla} \log \pi(\boldsymbol{x} \mid \boldsymbol{y}) = -(\boldsymbol{Q} + \operatorname{diag}(\boldsymbol{c}))\boldsymbol{x} + \boldsymbol{Q}\boldsymbol{\mu} + \boldsymbol{b}.$$

To evaluate the gradient involves potentially costly matrix-vector products like $\boldsymbol{Qx}$. However, this is computationally fast as it only requires $\mathcal{O}(n)$ flops for common GMRFs, see Algorithm B.1 in the Appendix. In our experience, line-search methods for optimizing $\log \pi(\boldsymbol{x}|\boldsymbol{y})$ should be preferred for huge GMRFs such as GMRFs for spatiotemporal problems, while the Newton-Raphson approach is preferred for smaller GMRFs. However, it is important to make all optimization methods robust as they need to be fail-safe when they are to be used within an MCMC algorithm. More details on robust strategies and constrained optimization for constrained GMRFs are available in the documentation of the `GMRFLib`-library described in Appendix B.

The Taylor expansion of $\log \pi(y_i|x_i)$ is most accurate at the point around which we expand. However, concerning an approximation to $\pi(\boldsymbol{x}|\boldsymbol{y})$ it is more important that the error is small in the region where the major part of the probability mass is. This motivates using numerical approximations to obtain the terms (a_i, b_i, c_i) in favor of analytical expressions of the Taylor expansion of $\log \pi(y_i|x_i)$. Let $f(\eta) = \log \pi(y_i|\eta)$ and η_0 be the point around which to construct an approximation. For $\delta > 0$ we use

$$\begin{aligned} c_i &= -\frac{f(\eta_0+\delta) - 2f(\eta_0) + f(\eta_0-\delta)}{\delta^2} \\ b_i &= \frac{f(\eta_0+\delta) - f(\eta_0-\delta)}{2\delta} - \eta_0 c_i \end{aligned}$$

and similarly for a_i. The value of δ should not be too small, for example, δ between 10^{-2} and 10^{-4}. This depends of course on the scale of x_i. These values of (a_i, b_i, c_i) ensure that the error in the approximation is zero not only at η_0, but also at $\eta_0 \pm \delta$. See also the discussion and the examples in Rue (2001).

In those cases where the GMRF approximation is not sufficiently accurate, we can go beyond the Gaussian and construct non-Gaussian approximations. We will return to this issue in Section 5.2.

Example: Revisiting Tokyo rainfall data

As a simple example, we now revisit the Tokyo rainfall data from Section 4.3.4. Assuming a logistic regression model, the likelihood for calendar

Number of iterations	Acceptance rates	Iterations per second
1	67.6%	64.8
2	82.3%	45.4
3	83.1%	34.8
until convergence	83.2%	33.1

Table 4.3 *Acceptance rates and number of iterations per second for Tokyo rainfall data*

day i equals

$$\pi(y_i \mid x_i) \propto \exp\left(x_i \sum_{j=1}^{m_i} y_{ij} - m_i \log(1 + \exp(x_i))\right).$$

We construct a GMRF approximation for the full conditional of $\boldsymbol{x}$ following the procedure in Section 4.4.1.

To illustrate the performance of the GMRF approximation, we have computed the acceptance rate and the speed of the MCMC algorithm for different values of the number of Newton-Raphson iterations to construct the GMRF approximation. When we stop the Newton-Raphson iterations before the mode is found, we obtain *an approximation* to the *GMRF approximation*. We have used an independence sampler for $(\kappa, \boldsymbol{x})$, which is described in Section 5.2.3, by first sampling κ from an approximation to the marginal posterior $\pi(\kappa|\boldsymbol{y})$, and then sampling from an approximation to $\pi(\boldsymbol{x}|\kappa, \boldsymbol{y})$. Finally, we accept/reject the two jointly. These number are reported in Table 4.3. It can be seen that one iteration already results in a very decent acceptance rate of 67.6%. For two iterations, the rate increases to 82.3% but the algorithm runs slower. The acceptance rates are only marginally larger if the proposal is defined exactly at the posterior mode. In this example, it seems not necessary to use more than two iterations. For comparison, we note that the auxiliary variable algorithms are considerably faster, with 153.6 (221.7) iterations per second on the same computer in the logit (probit) case.

4.4.2 Example: Joint disease mapping

In this final case study, we present a spatial analysis of oral cavity and lung cancer mortality rates in Germany, 1986–1990 (Held et al., 2004). The data y_{ij} are the number of cases during the 5-year period in area i for cancer type j, where $j = 1$ denotes oral cavity cancer while $j = 2$ denotes lung cancer. The number of cases in region i depends also on the number of people in that region, and their age distribution. The expected number of cases of disease j in region i is calculated based on

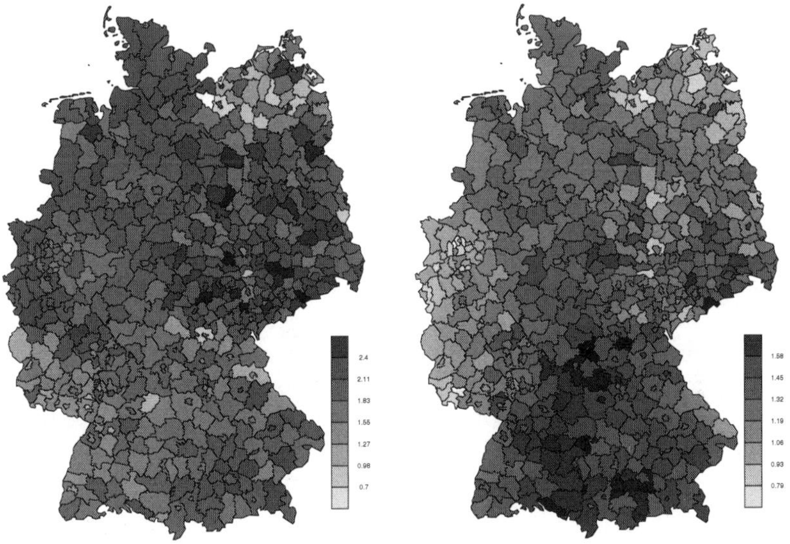

Figure 4.15 *The standardized mortality ratios for oral cavity cancer (left) and lung cancer (right) in Germany, 1986-1990.*

this information such that

$$\sum_i e_{ij} = \sum_i y_{ij}, \quad j = 1, 2$$

is fulfilled. Hence, we consider only the *relative risk*, not the absolute risk.

The common approach is to assume that the observed counts y_{ij} are conditionally independent Poisson observations,

$$y_{ij} \sim \mathcal{P}(e_{ij} \exp(\eta_{ij})),$$

where η_{ij} denotes the log relative risk in area i for disease j.

The standardized mortality ratios (SMRs) y_{ij}/e_{ij} are displayed in Figure 4.15 for both diseases. For some background information on why the SMRs are not suitable for estimating the relative risk, see, for example, Mollié (1996). We will first consider a model for a separate analysis of disease risk for each disease, then we will discuss a model for a joint analysis of both diseases.

Separate analysis of disease risk

We suppress the second index for simplicity. One of the most commonly used methods for a separate spatial analysis assumes that the log-relative risk $\boldsymbol{\eta}$ can be decomposed into

$$\boldsymbol{\eta} = \mu\mathbf{1} + \boldsymbol{u} + \boldsymbol{v}, \tag{4.28}$$

where μ is an intercept, $\boldsymbol{u}$ a *spatially structured* component, and $\boldsymbol{v}$ an *unstructured* component (Besag et al., 1991, Mollié, 1996). The intercept is often assumed to be zero mean normal with precision κ_μ. The unstructured component is typically modeled with independent zero mean normal variables with precision $\kappa_{\boldsymbol{v}}$, say. The spatially structured component $\boldsymbol{u}$ is typically an IGMRF of first order of the form (3.30) with precision $\kappa_{\boldsymbol{u}}$. Here we use the simple form without additional weights. A sum-to-zero restriction is placed on $\boldsymbol{u}$ to ensure identifiability of μ. If the prior on μ has zero precision, an equivalent model is to drop μ and the sum-to-zero restriction, but such a formulation is not so straightforward to generalize to a joint analysis of disease risk.

Let $\boldsymbol{\kappa} = (\kappa_{\boldsymbol{u}}, \kappa_{\boldsymbol{v}})^T$ denote the two unknown precisions while κ_μ is assumed to be fixed to some value, we simply set $\kappa_\mu = 0$. The posterior distribution is now

$$\begin{aligned}
\pi(\mu, \boldsymbol{u}, \boldsymbol{v}, \boldsymbol{\kappa} \mid \boldsymbol{y}) &\propto \exp\left(-\frac{\kappa_\mu}{2}\mu^2\right) \kappa_{\boldsymbol{v}}^{n/2} \exp\left(-\frac{\kappa_{\boldsymbol{v}}}{2}\sum_i v_i^2\right) \\
&\times \kappa_{\boldsymbol{u}}^{(n-1)/2} \exp\left(-\frac{\kappa_{\boldsymbol{u}}}{2}\sum_{i\sim j}(u_i - u_j)^2\right) \\
&\times \exp\left(\sum_i y_i(\mu + u_i + v_i) - e_i \exp(\mu + u_i + v_i)\right) \\
&\times \pi(\boldsymbol{\kappa}).
\end{aligned}$$

For $\pi(\boldsymbol{\kappa})$ we choose independent $\mathcal{G}(1, 0.01)$ priors for each of the two precisions. Note that the prior for $\boldsymbol{x} = (\mu, \boldsymbol{u}^T, \boldsymbol{v}^T)^T$ conditioned on $\boldsymbol{\kappa}$ is a GMRF. However, we see that y_i depends on the sum of three components of $\boldsymbol{x}$, due to (4.28). This is an implementation nuisance and can be solved here by reparameterization using $\boldsymbol{\eta}$ instead of $\boldsymbol{v}$,

$$\boldsymbol{\eta} \mid \boldsymbol{u}, \boldsymbol{\kappa}, \mu \sim \mathcal{N}(\mu\mathbf{1} + \boldsymbol{u},\ \kappa_{\boldsymbol{v}}\boldsymbol{I}).$$

In the new parameterization $\boldsymbol{x} = (\mu, \boldsymbol{u}^T, \boldsymbol{\eta}^T)^T$ and the posterior distribution is now

$$\begin{aligned}
\pi(\boldsymbol{x}, \boldsymbol{\kappa} \mid \boldsymbol{y}) &\propto \kappa_{\boldsymbol{v}}^{n/2}\kappa_{\boldsymbol{u}}^{(n-1)/2} \exp\left(-\frac{1}{2}\boldsymbol{x}^T\boldsymbol{Q}\boldsymbol{x}\right) \\
&\times \exp\left(\sum_i y_i\eta_i - e_i\exp(\eta_i)\right) \pi(\boldsymbol{\kappa}),
\end{aligned}$$

where

$$\boldsymbol{Q} = \begin{pmatrix} \kappa_\mu + n\kappa_{\boldsymbol{v}} & \kappa_{\boldsymbol{v}}\boldsymbol{1}^T & -\kappa_{\boldsymbol{v}}\boldsymbol{1}^T \\ \kappa_{\boldsymbol{v}}\boldsymbol{1} & \kappa_{\boldsymbol{u}}\boldsymbol{R} + \kappa_{\boldsymbol{v}}\boldsymbol{I} & -\kappa_{\boldsymbol{v}}\boldsymbol{I} \\ -\kappa_{\boldsymbol{v}}\boldsymbol{1} & -\kappa_{\boldsymbol{v}}\boldsymbol{I} & \kappa_{\boldsymbol{v}}\boldsymbol{I} \end{pmatrix} \tag{4.29}$$

and $\boldsymbol{R}$ is the structure matrix for the IGMRF of first order

$$R_{ij} = \begin{cases} m_i & \text{if } i = j \\ -1 & \text{if } i \sim j \\ 0 & \text{otherwise,} \end{cases} \tag{4.30}$$

where m_i is the number of neighbors to i. Note that $\boldsymbol{Q}$ is a $2n+1 \times 2n+1$ matrix with $6n + \sum_i m_i$ nonzero off-diagonal terms. In this example $n = 544$. Further, the spatially structured term $\boldsymbol{u}$ has a sum-to-zero constraint, $\boldsymbol{1}^T\boldsymbol{u} = 0$.

We now apply the one-block algorithm and generate a joint proposal using

$$\begin{aligned} \kappa_{\boldsymbol{u}}^* &\sim q(\kappa_{\boldsymbol{u}}^* \mid \kappa_{\boldsymbol{u}}) \\ \kappa_{\boldsymbol{v}}^* &\sim q(\kappa_{\boldsymbol{v}}^* \mid \kappa_{\boldsymbol{v}}) \\ \boldsymbol{x}^* &\sim \widetilde{\pi}(\boldsymbol{x} \mid \boldsymbol{\kappa}^*, \boldsymbol{y}) \end{aligned}$$

and then accept/reject $(\boldsymbol{\kappa}^*, \boldsymbol{x}^*)$ jointly. We use the simple choice (4.13) to update the precisions but this can be improved upon. The GMRF approximation is quite accurate in this case, and the factor F in (4.13) is tuned to obtain an acceptance rate between 30 and 40%. The algorithm does about 14 iterations per second on a 2.6-MHz processor. Since we approximately integrate out $\boldsymbol{x}$, the MCMC algorithm is essentially a simple Metropolis-Hastings algorithm for a two-dimensional density. We refer to Knorr-Held and Rue (2002) for a thorough discussion of constructing MCMC algorithms doing block updating for these models.

Figure 4.16 displays the nonzero terms of the precision matrix (4.29) in (a), after reordering to reduce the number of fill-ins in (b) and the Cholesky triangle in (c). The structure of (4.29) is clearly recognized. The estimated relative risks (posterior median) for the spatial component $\exp(\boldsymbol{u})$ are displayed in Figure 4.17 where (a) displays the results for lung cancer and (b) for oral cavity cancer.

If we choose a diffuse prior for μ using $\kappa_\mu = 0$, some technical issues appear using the GMRF approximation. Note that the full conditional of $\boldsymbol{x}$ and the GMRF approximation are both proper due to the constraint. However, since we are using Algorithm 2.6 to correct for the constraint $\boldsymbol{1}^T\boldsymbol{u} = 0$, we must require the GMRF approximation to be proper also *without* the constraint. This can be accomplished by modifying the GMRF approximation, either using a small positive value for κ_μ or to add a small positive constant to the diagonal of $\boldsymbol{R}$. The last option is

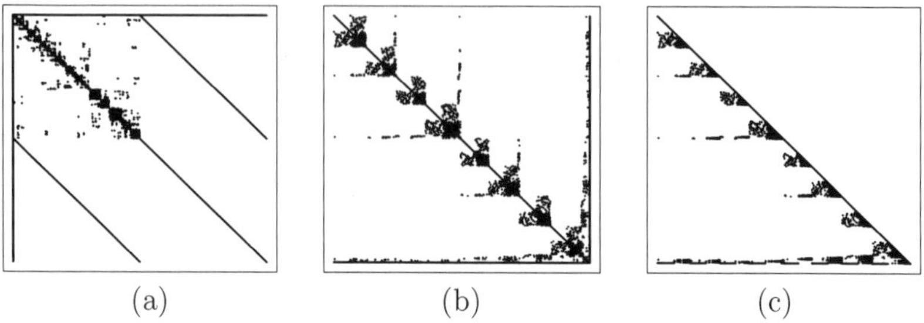

Figure 4.16 *(a) The precision matrix (4.29) in the original ordering, and (b) the precision matrix after appropriate reordering to reduce to number of nonzero terms in the Cholesky triangle shown in (c). Only the nonzero terms are shown and those are indicated by a dot.*

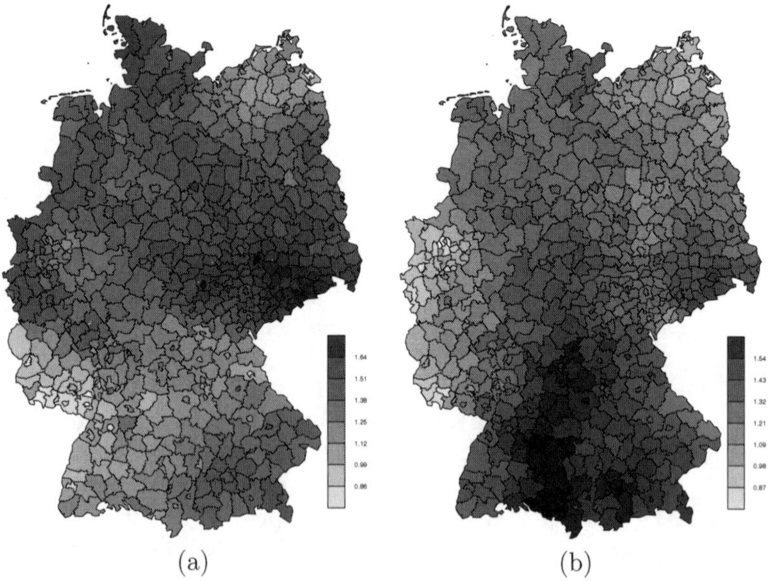

Figure 4.17 *Estimated relative risks (posterior median) of the spatial component* $\exp(\boldsymbol{u})$ *for (a) lung cancer and (b) oral cavity cancer.*

justified by (3.37). The acceptance probability is of course evaluated using this modified GMRF approximation, so the stationary limit of the MCMC algorithm is unaltered.

Joint analysis of disease risk

A natural extension of a separate analysis is to consider a joint analysis of two or more diseases (Held et al., 2004, Knorr-Held and Best, 2001,

Knorr-Held and Rue, 2002). In this example we will consider a joint analysis of oral cavity cancer and lung cancer shown in Figure 4.15. We assume that there is a latent spatial component $\boldsymbol{u}_1$, shared by both diseases and again modeled through an IGMRF. An additional unknown scale parameter $\delta > 0$ is included to allow for a different *risk gradient* of the shared component for the two diseases. A further spatial component $\boldsymbol{u}_2$ may enter for one of the diseases, which we again model through an IGMRF. In our setting, one of the diseases is oral cavity cancer, while the second one is lung cancer. Both are known to be related to tobacco smoking, but only oral cancer is known to be related to alcohol consumption. Considering the latent components as unobserved covariates, representing the two main risk factors tobacco ($\boldsymbol{u}_1$) and alcohol ($\boldsymbol{u}_2$), it is natural to include $\boldsymbol{u}_2$ only for oral cancer. More specifically we assume that

$$\begin{aligned}\boldsymbol{\eta}_1 \mid \boldsymbol{u}_1, \boldsymbol{u}_2, \boldsymbol{\mu}, \boldsymbol{\kappa} &\sim \mathcal{N}(\mu_1 \mathbf{1} + \delta \boldsymbol{u}_1 + \boldsymbol{u}_2,\ \kappa_{\boldsymbol{\eta}_1}^{-1} \boldsymbol{I}) \\ \boldsymbol{\eta}_2 \mid \boldsymbol{u}_1, \boldsymbol{u}_2, \boldsymbol{\mu}, \boldsymbol{\kappa} &\sim \mathcal{N}(\mu_2 \mathbf{1} + \delta^{-1} \boldsymbol{u}_1,\ \kappa_{\boldsymbol{\eta}_2}^{-1} \boldsymbol{I}),\end{aligned}$$

where $\kappa_{\boldsymbol{\eta}_1}$ and $\kappa_{\boldsymbol{\eta}_2}$ are the unknown precisions of $\boldsymbol{\eta}_1$ and $\boldsymbol{\eta}_2$, respectively. Additionally, we impose sum-to-zero constraints for both $\boldsymbol{u}_1$ and $\boldsymbol{u}_2$.

Assuming $\boldsymbol{\mu}$ is zero mean normal with covariance matrix $\kappa_{\boldsymbol{\mu}}^{-1}\boldsymbol{I}$, then $\boldsymbol{x} = (\boldsymbol{\mu}^T, \boldsymbol{u}_1^T, \boldsymbol{u}_2^T, \boldsymbol{\eta}_1^T, \boldsymbol{\eta}_2^T)^T$ is *a priori* a GMRF. The posterior distribution is

$$\begin{aligned}\pi(\boldsymbol{x}, \boldsymbol{\kappa}, \delta \mid \boldsymbol{y}) &\propto \kappa_{\boldsymbol{\eta}_1}^{n/2} \kappa_{\boldsymbol{\eta}_2}^{n/2} \kappa_{\boldsymbol{u}_1}^{(n-1)/2} \kappa_{\boldsymbol{u}_2}^{(n-1)/2} \exp\left(-\frac{1}{2}\boldsymbol{x}^T \boldsymbol{Q} \boldsymbol{x}\right) \\ &\times \exp\left(\sum_{j=1}^{2}\sum_{i=1}^{n} y_{ij}\eta_{ij} - e_{ij}\exp(\eta_{ij})\right) \\ &\times \pi(\boldsymbol{\kappa})\, \pi(\delta),\end{aligned}$$

where

$$\boldsymbol{Q} = \begin{pmatrix} \boldsymbol{Q}_{\boldsymbol{\mu}\boldsymbol{\mu}} & \boldsymbol{Q}_{\boldsymbol{\mu}\boldsymbol{u}_1} & \boldsymbol{Q}_{\boldsymbol{\mu}\boldsymbol{u}_2} & \boldsymbol{Q}_{\boldsymbol{\mu}\boldsymbol{\eta}_1} & \boldsymbol{Q}_{\boldsymbol{\mu}\boldsymbol{\eta}_2} \\ & \boldsymbol{Q}_{\boldsymbol{u}_1\boldsymbol{u}_1} & \boldsymbol{Q}_{\boldsymbol{u}_1\boldsymbol{u}_2} & \boldsymbol{Q}_{\boldsymbol{u}_1\boldsymbol{\eta}_1} & \boldsymbol{Q}_{\boldsymbol{u}_1\boldsymbol{\eta}_2} \\ & & \boldsymbol{Q}_{\boldsymbol{u}_2\boldsymbol{u}_2} & \boldsymbol{Q}_{\boldsymbol{u}_2\boldsymbol{\eta}_1} & \boldsymbol{Q}_{\boldsymbol{u}_2\boldsymbol{\eta}_2} \\ & \text{sym.} & & \boldsymbol{Q}_{\boldsymbol{\eta}_1\boldsymbol{\eta}_1} & \boldsymbol{Q}_{\boldsymbol{\eta}_1\boldsymbol{\eta}_2} \\ & & & & \boldsymbol{Q}_{\boldsymbol{\eta}_2\boldsymbol{\eta}_2} \end{pmatrix}, \tag{4.31}$$

$\mathbf{1}^T\boldsymbol{u}_1 = 0$ and $\mathbf{1}^T\boldsymbol{u}_2 = 0$. Defining the two $n \times 2$ matrices $\boldsymbol{C}_1 = (\mathbf{1}\ \mathbf{0})$ and $\boldsymbol{C}_2 = (\mathbf{0}\ \mathbf{1})$, the elements of $\boldsymbol{Q}$ are as follows:

$$\begin{aligned}\boldsymbol{Q}_{\boldsymbol{\mu}\boldsymbol{\mu}} &= \kappa_{\boldsymbol{\mu}}\boldsymbol{I} + \kappa_{\boldsymbol{\eta}_1}\boldsymbol{C}_1^T\boldsymbol{C}_1 + \kappa_{\boldsymbol{\eta}_2}\boldsymbol{C}_2^T\boldsymbol{C}_2 \\ \boldsymbol{Q}_{\boldsymbol{\mu}\boldsymbol{u}_1} &= \kappa_{\boldsymbol{\eta}_1}\delta\,\boldsymbol{C}_1^T \\ \boldsymbol{Q}_{\boldsymbol{\mu}\boldsymbol{u}_2} &= \kappa_{\boldsymbol{\eta}_1}\boldsymbol{C}_1^T + \kappa_{\boldsymbol{\eta}_2}\delta^{-1}\boldsymbol{C}_2^T\end{aligned}$$

$$
\begin{aligned}
\boldsymbol{Q}_{\mu\eta_1} &= -\kappa_{\eta_1}\boldsymbol{C}_1^T \\
\boldsymbol{Q}_{\mu\eta_2} &= -\kappa_{\eta_2}\boldsymbol{C}_2^T \\
\boldsymbol{Q}_{\boldsymbol{u}_1\boldsymbol{u}_1} &= \kappa_{\boldsymbol{u}_1}\boldsymbol{R} + \kappa_{\boldsymbol{\eta}_1}\delta^2\boldsymbol{I} \\
\boldsymbol{Q}_{\boldsymbol{u}_1\boldsymbol{u}_2} &= \kappa_{\boldsymbol{\eta}_1}\delta\,\boldsymbol{I} \\
\boldsymbol{Q}_{\boldsymbol{u}_1\boldsymbol{\eta}_1} &= -\kappa_{\boldsymbol{\eta}_1}\delta\,\boldsymbol{I} \\
\boldsymbol{Q}_{\boldsymbol{u}_1\boldsymbol{\eta}_2} &= \boldsymbol{0} \\
\boldsymbol{Q}_{\boldsymbol{u}_2\boldsymbol{u}_2} &= \kappa_{\boldsymbol{u}_2}\boldsymbol{R} + \kappa_{\boldsymbol{\eta}_1}\boldsymbol{I} + \kappa_{\boldsymbol{\eta}_2}\delta^{-2}\boldsymbol{I} \\
\boldsymbol{Q}_{\boldsymbol{u}_2\boldsymbol{\eta}_1} &= -\kappa_{\boldsymbol{\eta}_1}\boldsymbol{I} \\
\boldsymbol{Q}_{\boldsymbol{u}_2\boldsymbol{\eta}_2} &= -\kappa_{\boldsymbol{\eta}_2}\delta^{-1}\boldsymbol{I} \\
\boldsymbol{Q}_{\boldsymbol{\eta}_1\boldsymbol{\eta}_1} &= \kappa_{\boldsymbol{\eta}_1}\boldsymbol{I} \\
\boldsymbol{Q}_{\boldsymbol{\eta}_1\boldsymbol{\eta}_2} &= \boldsymbol{0} \\
\boldsymbol{Q}_{\boldsymbol{\eta}_2\boldsymbol{\eta}_2} &= \kappa_{\boldsymbol{\eta}_2}\boldsymbol{I}.
\end{aligned}
$$

Here, $\boldsymbol{R}$ is the structure matrix of the IGMRF (4.30). We assign a $\mathcal{N}(0, 0.17^2)$ prior on $\log\delta$ and independent $\mathcal{G}(1.0, 0.01)$ priors on all precisions.

There are now various ways to update $(\boldsymbol{x}, \boldsymbol{\kappa}, \delta)$. One option is to update all in one block, by first updating the hyperparameters $(\boldsymbol{\kappa}, \delta)$, then sample a proposal for $\boldsymbol{x}$ using the GMRF approximation and then accept/reject jointly. Although this is feasible, in this example it is both faster and sufficient to use subblocks. Those can be chosen as

$$
(\kappa_{\boldsymbol{\eta}_1}, \mu_1, \boldsymbol{\eta}_1),\ (\kappa_{\boldsymbol{\eta}_2}, \mu_2, \boldsymbol{\eta}_2),\ (\delta, \kappa_{\boldsymbol{u}_1}, \boldsymbol{u}_1),\ (\delta, \kappa_{\boldsymbol{u}_2}, \boldsymbol{u}_2),
$$

but other choices are possible. However, it is important to update each field jointly with its hyperparameter. Note that some variables can occur in more than one subblock, like δ in this example. We might also choose to merge the two last subblocks into

$$
(\kappa_{\boldsymbol{u}_1}, \kappa_{\boldsymbol{u}_2}, \delta, \boldsymbol{\mu}, \boldsymbol{u}_1, \boldsymbol{u}_2).
$$

This is justified as the full conditional of $(\boldsymbol{\mu}^T, \boldsymbol{u}_1^T, \boldsymbol{u}_2^T)^T$ is a GMRF.

Figure 4.18 displays the nonzero terms of the precision matrix (4.31) in (a), after reordering to reduce the number of fill-ins in (b) and the Cholesky triangle $\boldsymbol{L}$ in (c). It took 0.03 seconds to reorder and factorize the matrix on a 2.6-MHz computer. The number of nonzero terms in $\boldsymbol{L}$ is 24 987 including 15 081 fill-ins.

The estimated shared component $\boldsymbol{u}_1$ and the component $\boldsymbol{u}_2$ only relevant for oral cancer are displayed in Figure 4.19. The two estimates display very different spatial patterns and reflect quite nicely known geographical differences in tobacco ($\boldsymbol{u}_1$) and alcohol ($\boldsymbol{u}_2$) consumptions. The posterior median for δ is estimated to be 0.70 with a 95% credible

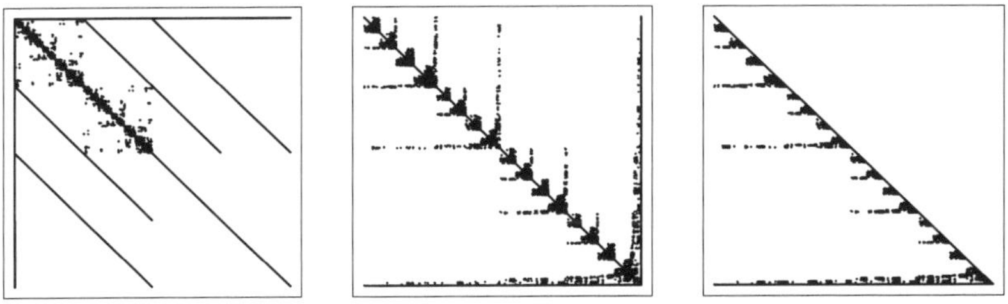

(a) (b) (c)

Figure 4.18 *(a) The precision matrix (4.31) in the original ordering, and (b) the precision matrix after appropriate reordering to reduce to number of nonzero terms in the Cholesky triangle shown in (c). Only the nonzero terms are shown and those are indicated by a dot.*

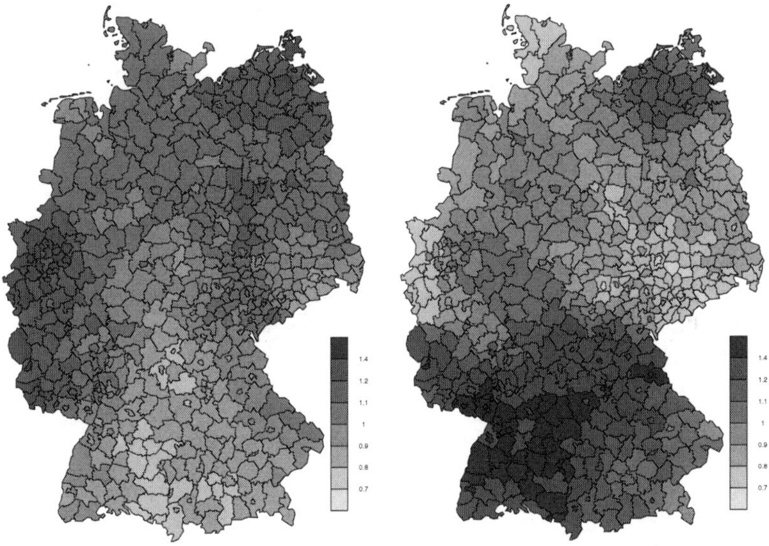

(a) (b)

Figure 4.19 *Estimated relative risks (posterior median) for (a) the shared component* $\exp(\boldsymbol{u}_1)$ *(related to tobacco) and (b) the oral-specific component* $\exp(\boldsymbol{u}_2)$ *(related to alcohol).*

interval [0.52, 0.88], suggesting that the shared component carries more weight for lung cancer than for oral cancer. For more details on this particular application we refer to Held et al. (2004), who also discuss the connection to ecological regression models and generalizations to more than two diseases.

4.5 Bibliographic notes

Our blocking strategy in hierarchical GMRF models is from Knorr-Held and Rue (2002). Pitt and Shephard (1999) discuss in great detail convergence and reparameterization issues for the first-order autoregressive process with normal observations also valid for the limiting RW1 model. Papaspiliopoulos et al. (2003) discuss reparameterization issues, and convergence for the two-block Gibbs sampler in normal hierarchical models extending previous results by Gelfand et al. (1995). Wilkinson (2003) comments on reparameterization issues and the joint update approach taken here. Gamerman et al. (2003) compare various block algorithms while Wilkinson and Yeung (2004) use the one-block approach for a normal response model. For theoretical results regarding blocking in MCMC, see Liu et al. (1994) and Roberts and Sahu (1997). Steinsland and Rue (2003) present an alternative approach that updates the GMRF in (4.8) as a sequence of overlapping blocks, a major benefit for large GMRFs. Barone and Frigessi (1989) studied overrelaxation MCMC methods for simulating from a GMRF, Barone et al. (2001) study the case of general overrelaxation combined with blocking while Barone et al. (2002) study the combination of coupling and overrelaxation.

The program `BayesX` is a (open source) software tool for performing complex Bayesian inference using GMRFs (among others) (Brezger et al., 2003). `BayesX` uses numerical methods for sparse matrices and block updates large blocks of the GMRF using the GMRF approximation for nonnormal data.

Other distributions that can be written as scale mixture of normals are the symmetric stable and the symmetric gamma distributions, see Devroye (1986) for the corresponding distribution of the mixing variables λ in each case. For more information on scale mixtures of normals, see Andrews and Mallows (1974), Barndorff-Nielsen et al. (1982), Kelker (1971) and Devroye (1986).

The hierarchical-t formulation is used particularly in state-space models (Carlin et al., 1992, Carter and Kohn, 1996) but also in spatial models, for example, in agricultural field experiments (Besag and Higdon, 1999). Additionally, one may also add a prior on the degrees of freedom ν (Besag and Higdon, 1999).

The use of GMRF priors in additive models has been proposed in

Fahrmeir and Lang (2001b), see also Fahrmeir and Knorr-Held (2000). Biller and Fahrmeir (1997) use priors as those discussed in Section 3.5. Similarly, GMRF priors can be used to model time-changing covariate effects and are discussed in Harvey (1989) and Fahrmeir and Tutz (2001), see also Fahrmeir and Knorr-Held (2000). Spatially varying covariate effects as proposed in Assunção et al. (1998) and further developed in Assunção et al. (2002) and Gamerman et al. (2003).

It is well known that the auxiliary variable approach is also useful for multicategorical response data. Consider first the case, where the response categories are ordered. Albert and Chib (1993) and Albert and Chib (2001) have shown how to use the auxiliary variable approach for the *cumulative probit* and *sequential probit* model. Similarly, it is straightforward to adopt the auxiliary variable approach for logistic regression to the cumulative and sequential model. For an introduction to these models see Fahrmeir and Tutz (2001, Chapter 3). For an application of such models with latent IGMRFs, but without auxiliary variables, see Fahrmeir and Knorr-Held (2000) and Knorr-Held et al. (2002). Turning to models for unordered response categories, the auxiliary variable approach can also be used in the multinomial probit model, as noted by Albert and Chib (1993), and in the multinomial logit model, as described in Holmes and Held (2003). For an application of the multinomial probit model using GMRF priors and auxiliary variables see Fahrmeir and Lang (2001c). The auxiliary variable approach does also extend to a certain class of non-Gaussian (intrinsic) MRFs such that the full conditional for the MRF is a GMRF, see Geman and Yang (1995).

The use of the algorithms to construct appropriate GMRF approximations in Section 4.4.1 was first advocated by Gamerman (1997) in the context of generalized linear mixed models. Follestad and Rue (2003) and Rue and Follestad (2003) discuss the construction of GMRF approximations using *soft constraints* (see Section 2.3.3) where aggregated Poisson depends on the sum of all relative risks in a region.

Various other examples of hierarchical models with latent GMRF components can also be found in Banerjee et al. (2004). Sun et al. (1999) considers propriety of the posterior distribution of hierarchical models using IGMRFs. Ferreira and De Oliveira (2004) discuss default/reference priors for parameters of GMRFs for use in an 'objective Bayesian analysis'.

CHAPTER 5

Approximation techniques

This chapter is reserved for the presentation of two recent developments regarding GMRFs. At the time of writing these new developments have not been fully explored but extend the range of applications regarding GMRFs and bring GMRFs into new areas.

Section 5.1 provides a link between GMRFs and Gaussian fields used in geostatistics. Commonly used Gaussian fields on regular lattices can be well approximated using GMRF models with a small neighborhood. The benefit of using GMRF approximations instead of Gaussian fields is purely computational, as efficient computation of GMRFs utilize algorithms for sparse matrices as discussed in Chapter 2.

The second topic is motivated by the extensive use of the GMRF approximation to the full conditional of the prior GMRF $\boldsymbol{x}$ in Chapter 4. The GMRF approximation is found by Taylor expanding to second order the nonquadratic terms around the mode. In those cases where this approximation is not sufficiently accurate, we might wish to 'go beyond the Gaussian' and construct non-Gaussian approximations. However, to apply non-Gaussian approximations in the setting of Chapter 4, we need to be able to sample exactly from the approximation and compute the normalizing constant. Despite these rather strict requirements, we will present a class of non-Gaussian approximations that satisfy these requirements and are adaptive in the sense that the approximation adapts itself to the particular full conditional for $\boldsymbol{x}$ under study.

5.1 GMRFs as approximations to Gaussian fields

This section discusses the link between Gaussian fields (GFs) used in geostatistics and GMRFs. We will restrict ourselves to isotropic Gaussian fields on regular lattices $\mathcal{I}_{\boldsymbol{n}}$. We will demonstrate that GMRFs with a local neighborhood can well approximate isotropic Gaussian fields with commonly used covariance functions, in the sense that each element in the covariance matrix of the GMRF is close to the corresponding element of the covariance matrix of the Gaussian field. This allows us to use GMRFs as approximations to Gaussian fields on regular lattices. The advantage is purely computational, but the speedup can be $\mathcal{O}(n^{3/2})$. This is our approach to solve (partially) what Banerjee et al. (2004) call the *big n problem*. Further, the Markov property of the GMRFs can be valuable for applying these models as a component in a larger complex model, especially if simulation-based methods for inference are used.

5.1.1 Gaussian fields

Let $\{z(s),\ s \in D\}$ be a stochastic process where $D \subset \mathbb{R}^d$ and $s \in D$ represents the location. In most applications, d is either 1, 2, or 3.

Definition 5.1 (Gaussian field) *The process* $\{z(s),\ s \in D\}$ *is a Gaussian field if for any* $k \geq 1$ *and any locations* $s_1, \ldots, s_k \in D$, $(z(s_1), \ldots, z(s_k))^T$ *is normally distributed. The* mean function *and* covariance function *(CF) of* z *are*

$$\mu(s) = E(z(s)), \qquad C(s, t) = Cov(z(s), z(t)),$$

which are both assumed to exist for all s *and* t. *The Gaussian field is* stationary *if* $\mu(s) = \mu$ *for all* $s \in D$ *and if the covariance function only depends on* $s - t$. *A stationary Gaussian field is called* isotropic *if the covariance function only depends on the Euclidean distance between* s *and* t, *i.e.,* $C(s, t) = C(h)$ *with* $h = \sqrt{||s - t||}$.

For any finite set of locations, a CF must necessarily induce a positive definite covariance matrix, i.e.,

$$\sum_i \sum_j a_i a_j C(s_i, s_j) > 0$$

must hold for any $k \geq 1$, any $s_1, \ldots, s_k$, and any (real) coefficients $a_1, \ldots, a_n$. If this is the case, then the CF is called *positive definite*. All continuous CFs on $\mathbb{R}^d$ can be represented as Fourier transforms of a finite measure, a rather deep result known as *Bockner's theorem*, see, for example, Cressie (1993). Bockner's theorem is often used as a remedy to construct CF or to verify that a (possible) CF is positive definite.

In most applications one of the following isotropic CFs is used in geostatistics:

Exponential	$C(h) = \exp(-3h)$	
Gaussian	$C(h) = \exp(-3h^2)$	
Powered exponential	$C(h) = \exp(-3h^\alpha), \quad 0 < \alpha \leq 2$	(5.1)
Matérn	$C(h) = \frac{1}{\Gamma(\nu)2^{\nu-1}}(s_\nu h)^\nu K_\nu(s_\nu h).$	(5.2)

Here K_ν is the modified Bessel function of the second kind and order $\nu > 0$, and s_ν is a function of ν such that the covariance function is scaled to $C(1) = 0.05$. For similar reasons the multiplicative factor 3 enters in the exponent of the exponential, Gaussian and powered exponential CF,

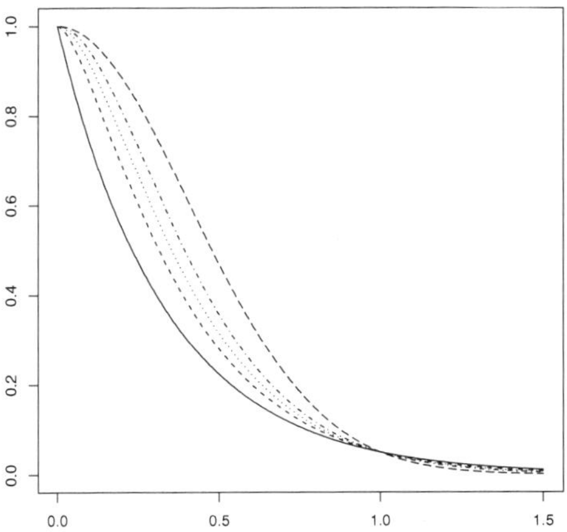

Figure 5.1 *The Matérn CF with range* $r = 1$, *and* $\nu = 1/2, 1, 3/2, 5/2$, *and* 100 *(from left to right). The case* $\nu = 1/2$ *corresponds to the exponential CF while* $\nu = 100$ *is essentially the Gaussian CF.*

where now $C(1) = 0.04979 \approx 0.05$. Note that $C(0) = 1$ in all four cases, so the CFs are also *correlation* functions. Of course, if we multiply $C(h)$ with σ^2 then the variance becomes σ^2.

The powered exponential CF includes both the exponential ($\alpha = 1$) and the Gaussian ($\alpha = 2$), and so does the often-recommended Matérn CF with $\nu = 1/2$ and $\nu \to \infty$, respectively. The Matérn CF is displayed in Figure 5.1 for $\nu = 1/2, 1, 3/2, 5/2$, and 100.

A further parameter r, called the *range*, is often introduced to scale the Euclidean distance, so the CF is $C(h/r)$. The range parameter can be interpreted as the (minimum) distance h for which the correlation function $C(h) = 0.05$. Hence, two locations more than distance r apart are essentially uncorrelated and hence also nearly independent.

Nonisotropic CFs can be constructed from isotropic CFs by replacing the Euclidean distance h between locations $\boldsymbol{s}$ and $\boldsymbol{t}$ with

$$h' = \sqrt{(\boldsymbol{t} - \boldsymbol{s})^T \boldsymbol{A} (\boldsymbol{t} - \boldsymbol{s})},$$

where $\boldsymbol{A}$ is SPD. This gives elliptical contours of the covariance function, whereas isotropic ones always have circular contours. The Euclidean distance is obtained if $\boldsymbol{A} = \boldsymbol{I}$.

5.1.2 Fitting GMRFs to Gaussian fields

We will now formulate how to fit GMRFs to Gaussian fields and then discuss in detail some practical and technical issues. We assume the Gaussian field is isotropic with zero mean and restricted to the torus $\mathcal{T}_{\boldsymbol{n}}$. The torus is preferred over the lattice for computational reasons *and* to make the GMRF stationary (see Section 2.6.3).

Let $\boldsymbol{z}$ denote a zero mean Gaussian field on $\mathcal{T}_{\boldsymbol{n}}$. We denote by $d((i,j),(i',j'))$ the Euclidean distance between z_{ij} and $z_{i'j'}$ on the torus. The covariance between z_{ij} and $z_{i'j'}$ is

$$\mathrm{Cov}(z_{ij}, z_{i'j'}) = C(d((i,j),(i',j'))/r),$$

where $C(\cdot/r)$ is one of the isotropic CFs presented in Section 5.1.1 and r is the range parameter. A value of $r = 10$ means that the range is 10 pixels. Denote by $\boldsymbol{\Sigma}$ the covariance matrix of $\boldsymbol{z}$.

Let $\boldsymbol{x}$ be a GMRF defined on $\mathcal{T}_{\boldsymbol{n}}$ with precision matrix $\boldsymbol{Q}(\boldsymbol{\theta})$ depending on some parameter vector $\boldsymbol{\theta} = (\theta_1, \theta_2, \ldots)^T$. Let $\boldsymbol{\Sigma}(\boldsymbol{\theta}) = \boldsymbol{Q}(\boldsymbol{\theta})^{-1}$ denote the covariance matrix of $\boldsymbol{x}$. Let $\pi(\boldsymbol{x};\boldsymbol{\theta})$ be the density of the GMRF and $\pi(\boldsymbol{z})$ the density of the Gaussian field we want to fit. The best fit is obtained using the 'optimal' parameter value

$$\boldsymbol{\theta}^* = \arg \min_{\boldsymbol{\theta} \in \boldsymbol{\Theta}_\infty^+} \mathcal{D}(\pi(\boldsymbol{x};\boldsymbol{\theta}), \pi(\boldsymbol{z})). \tag{5.3}$$

Here, $\mathcal{D}(\cdot,\cdot)$ is a metric or measure of discrepancy between the two densities and $\boldsymbol{\Theta}_\infty^+$ is the space of valid values of $\boldsymbol{\theta}$ as defined in (2.71). In practice, we face a trade-off problem between sparseness of $\boldsymbol{Q}(\boldsymbol{\theta})$ and how close the two densities are. If $\boldsymbol{\Sigma}(\boldsymbol{\theta}) = \boldsymbol{\Sigma}$ then the two densities are identical but $\boldsymbol{Q}(\boldsymbol{\theta})$ is typically a completely dense matrix. If $\boldsymbol{Q}(\boldsymbol{\theta})$ is (very) sparse, the fit may not be sufficiently accurate.

The fitted parameters may also depend on the size of the torus, which means that we may need a different set of parameters for each size. However, it will be demonstrated in Section 5.1.3 that the fitted parameters are nearly invariant to the size of the torus if the size is large enough compared to the range.

In order to obtain a working algorithm, we need to be more specific about the choice of the neighborhood, the parameterization, the distance measure and the numerical algorithms to compute the best fit.

Choosing the torus $\mathcal{T}_{\boldsymbol{n}}$ instead of the lattice $\mathcal{I}_{\boldsymbol{n}}$

The covariance matrix $\boldsymbol{\Sigma}$ and the precision matrix $\boldsymbol{Q}(\boldsymbol{\theta})$ are block-circulant matrices, see Section 2.6.3. We denote their bases by $\boldsymbol{\sigma}$ and $\boldsymbol{q}(\boldsymbol{\theta})$, respectively. Denote by $\boldsymbol{\sigma}(\boldsymbol{\theta})$ the base of $\boldsymbol{Q}(\boldsymbol{\theta})^{-1}$.

Numerical algorithms for block-circulant matrices are discussed in Section 2.6.2 and are all based on the discrete Fourier transform. For this reason, the size of the torus is selected as products of small primes. However, delicate technical issues appear if we define a Gaussian field on $\mathcal{T}_n$ using a CF valid on $\mathbb{R}^2$, as it might not be a valid CF on the torus. Wood (1995) discusses this issue in the one-dimensional case where a line is wrapped onto a circle and proves that the exponential CF will still be valid while the Gaussian CF will never be valid no matter the size of the circle compared to the range. However, considering only the CF restricted to a lattice, Dietrich and Newsam (1997) prove essentially, that if the size of the lattice is large enough compared to the range, then the (block) circulant matrix will be SPD. See also Grenander and Szegö (1984) and Gray (2002) for conditions when the (block) Toeplitz matrix and its circulant embedding are asymptotically equivalent. The consequence is simply to choose the size of the torus large enough compared to the range.

Choice of neighborhood and parameterization

Regarding the neighborhood and parameterization, we will use a square window of size $(2m+1) \times (2m+1)$ centered at each (i, j), where $m = 2$ or 3. The reason to use a square neighborhood is purely numerical (see Section 2), as if $x_{i,j+2}$ is a neighbor to x_{ij} then the additional cost is negligible if we let also $x_{i+2,j+2}$ be a neighbor of x_{ij}. Hence, by choosing the square neighborhood we maximize the number of parameters, keeping the computational costs nearly constant. Further, since the CF is isotropic, we must impose similar symmetries regarding the precision matrix. For $m = 2$ this implies that $\mathrm{E}(x_{ij} | \boldsymbol{x}_{-ij})$ is parameterized with 6 parameters

$$-\frac{1}{\theta_1}\left(\theta_2 \begin{smallmatrix} \circ & \circ & \circ & \circ & \circ \\ \circ & \circ & \bullet & \circ & \circ \\ \circ & \bullet & \circ & \bullet & \circ \\ \circ & \circ & \bullet & \circ & \circ \\ \circ & \circ & \circ & \circ & \circ \end{smallmatrix} + \theta_3 \begin{smallmatrix} \circ & \circ & \circ & \circ & \circ \\ \circ & \bullet & \circ & \bullet & \circ \\ \circ & \circ & \circ & \circ & \circ \\ \circ & \bullet & \circ & \bullet & \circ \\ \circ & \circ & \circ & \circ & \circ \end{smallmatrix} + \theta_4 \begin{smallmatrix} \circ & \circ & \bullet & \circ & \circ \\ \circ & \circ & \circ & \circ & \circ \\ \bullet & \circ & \circ & \circ & \bullet \\ \circ & \circ & \circ & \circ & \circ \\ \circ & \circ & \bullet & \circ & \circ \end{smallmatrix} + \theta_5 \begin{smallmatrix} \circ & \bullet & \circ & \bullet & \circ \\ \bullet & \circ & \circ & \circ & \bullet \\ \circ & \circ & \circ & \circ & \circ \\ \bullet & \circ & \circ & \circ & \bullet \\ \circ & \bullet & \circ & \bullet & \circ \end{smallmatrix} + \theta_6 \begin{smallmatrix} \bullet & \circ & \circ & \circ & \bullet \\ \circ & \circ & \circ & \circ & \circ \\ \circ & \circ & \circ & \circ & \circ \\ \circ & \circ & \circ & \circ & \circ \\ \bullet & \circ & \circ & \circ & \bullet \end{smallmatrix}\right) \tag{5.4}$$

using the same notation as in Section 3.4.2, while

$$\mathrm{Prec}(x_{ij} \mid \boldsymbol{x}_{-ij}) = \theta_1.$$

However, we have to ensure that the corresponding precision matrix is SPD, see the discussion in Section 2.7.

We will later display the coefficients as

$$\theta_1 \begin{bmatrix} & & \theta_6/\theta_1 \\ & \theta_3/\theta_1 & \theta_5/\theta_1 \\ 1 & \theta_2/\theta_1 & \theta_4/\theta_1 \end{bmatrix} \tag{5.5}$$

representing the lower part of the upper right quadrant. For $m = 3$ we have 10 available parameters:

$$\theta_1 \begin{bmatrix} & & & \theta_{10}/\theta_1 \\ & & \theta_6/\theta_1 & \theta_9/\theta_1 \\ & \theta_3/\theta_1 & \theta_5/\theta_1 & \theta_8/\theta_1 \\ 1 & \theta_2/\theta_1 & \theta_4/\theta_1 & \theta_7/\theta_1 \end{bmatrix} \tag{5.6}$$

with obvious notation.

Choice of metric between the densities

When considering the choice of metric between the two densities, we note that we can restrict ourselves to the case where both densities have unit marginal precision, so $\text{Prec}(x_{ij}) = \text{Prec}(z_{ij}) = 1$ for all ij. We can obtain this situation if the CF satisfies $C(0) = 1$ and $\boldsymbol{\sigma}(\boldsymbol{\theta})$ is scaled so that element 00 is 1. The scaling makes one of the parameters redundant, hence we set $\theta_1 = 1$ without loss of generality. After the best fit is found, we compute the value of θ_1 giving unit marginal precision and scale the solution accordingly. The coefficients found in the minimization are those within the brackets in (5.5) and (5.6).

As both densities have unit marginal precision, the covariance matrix equals the correlation matrix. Let $\boldsymbol{\rho}$ be the base of the correlation matrix of the Gaussian field and $\boldsymbol{\rho}(\boldsymbol{\theta})$ the base of the correlation matrix of the GMRF. Since $\boldsymbol{\rho}$ uniquely determines the density of $\boldsymbol{z}$ and $\boldsymbol{\rho}(\boldsymbol{\theta})$ uniquely determines the density of $\boldsymbol{x}$, we can define the norm between the densities in (5.3) by a norm between the corresponding correlation matrices. We use the weighted 2-norm,

$$\|\boldsymbol{\rho} - \boldsymbol{\rho}(\boldsymbol{\theta})\|_{\boldsymbol{w}}^2 = \sum_{ij} (\rho_{ij} - \rho_{ij}(\boldsymbol{\theta}))^2 w_{ij}$$

with positive weights $w_{ij} > 0$ for all ij. The CF is isotropic and its value at (i,j) only depends on the distance to $(0,0)$. It is therefore natural to choose $w_{ij} \propto 1/d((i,j),(0,0)))$ for $ij \neq 00$. However, we will put slightly more weight on lags with distance close to the range, and use

$$w_{ij} \propto \begin{cases} 1 & \text{if } ij = 00 \\ \frac{1+r/d((i,j),(0,0))}{d((i,j),(0,0))} & \text{otherwise.} \end{cases}$$

The coefficients giving the best fit are then found as

$$\boldsymbol{\theta}^* = \arg \min_{\boldsymbol{\theta} \in \boldsymbol{\Theta}_\infty^+} \|\boldsymbol{\rho} - \boldsymbol{\rho}(\boldsymbol{\theta})\|_{\boldsymbol{w}}^2, \tag{5.7}$$

which has to be computed numerically.

Numerical optimization: computing the objective function and its gradient

Numerical optimization methods for solving (5.7) needs a function that evaluates the objective function

$$U(\boldsymbol{\theta}) = \|\boldsymbol{\rho} - \boldsymbol{\rho}(\boldsymbol{\theta})\|_w^2$$

and if possible, the gradient

$$(\frac{\partial}{\partial \theta_2}, \frac{\partial}{\partial \theta_3}, \ldots)^T U(\boldsymbol{\theta}). \tag{5.8}$$

In (5.8), recall that we have fixed $\theta_1 = 1$. Both the objective function and its gradient ignore the constraint that $\boldsymbol{\theta} \in \boldsymbol{\Theta}_\infty^+$. We will return to this issue shortly.

The objective function and its gradient can be computed in $\mathcal{O}(n \log n)$ flops using algorithms for block-circulant matrices based on the discrete Fourier transform as discussed in Section 2.6.2. The details are as follows. The base $\boldsymbol{\rho}$ is determined by the target correlation function of the GF. The base $\boldsymbol{\rho}(\boldsymbol{\theta})$ is computed using (2.50),

$$\boldsymbol{\sigma}(\boldsymbol{\theta}) = \frac{1}{n_1 n_2} \text{IDFT2}(\text{DFT2}(\boldsymbol{q}(\boldsymbol{\theta}) \oslash (-1))),$$

which we need to scale to obtain unit precision:

$$\boldsymbol{\rho}(\boldsymbol{\theta}) = \boldsymbol{\sigma}(\boldsymbol{\theta}) / \sigma_{00}(\boldsymbol{\theta}). \tag{5.9}$$

Here, '$\oslash$' is defined in Section 2.1.1 and $\sigma_{00}(\boldsymbol{\theta})$ is the element 00 of $\boldsymbol{\sigma}(\boldsymbol{\theta})$. The objective function is then

$$U(\boldsymbol{\theta}) = \sum_{ij} \left(\rho_{ij} - \rho_{ij}(\boldsymbol{\theta})\right)^2 w_{ij}. \tag{5.10}$$

Because the CF is defined on a torus and assumed to be isotropic, we only need to sum over 1/8 of all indices in (5.10).

The computation of the gradient (5.8) is somewhat more involved. Let $\boldsymbol{A}$ be a nonsingular matrix depending on a parameter α. Since $\boldsymbol{A}\boldsymbol{A}^{-1} = \boldsymbol{I}$ it follows from the product rule that

$$\left(\frac{\partial}{\partial \alpha}\boldsymbol{A}\right)\boldsymbol{A}^{-1} + \boldsymbol{A}\left(\frac{\partial}{\partial \alpha}(\boldsymbol{A}^{-1})\right) = \boldsymbol{0},$$

where

$$\frac{\partial}{\partial \alpha}\boldsymbol{A} = \left(\tfrac{\partial}{\partial \alpha} A_{ij}\right).$$

So

$$\frac{\partial}{\partial \theta_k}\boldsymbol{\Sigma}(\boldsymbol{\theta}) = -\boldsymbol{\Sigma}(\boldsymbol{\theta})\left(\frac{\partial}{\partial \theta_k}\boldsymbol{Q}(\boldsymbol{\theta})\right)\boldsymbol{\Sigma}(\boldsymbol{\theta})$$

is a block-circulant matrix with base $\frac{\partial}{\partial\theta_k}\boldsymbol{\sigma}(\boldsymbol{\theta})$ equal to

$$\frac{1}{n_1 n_2}\text{IDFT2}\left(-\left(\text{DFT2}(\boldsymbol{q}(\boldsymbol{\theta}))\mathbin{\bigcirc\!\!\!\!\wedge} -2\right)\odot \text{DFT2}\left(\frac{\partial}{\partial\theta_k}\boldsymbol{q}(\boldsymbol{\theta})\right)\right). \tag{5.11}$$

Recall that '$\odot$' is elementwise multiplication, see Section 2.1.1. Eq. (5.11) is easily derived using the techniques presented in Section 2.6.2. Note that the base $\partial q(\boldsymbol{\theta})/\partial\theta_k$ only contains zeros and ones. More specifically, for $m = 2$ the base contain four 1's for $k \in \{2, 3, 4, 6\}$ and eight 1's for $k = 5$, see (5.4). From (5.9) we obtain

$$\frac{\partial}{\partial\theta_k}\boldsymbol{\rho}(\boldsymbol{\theta}) = \frac{1}{\sigma_{00}(\boldsymbol{\theta})}\left(\frac{\partial}{\partial\theta_k}\boldsymbol{\sigma}(\boldsymbol{\theta})\right) - \frac{1}{\sigma_{00}(\boldsymbol{\theta})^2}\frac{\partial\sigma_{00}(\boldsymbol{\theta})}{\partial\theta_k}\boldsymbol{\sigma}(\boldsymbol{\theta});$$

hence

$$\frac{\partial}{\partial\theta_k}U(\boldsymbol{\theta}) = -2\sum_{ij} w_{ij}\left(\rho_{ij} - \rho_{ij}(\boldsymbol{\theta})\right)\frac{\partial\rho_{ij}(\boldsymbol{\theta})}{\partial\theta_k}.$$

We perform the optimization in `R` (Ihaka and Gentleman, 1996) using the routine `optim` and the option `BFGS`. `optim` is a quasi-Newton method also known as a variable metric algorithm. The algorithm uses function values and gradients to build up a picture of the surface to be optimized. A further advantage is that the implementation allows for returning a nonvalid function value if the parameters are outside the valid area, i.e., if some of the eigenvalues of $\boldsymbol{Q}(\boldsymbol{\theta})$ are negative.

As discussed in Section 2.7, it is hard to determine $\boldsymbol{\Theta}^+_\infty$. Our strategy is to replace $\boldsymbol{\Theta}^+_\infty$ by $\boldsymbol{\Theta}^+_{\boldsymbol{n}}$ where $\boldsymbol{n}$ is the size of the torus. After the solution is found, we will take a large $\boldsymbol{n}'$ and verify that $\boldsymbol{\theta}^* \in \boldsymbol{\Theta}^+_{\boldsymbol{n}'}$. If $\boldsymbol{\theta}^*$ passes this test (which is nearly always the case), we accept $\boldsymbol{\theta}^*$, otherwise we rerun the optimization with a larger $\boldsymbol{n}$.

5.1.3 *Results*

We will now present some typical results showing how well the fitted GMRFs approximate the CFs in Section 5.1. We concentrate on the exponential and Gaussian CF using both a 5×5 and 7×7 neighborhood with range 30 and 50. The size of the torus is taken as 512×512.

Figure 5.2(a) and Figure 5.2(c) shows the fit obtained for the exponential CF with range 30 using a 5×5 and 7×7 neighborhood, respectively. Figure 5.2(b) and Figure 5.2(d) show similar results for the Gaussian CF with range 50. The fitted CF is drawn with a solid line, while the target CF is drawn with a dashed line, while the difference between the two is shown in Figure 5.3.

The approximation obtained is quite accurate. For the exponential CF the absolute difference is less than 0.01 using a 5×5 neighborhood, while it is less than 0.005 using a 7×7 neighborhood. The Gaussian CF

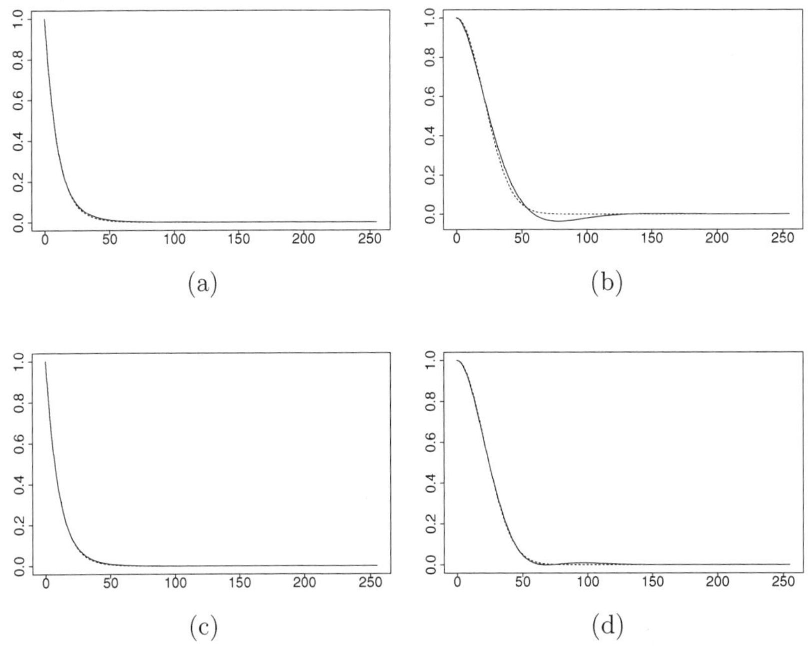

Figure 5.2 *The figures display the correlation function (CF) for the fitted GMRF (solid line) and the target CF (dashed line) with the following parameters: (a) exponential CF with range* 30 *and a* 5×5 *neighborhood, (b) Gaussian CF with range* 50 *and a* 5×5 *neighborhood, (c) exponential CF with range* 30 *and a* 7×7 *neighborhood, and (d) Gaussian CF with range* 50 *and a* 7×7 *neighborhood.*

is more difficult to fit, which is due to the CF type, not the increase in the range. In order to fit the CF accurately for small lags, the fitted correlation needs to be negative for larger lags. However, the absolute difference is still reasonably small and about 0.04 and 0.008 for the 5×5 and 7×7 neighborhood, respectively. The improvement by enlarging the neighborhood is larger for the Gaussian CF than for the exponential CF.

The results obtained in Figure 5.2 are quite typical for different range parameters and other CFs. For other values of the range, the shape of the fitted CF is about the same and only the horizontal scale is different (due to the different range). We do not present the fits using the powered exponential CF (5.1) for $1 \leq \alpha \leq 2$, and the Matérn CF (5.2), but they are also quite good. The errors are typically between those obtained for the exponential and the Gaussian CF.

For the CFs shown in Figure 5.3 the GMRF coefficients (compare (5.5)

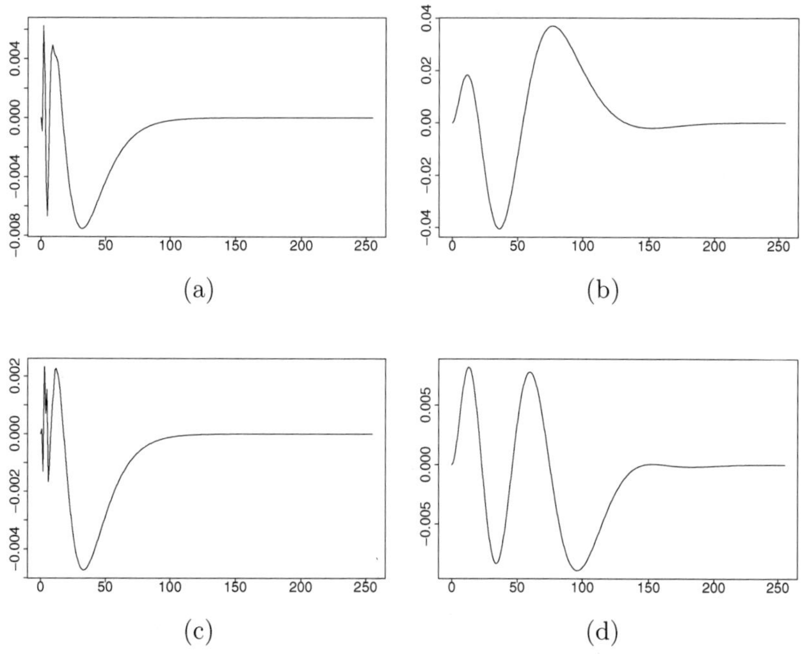

Figure 5.3 *The figures display the difference between the target correlation function (CF) and the fitted CF for the corresponding fits displayed in Figure 5.2. The difference goes to zero for lags larger than shown in the figures.*

and (5.6)) are

$$26.685 \begin{bmatrix} & & 0.091 \\ & 0.304 & -0.191 \\ 1 & -0.537 & 0.275 \end{bmatrix} \tag{5.12}$$

and

$$14.876 \begin{bmatrix} & & & 0.063 \\ & & 0.085 & -0.100 \\ & -0.015 & -0.002 & 0.071 \\ 1 & -0.280 & -0.005 & -0.033 \end{bmatrix}$$

for the exponential CF and

$$118\,668.081 \begin{bmatrix} & & -0.083 \\ & 0.000 & 0.166 \\ 1 & -0.333 & -0.165 \end{bmatrix} \tag{5.13}$$

and

$$73\,382.052 \begin{bmatrix} & & & 0.009 \\ & & 0.173 & -0.031 \\ & -0.263 & -0.002 & -0.057 \\ 1 & -0.122 & 0.029 & 0.106 \end{bmatrix}$$

for the Gaussian CF. We have truncated the values of θ_i/θ_1 showing only 3 digits.

Note that the coefficients are all far from producing a diagonal-dominant precision matrix. This supports the claim made in Section 2.7.2 that imposing diagonal dominance can be severely restrictive for larger neighborhoods. Further, we do not find the magnitude of the coefficients and (some of) their signs particularly intuitive. Although the coefficients have a nice conditional interpretation (see Theorem 2.3), it seems hard to extrapolate these to knowledge of the covariance matrix without doing the matrix inversion. A way to interpret the coefficients of the fitted GMRF is to consider them purely as a mapping of the parameters of the CF. The parameters in the CF are easier to interpret.

The fitted GMRF is sensitive wrt small changes in the coefficients. To illustrate this issue, we took the coefficients in (5.13) and use 3, 5, 7, and 9 significant digits to represent θ_i/θ_1, for $i = 2, \ldots, 5$. The scaling factor θ_0 is not truncated. Using these truncated coefficients we computed the CF and compared it with the Gaussian CF with range 50. The result is shown in Figure 5.4. Using only 3 significant (panel a) digits makes $\boldsymbol{Q}(\boldsymbol{\theta})$ not SPD and these parameters are not valid. With 5 significant digits (panel b) the range is far too small and the marginal precision is 276, far too large. Increasing the number of digits to 7 (panel c) decreases the marginal precision to 17, while for 9 digits (panel d) the marginal precision is now 0.95 and the fit has considerably improved. We need 12 digits to reproduce Figure 5.2(b). There is somewhat less sensitivity of the coefficients for the exponential CF, but still we need 8 significant digits to reproduce Figure 5.2(a).

The coefficients obtained as the solution of (5.7) also depend on the size of the torus. How strong this dependency is can be investigated by using the coefficients computed for one size to compute the CF and the marginal precision using a torus of a different size. If the CF and the marginal precision are close to the target CF and the marginal precision is near 1, then we may use the computed coefficients on toruses of a different size. This is an advantage as we only need one set of coefficients for each CF. It turns out that as long as the size of the torus is large compared to the range r, the effect of the size of the torus is (very) small.

We illustrate this result using the coefficients in (5.13) corresponding to a Gaussian CF with range 50. We applied these coefficients to toruses of size $n \times n$ with n between 32 and 512. For each size n we computed

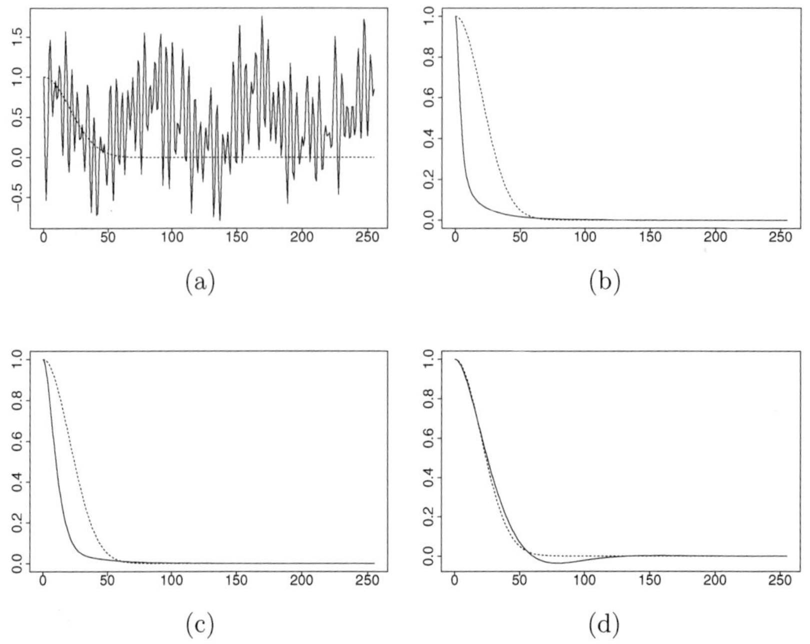

Figure 5.4 *The figures display the correlation function (CF) for the fitted GMRF (solid line) and the Gaussian CF (dashed line) with range* 50*. The figures are produced using the coefficients as computed for Figure 5.2(b) using only (a)* 3 *significant digits, (b)* 5 *significant digits, (c)* 7 *significant digits and (d)* 9 *significant digits, for* θ_i/θ_1, $i = 2, \ldots, 5$ *while* θ_0 *was not truncated. For (a) the parameters are outside the valid parameter space so* $\boldsymbol{Q}(\boldsymbol{\theta})$ *not SPD. The marginal precisions based on the coefficients in (b), (c), and (d) is* 276*,* 17*, and* 0.95*, respectively.*

the marginal precision, shown in Figure 5.5. The marginal precision is less than one for small n, becomes larger for larger n, before it stabilizes at unity for $n > 300$. This corresponds to a size that is 6 times the range. The marginal precision is not equal to one if the CF does not reach its stable value at zero at lag $n_1/2$. Due to the cyclic boundary conditions the CF is cyclic as well. Repeating this exercise using the coefficients (5.12) obtained for the exponential CF with range 30 shows that in this case, too, the dimension of the torus must be about 6 times the range.

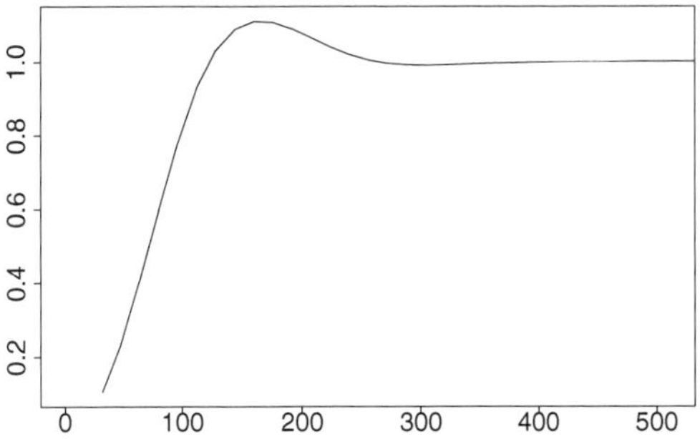

Figure 5.5 *The marginal precision with the coefficients in (5.13) corresponding to a Gaussian CF with range* 50*, on a torus with size from* 32 *to* 512*. The marginal precision is about* 1 *for dimension larger than* 300*, i.e.,* 5 *times the range in this case.*

5.1.4 Regular lattices and boundary conditions

In applications where we use zero mean Gaussian fields restricted to a regular lattice, cyclic boundary conditions are rarely justified from a modeling point of view. This means that we must consider approximations on the lattice $\mathcal{I}_{\boldsymbol{n}}$ rather than the torus $\mathcal{T}_{\boldsymbol{n}}$.

However, the full conditional $\pi(x_{ij}|\boldsymbol{x}_{-ij})$ of a GMRF approximation will have a complicated structure with coefficients depending on ij and in particular on the distance from the boundary, in order to fulfill stationarity and isotropy.

Consider a $(2m+1)\times(2m+1)$ neighborhood. A naive approach is to specify the full conditionals as

$$\mathrm{E}(x_{ij} \mid \boldsymbol{x}_{-ij}) \quad = \quad -\frac{1}{\theta_{00}}\sum_{kl}\theta_{kl}x_{i+k,j+l} \tag{5.14}$$

$$\mathrm{Prec}(x_{ij} \mid \boldsymbol{x}_{-ij}) \quad = \quad \theta_{00}, \tag{5.15}$$

using the same coefficients as obtained through (5.7) and simply setting $x_{i+k,j+l} = 0$ (the mean value) (5.14) for $(i+l, j+l) \notin \mathcal{I}_{\boldsymbol{n}}$. Note that the coefficients in (5.14) equal those in (5.5) and in (5.6) with a slight change in notation.

The full conditional $\pi(x_{ij}|\boldsymbol{x}_{-ij})$ now has constant coefficients but the marginal precision and the CF will no longer be constant but varies with ij and depends on the distance from the boundary. This effect is not desirable and we will now discuss strategies to avoid these features.

Our first concern, however, is to make sure that it is valid to take the coefficients on $\mathcal{T}_{\boldsymbol{n}}$ and use them to construct a GMRF on $\mathcal{I}_{\boldsymbol{n}}$ with

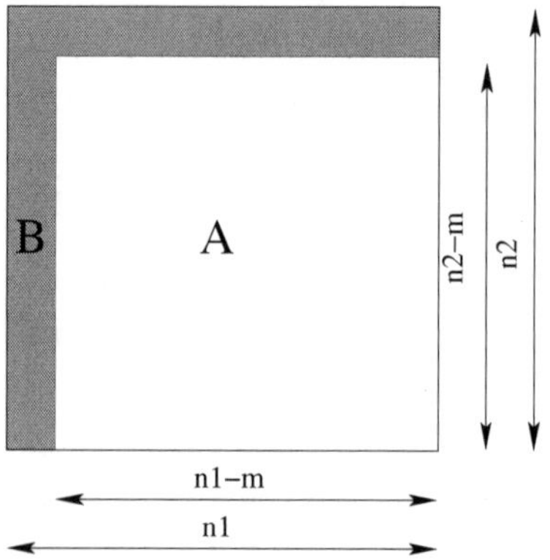

Figure 5.6 *The figure displays an unwrapped $n_1 \times n_2$ torus. The set B have thickness m and A is an $(n_1 - m) \times (n_2 - m)$ lattice.*

full conditionals as in (5.14) and (5.15). In other words, under what conditions will $\boldsymbol{Q}(\boldsymbol{\theta})$, defined by (5.14) and (5.15), be SPD? The solution is trivial as soon as we get the right view.

Consider now Figure 5.6 where we have unwrapped the torus $\mathcal{T}_{\boldsymbol{n}}$. Define the set B with thickness m shown in gray and let A denote the remaining sites. Consider the density of $\boldsymbol{x}_A | \boldsymbol{x}_B$. Since B is thick enough to destroy the effect of the cyclic boundary conditions, $\boldsymbol{Q}_{AA}$ equals the precision matrix obtained when using the coefficients $\boldsymbol{\theta}$ on an $(n_1 - m) \times (n_2 - m)$ lattice. Since all principal submatrices of $\boldsymbol{Q}$ are SPD (see Section 2.1.6), then $\boldsymbol{Q}_{AA}$ is SPD. Hence we can use the coefficients computed in Section 5.1.2 on $\mathcal{T}_n$ to define GMRFs through the full conditionals (5.14) and (5.15) on lattices not larger than $(n_1 - m) \times (n_2 - m)$, see also (2.75).

To illustrate the boundary effect of using the full conditionals (5.14) and (5.15), we use the 5×5 coefficients corresponding to Gaussian and exponential CFs with range 50 and define a GMRF on a 200×200 lattice. We computed the marginal variance for x_{ii}, as a function of i using the algorithms in Section 2.3.1. Figure 5.7 shows the result.

The marginal variance at the four corners is 0.005^2 (Gaussian) and 0.10^2 (exponential), which is small compared to the target unit variance. Increasing the distance from the border the variance increases, and reaches unity at a distance of around 75 (Gaussian) and 50 (exponential). This corresponds to 1.5 (Gaussian) and 1.0 (exponential) times the range. The different behavior of the Gaussian and exponential CF is mainly due to the different smoothness properties of the CF at lag 0.

The consequence of Figure 5.7 is to enlarge the lattice to remove the effect of x_{ij} being zero outside the lattice. As an example, if the problem

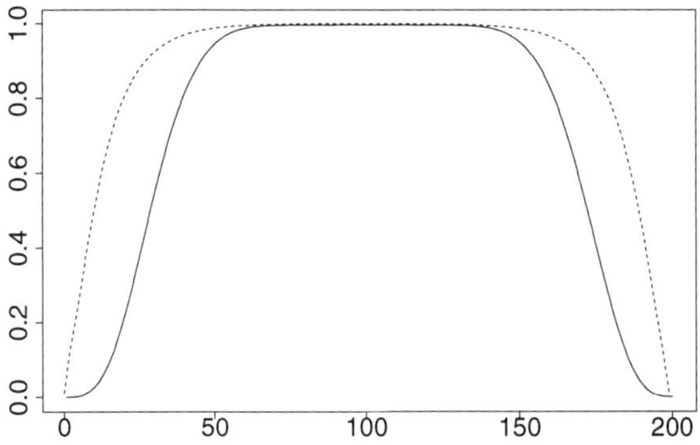

Figure 5.7 *The variance of* x_{ii}, *for* $i = 1, \ldots, 200$, *when using the* 5×5 *coefficients corresponding to Gaussian (solid line) and exponential (dashed line) CFs with range* 50, *on a* 200×200 *lattice.*

at hand requests a 200×200 lattice and we expect a maximum range of about 50, then using GMRF approximations, we should enlarge the region of interest and use a lattice with dimension between 300×300 and 400×400.

An alternative approach is to scale the coefficients near the boundary, in 'some appropriate fashion'. However, if the precision matrix is not diagonal dominant, it is not clear how to modify the coefficients near and at the boundary to ensure that the modified precision matrix remains SPD.

We will now discuss an alternative approach that apparently solves this problem by an embedding technique. This approach is technically more involved but the computational complexity to factorize precision matrix is $\mathcal{O}(n^{3/2} \log n)$, where $n = n_1 n_2$. Let $\boldsymbol{x}$ be a zero mean GMRF defined through its full conditionals (5.14) and (5.15) on the *infinite* lattice $\mathcal{I}_\infty$. Let

$$\gamma_{ij} = \mathrm{E}(x_{ij} x_{00})$$

be the covariance at lag ij, which is related to the coefficients $\boldsymbol{\theta}$ of the GMRF by

$$\gamma_{ij} = \frac{1}{4\pi^2} \int_{-\pi}^{\pi} \int_{-\pi}^{\pi} \frac{\cos(\omega_1 i + \omega_2 j)}{\sum_{kl} \theta_{kl} \cos(\omega_1 k + \omega_2 j)} \, d\omega_1 d\omega_2, \tag{5.16}$$

see Section 2.6.5. Figure 5.8 illustrates an $(n_1 + 2m) \times (n_2 + 2m)$ lattice, which is to be considered as a subset of $\mathcal{I}_\infty$. Further, the thickness of B is m, such that

$$\boldsymbol{x}_A \perp \boldsymbol{x}_{(A \cup B)^c} \mid \boldsymbol{x}_B.$$

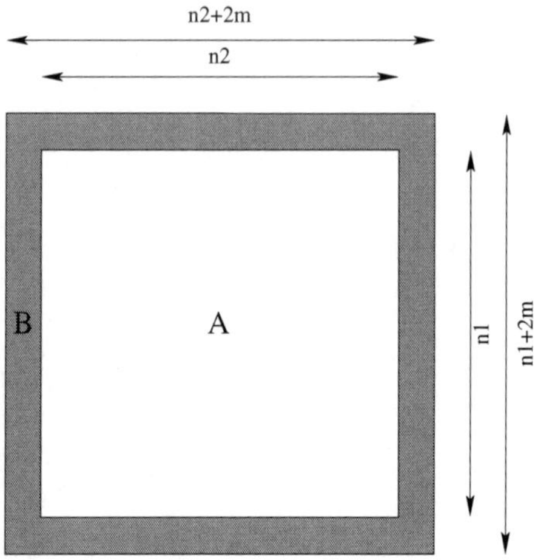

Figure 5.8 *An $n_1 \times n_2$ lattice (A) with an additional boundary (B) of thickness m.*

where $(A \cup B)^c$ is the complement of $A \cup B$. We factorize the density for $\boldsymbol{x}_{A\cup B}$ as

$$\pi(\boldsymbol{x}_{A\cup B}) = \pi(\boldsymbol{x}_B)\, \pi(\boldsymbol{x}_A \mid \boldsymbol{x}_B).$$

Here, $\boldsymbol{x}_B$ is Gaussian with covariance matrix $\boldsymbol{\Sigma}_{BB}$ with elements found from (5.16), while $\boldsymbol{x}_A|\boldsymbol{x}_B$ has density

$$\begin{aligned} \pi(\boldsymbol{x}_A \mid \boldsymbol{x}_B) &\propto \exp\left(-\frac{1}{2}\boldsymbol{x}_{A\cup B}^T \boldsymbol{Q} \boldsymbol{x}_{A\cup B}\right) \\ &\propto \exp\left(-\frac{1}{2}\boldsymbol{x}_A^T \boldsymbol{Q}_{AA} \boldsymbol{x}_A + \boldsymbol{x}_B^T \boldsymbol{Q}_{AB}^T \boldsymbol{x}_A\right), \end{aligned}$$

where $Q_{(i,j),(k,l)} = \theta_{i-k,j-l}$. The marginal density of $\boldsymbol{x}_A$ now has the correct covariance matrix with elements found from (5.16).

To simulate from $\pi(\boldsymbol{x}_{A\cup B})$ we first sample $\boldsymbol{x}_B$ from $\pi(\boldsymbol{x}_B)$ then $\boldsymbol{x}_A$ from $\pi(\boldsymbol{x}_A|\boldsymbol{x}_B)$. Conditional simulation requires, however, the joint precision matrix for $\boldsymbol{x}_{A\cup B}$:

$$\text{Prec}\begin{pmatrix}\boldsymbol{x}_A \\ \boldsymbol{x}_B\end{pmatrix} = \begin{pmatrix}\boldsymbol{Q}_{AA} & \boldsymbol{Q}_{AB} \\ \boldsymbol{Q}_{AB}^T & \boldsymbol{Q}_{AB}^T \boldsymbol{Q}_{AA}^{-1} \boldsymbol{Q}_{AB} + \boldsymbol{\Sigma}_{BB}^{-1}\end{pmatrix}. \tag{5.17}$$

Note that B has size n_B of order $\mathcal{O}(n^{1/2})$. To factorize the (dense) $n_B \times n_B$ matrix

$$\boldsymbol{Q}_{AB}^T \boldsymbol{Q}_{AA}^{-1} \boldsymbol{Q}_{AB} + \boldsymbol{\Sigma}_{BB}^{-1}, \tag{5.18}$$

we need $\mathcal{O}(n_B^3) = \mathcal{O}(n^{3/2})$ flops. This is the same cost needed to factorize $\boldsymbol{Q}_{AA}$. However, to compute (5.18), we need $\boldsymbol{Q}_{AB}^T \boldsymbol{Q}_{AA}^{-1} \boldsymbol{Q}_{AB}$. Let $\boldsymbol{L}_A$ be the Cholesky triangle of $\boldsymbol{Q}_{AA}$, then $\boldsymbol{Q}_{AB}^T \boldsymbol{Q}_{AA}^{-1} \boldsymbol{Q}_{AB} = \boldsymbol{G}^T \boldsymbol{G}$ where

$\boldsymbol{L}_A\boldsymbol{G} = \boldsymbol{Q}_{AB}$. We can compute $\boldsymbol{G}$ by solving n_B linear systems each of cost $\mathcal{O}(n \log n)$, hence the total cost is $\mathcal{O}(n^{3/2} \log n)$. In total, (5.17) can be computed and factorized using $\mathcal{O}(n^{3/2} \log n)$ flops.

To compute the covariances of the GMRF on $\mathcal{I}_\infty$ to find $\boldsymbol{\Sigma}_{BB}$, there is no need to use (5.16), which involves numerical integration of an awkward integrand. As $\boldsymbol{Q}_{\boldsymbol{n}}$ is a block Toeplitz matrix we can construct its cyclic approximation $\boldsymbol{C}_{\boldsymbol{n}}$. Since $\boldsymbol{Q}_{\boldsymbol{n}}$ and $\boldsymbol{C}_{\boldsymbol{n}}$ are asymptotically equivalent, so are $\boldsymbol{Q}_{\boldsymbol{n}}^{-1}$ and $\boldsymbol{C}_{\boldsymbol{n}}^{-1}$ under mild conditions, see Gray (2002, Theorem 2.1). The consequence, is that $\{\gamma_{ij}\}$ can be taken as the base of $\boldsymbol{C}_{\boldsymbol{n}}^{-1}$ for a large value of $\boldsymbol{n}$.

The embedding approach extends directly to regions of nonsquare shape S as long as we consider $S \cap \mathcal{I}_{\boldsymbol{n}}$, and can embed it with a boundary region with thickness m.

5.1.5 Example: Swiss rainfall data

In applications we are often faced with the problem of fitting a stationary Gaussian field to some observed data. The observed data are most often of the form

$$(\boldsymbol{s}_1, x_1), (\boldsymbol{s}_2, x_2), \ldots, (\boldsymbol{s}_d, x_d),$$

where $\boldsymbol{s}_i$ is the spatial location for the ith observation x_i, and d the number of observations. This is a routine problem within geostatistics and different approaches exist, see, for example, Cressie (1993), Chilés and Delfiner (1999) and Diggle et al. (2003). We will not focus here on the fitting problem itself, but on the question of how we may take advantage of the GMRF approximations to Gaussian fields in this process.

Recall that we consider the GMRF approximations only as computationally more efficient representations of the corresponding GF. However, we do not recommend computing maximum likelihood estimates of the parameters in the GMRF directly from the observed data in geostatistical applications. Such an approach may give surprising results. This is related to model error and to the form of the sufficient statistics, see Rue and Tjelmeland (2002) for examples and further discussion.

To fix the discussion consider the data displayed in Figure 5.9, available in the `geoR`-library) (Ribeiro Jr. and Diggle, 2001). The data is the measured rainfall on May 8, 1986 from 467 locations in Switzerland where Figure 5.9 displays the observations at 100 of these locations. The motivating scientific problem is to construct a continuous spatial map of rainfall values from the observed data, but we have simplified the problem for presentation issues. Assume the spatial field is a GF with a common mean μ and a Matérn CF with unknown range, variance, and smoothness ν. Denote these parameters by $\boldsymbol{\theta}$. Estimating the parameters from the observed data using a Bayesian or a likelihood approach,

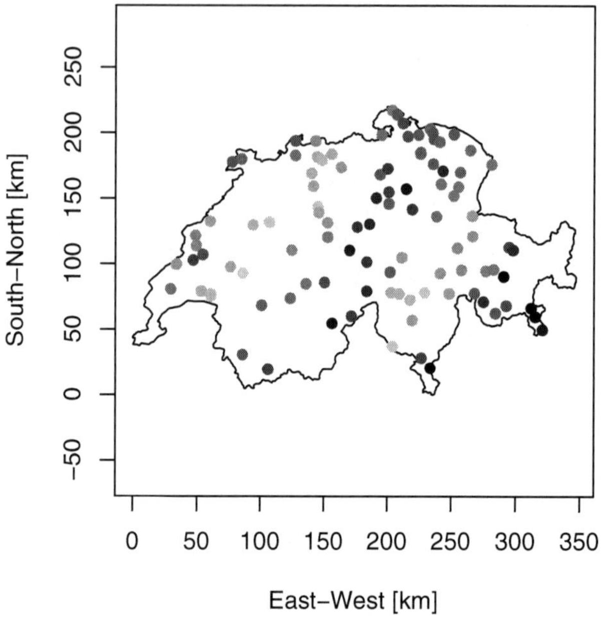

Figure 5.9 *Swiss rainfall data at* 100 *sample locations. The plot displays the locations and the observed values ranging from black to white.*

involves evaluating the log-likelihood,

$$-\frac{d}{2}\log 2\pi - \frac{1}{2}\log|\boldsymbol{\Sigma}(\boldsymbol{\theta})| - \frac{1}{2}(\boldsymbol{x}-\mu\mathbf{1})^T\boldsymbol{\Sigma}(\boldsymbol{\theta})^{-1}(\boldsymbol{x}-\mu\mathbf{1}), \qquad (5.19)$$

for several different values of the unknown parameters $\boldsymbol{\theta}$. Note that the $d \times d$ matrix $\boldsymbol{\Sigma}(\boldsymbol{\theta})$ is a dense matrix and due to the irregular locations $\{\boldsymbol{s}_i\}$, it has no special structure. Hence, a general Cholesky factorization algorithm must be used to factorize $\boldsymbol{\Sigma}(\boldsymbol{\theta})$, which costs $d^3/3$ flops.

If we estimate parameters using only $d = 100$ observations, then the computational cost is not that high and there is no need to involve GMRF approximations at this stage. However, if we use all the data $d = 467$ and factorizing $\boldsymbol{\Sigma}(\boldsymbol{\theta})$ will be costly. We may take advantage of the GMRF approximations to GFs. Assume for simplicity $\nu = 1/2$ so the CF is exponential. We have precomputed the best fits for values of the range $r = 1, 1.1, 1.2, \ldots$, and so on. Since GMRF approximations are for regular lattices only, we first need to replace the continuous area of interest with a fine grid of size n, say. The resolution of the grid has to be high enough so we can assign each observed data point $(\boldsymbol{s}_i, x_i)$ to the nearest grid point without (too much) error. Using some kind of interpolation is also possible. The boundary conditions should be treated as discussed in Section 5.1.4.

Let $\boldsymbol{x}$ denote the GMRF at the lattice with size n, and denote by D the set of sites that is observed and M those sites not observed. Note that $\boldsymbol{x} = \{\boldsymbol{x}_D, \boldsymbol{x}_M\}$. The likelihood for the observed data $\boldsymbol{x}_D$ is *not* directly available using GMRF approximations, but is available indirectly as

$$\pi(\boldsymbol{x}_D \mid \boldsymbol{\theta}) = \frac{\pi(\boldsymbol{x} \mid \boldsymbol{\theta})}{\pi(\boldsymbol{x}_M \mid \boldsymbol{x}_D, \boldsymbol{\theta})},$$

which holds for any $\boldsymbol{x}_M$. We may use $\boldsymbol{x}_M = \boldsymbol{0}$, say. Both terms on the rhs are easy to evaluate if $\boldsymbol{x}$ is a GMRF, and the cost is $\mathcal{O}(n^{3/2})$ flops. Comparing this cost with the cost of evaluating (5.19), we see that the breakpoint is $n \approx d^2$. If $n < d^2$ we will gain if we use GMRF approximations and if $n > d^2$ it will be better to use (5.19) directly. Choosing a resolution of the grid of 1×1 km^2, we would use (5.19) for $d = 100$, while for the full dataset, where $d = 467$, it would be computationally more efficient to use the GMRF approximations.

The next stage in the process is to provide spatial predictions for the whole area of interest with error bounds. Although it is also important to propagate the uncertainly in the parameter estimation into the prediction (Draper, 1995), we will for a short moment treat the estimated parameters as fixed and apply the simple plug-in approach. Maximum likelihood estimation using the exponential CF on the square-root transformed data give the estimates $\hat{r} = 125$ and $\hat{\sigma}^2 = 83$ and $\hat{\mu} = 21$, which we will use further.

Spatial predictions are usually performed by imposing a fine regular lattice of size n over the area of interest. Let $\boldsymbol{x}$ be the GF on this lattice. The best prediction is then

$$\mathrm{E}(\boldsymbol{x} \mid \text{observed data}, \hat{\boldsymbol{\theta}}) \tag{5.20}$$

and the associated uncertainty is determined through the conditional variance. To compute or estimate (5.20) we can make use of a GMRF approximation as the size of the grid is usually large. The locations of the observed data must again be moved to the nearest grid point using the GMRF approximations. The conditional variance is not directly available from the GMRF formulation, but is easily estimated using iid samples from the conditional distribution. Figure 5.10 shows the estimated conditional mean and standard deviation using a 200×200 lattice restricted to the area of interest.

To account for the uncertainty in the parameter estimates it will be required to use simulation-based inference using MCMC techniques. This also allows us to deal with non-Gaussian data, for example, binomial or Poisson observations. For a thorough discussion of this issue in relation to the Swiss rainfall data, see Diggle et al. (2003, Section 2.8). Simulating

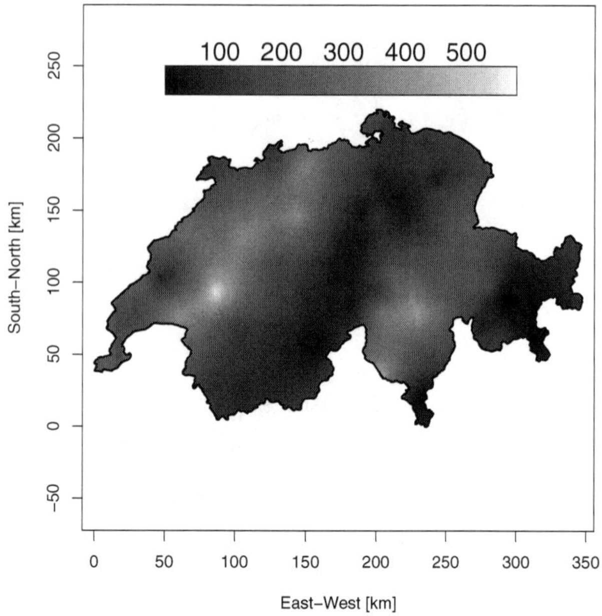

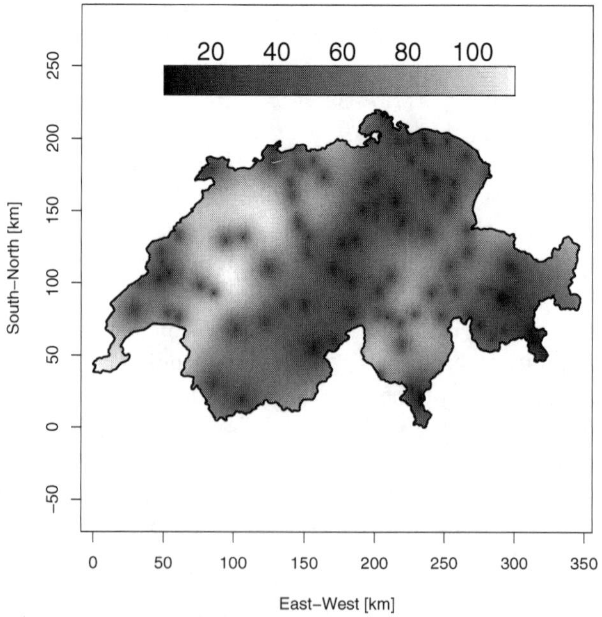

Figure 5.10 *Spatial interpolation of the Swiss rainfall data using a square-root transform and an exponential CF with range* 125. *The top figure shows the predictions and the bottom figure shows the prediction error (stdev), both computed using plug-in estimates.*

from the posterior density of the parameters of interest will explore their uncertainty, which can (and should) be accounted for in the predictions. Therefore, the predictive samples have to be generated as part of the MCMC algorithm. Averages over the samples will give an estimate of the expectation conditioned on the observed data. This is no different from averaging over predictions made using different parameter values. The GMRF approximations do only offer computational gain, but this can be important as it is typically computationally demanding to account for the parameter uncertainty using GFs.

5.2 Approximating hidden GMRFs

In this section we will discuss how to approximate a nonnormal density of the form

$$\pi(\boldsymbol{x} \mid \boldsymbol{y}) \propto \exp\left(-\frac{1}{2}\boldsymbol{x}^T \boldsymbol{Q}\boldsymbol{x} - \sum_i g_i(x_i, y_i)\right). \tag{5.21}$$

We assume that the terms $g_i(x_i, y_i)$ contain also nonquadratic terms of x_i, so that $\boldsymbol{x}|\boldsymbol{y}$ is not normal. If $\boldsymbol{x}$ is a GMRF wrt $\mathcal{G}$ that is partially observed through $\boldsymbol{y}$, then $\boldsymbol{x}|\boldsymbol{y}$ is called a *hidden GMRF* abbreviated as HGMRF. Note that (5.21) defines $\boldsymbol{x}$ as a Markov random field wrt (the same graph) $\mathcal{G}$, but it is not Gaussian.

Densities of the form (5.21) occurred frequently in Chapter 4 as full conditionals for a GMRF conditioned on nonnormal and independent observations $\{y_i\}$ where y_i only depends on x_i. For example, assume y_i is Poisson with mean $\exp(x_i)$, then the full conditional $\pi(\boldsymbol{x}|\boldsymbol{y})$ is of the form (5.21) with

$$g_i(x_i, y_i) = -x_i y_i + \exp(x_i).$$

Alternatively, if y_i is a Bernoulli variable with mean $\exp(x_i)/(1 + \exp(x_i))$, then $g_i(x_i, y_i)$ reads

$$-x_i y_i + \log\left(1 + \exp(x_i)\right)$$

using a logit link. In both cases, the $g_i(x_i, y_i)$ contain nonquadratic terms of x_i.

One approach taken in Chapter 4 was to construct a GMRF approximation to $\pi(\boldsymbol{x}|\boldsymbol{y})$, which we here denote as $\pi_G(\boldsymbol{x})$. The mean of the approximation equals the mode of $\pi(\boldsymbol{x}|\boldsymbol{y})$, $\boldsymbol{x}^*$, and the precision matrix is

$$\boldsymbol{Q} + \mathrm{diag}(\boldsymbol{c}).$$

Here, c_i is the coefficient in the second-order Taylor expansion of $g_i(x_i, y_i)$ at the mode x_i^*,

$$g_i(x_i, y_i) \approx a_i + b_i x_i + \frac{1}{2} c_i x_i^2 \tag{5.22}$$

where the coefficients a_i, b_i, and c_i depend on x_i^* and y_i. See also the discussion in Section 4.4. If n is large and/or the $g_i(x_i, y_i)$ terms are too influential, the GMRF approximation may not be sufficiently accurate. This will be evident in the MCMC algorithms as the acceptance rate for a proposal sampled from the GMRF approximation will become (very) low. In such cases it would be beneficial to construct a non-Gaussian approximation with improved accuracy compared to the GMRF approximation. An approach to construct improved approximations compared to the GMRF approximation is the topic in this section. We assume throughout that (5.21) is unimodal.

In Section 5.2.1 we will discuss how to construct approximations to a HGMRF, then apply these to a stochastic volatility model in Section 5.2.2 and to construct independence samplers for the Tokyo rainfall data in Section 5.2.3. More complex applications are found in Rue et al. (2004), which also combine these new approximations with the GMRF approximations to GFs in Section 5.1.

5.2.1 Constructing non-Gaussian approximations

Before we start discussing how an improved approximation to (5.21) can be constructed we should remind ourselves *why* we need an approximation and what properties we must require from one.

The improved approximation is needed to update $\boldsymbol{x}$ in one block, preferable jointly with its hyperparameters. For this reason, we need to be able to sample from the improved approximation directly and have access to the normalizing constant. For the normal distribution, this is feasible but there are not that many other candidates (for high dimension) around with these properties. We will outline an approach to construct improved approximations that have these properties, additional to being adaptive to the particular g_i's under study.

The first key observation is that we can convert a GMRF wrt $\mathcal{G}$ to a nonhomogeneous autoregressive model, defined backward in 'time', using the Cholesky triangle of the reordered precision matrix,

$$\boldsymbol{L}^T\boldsymbol{x} = \boldsymbol{z}. \tag{5.23}$$

Here, $\boldsymbol{z} \sim \mathcal{N}(\boldsymbol{0}, \boldsymbol{I})$ and $\boldsymbol{Q} = \boldsymbol{L}\boldsymbol{L}^T$. The representation is sequential backward in time, as (5.23) is equivalent to

$$\begin{aligned}
x_n &= \frac{1}{L_{nn}} z_n \\
x_{n-1} &= \frac{1}{L_{n-1,n-1}} \left(z_{n-1} - L_{n,n-1} x_n\right) \\
x_{n-2} &= \frac{1}{L_{n-2,n-2}} \left(z_{n-2} - L_{n,n-2} x_n - L_{n-1,n-2} x_{n-1}\right),
\end{aligned}$$

and so on. In short, (5.23) represents $\pi(\boldsymbol{x})$ as

$$\pi(\boldsymbol{x}) = \prod_{i=n}^{1} \pi(x_i \mid x_{i+1}, \ldots, x_n). \tag{5.24}$$

The autoregressive process is nonhomogeneous as the coefficients L_{ij} are not a function of $i - j$. The order of the process also varies and can be as large as $n - 1$. However, $\boldsymbol{L}$ is by construction sparse.

By using (5.24) we may rewrite (5.21) as

$$\begin{aligned}\pi(\boldsymbol{x} \mid \boldsymbol{y}) &= \frac{1}{Z} \prod_{i=n}^{1} \pi(x_i \mid x_{i+1}, \ldots, n) \exp(-g_i(x_i, y_i)) \\ &= \prod_{i=n}^{1} \pi(x_i \mid x_{i+1}, \ldots, x_n, y_1, \ldots, y_i),\end{aligned}$$

where Z is the normalizing constant and

$$\begin{aligned}\pi(x_i | x_{i+1}, \ldots, x_n, y_1, \ldots, y_i) &\propto \pi(x_i \mid x_{i+1}, \ldots, x_n) \\ &\times \exp\left(-g_i(x_i, y_i)\right) \\ &\times \int \exp\left(-\sum_{j=1}^{i-1} g_j(x_j, y_j)\right) \\ &\qquad \pi(x_1, \ldots, x_{i-1} \mid x_i, \ldots, x_n)\, dx_1 \cdots dx_{i-1}.\end{aligned} \tag{5.25}$$

The second key observation is to note that *if* we can approximate (5.25) by

$$\tilde{\pi}(x_i \mid x_{i+1}, \ldots, x_n, y_1, \ldots, y_i), \tag{5.26}$$

say, then we can approximate $\pi(\boldsymbol{x}|\boldsymbol{y})$ by

$$\tilde{\pi}(\boldsymbol{x} \mid \boldsymbol{y}) = \prod_{i=n}^{1} \tilde{\pi}(x_i \mid x_{i+1}, \ldots, x_n, y_1, \ldots, y_i). \tag{5.27}$$

Since (5.26) is univariate, we can construct the approximation so that the normalizing constant is known. As the approximation (5.27) is *defined* sequentially backward in 'time', it automatically satisfies the two requirements for an improved approximation:

1. We can sample from $\tilde{\pi}(\boldsymbol{x}|\boldsymbol{y})$ directly (and exact), by successively sampling

$$\begin{aligned} x_n &\sim \tilde{\pi}(x_n \mid y_1, \ldots, y_n) \\ x_{n-1} &\sim \tilde{\pi}(x_{n-1} \mid x_n, y_1, \ldots, y_{n-1}) \\ x_{n-2} &\sim \tilde{\pi}(x_{n-2} \mid x_{n-1}, x_n, y_1, \ldots, y_{n-2}), \end{aligned}$$

and so on, until we sample x_1.

2. The density $\tilde{\pi}(\boldsymbol{x}|\boldsymbol{y})$ is normalized since each term (5.26) is normalized.

Before we discuss how we can construct approximations to (5.25) by neglecting 'not that important' terms, we will simply shift the reference from $\pi(\boldsymbol{x})$ to the GMRF approximation of (5.21), $\pi_G(\boldsymbol{x})$, say, so that (5.25) reads

$$\begin{aligned} \pi(x_i|x_{i+1},\ldots,x_n,y_1,\ldots,y_i) &\propto \pi_G(x_i \mid x_{i+1},\ldots,x_n) \\ &\times \exp\left(-h_i(x_i,y_i)\right) \\ &\times \int \exp\left(-\sum_{j=1}^{i-1} h_j(x_j,y_j)\right) \\ &\quad \pi_G(x_1,\ldots,x_{i-1} \mid x_i,\ldots,x_n)\, dx_1 \cdots dx_{i-1}, \end{aligned} \tag{5.28}$$

where

$$h_i(x_i,y_i) = g_i(x_i,y_i) - (a_i + b_i x_i + \frac{1}{2}c_i x_i^2),$$

using the Taylor expansion in (5.22). Further, let $\boldsymbol{\mu}_G$ denote the mean in $\pi_G(\boldsymbol{x})$.

Starting from (5.28) we will construct three classes of approximations, which we denote by $A1$, $A2$, and $A3$.

Approximation A1

Approximation $A1$ is found by removing all terms in (5.28) apart from $\pi_G(x_i|x_{i+1},\ldots,x_n)$, so

$$\pi_{A1}(x_i \mid x_{i+1},\ldots,x_n,y_1,\ldots,y_i) = \pi_G(x_i \mid x_{i+1},\ldots,x_n). \tag{5.29}$$

Using (5.27), we obtain

$$\begin{aligned} \pi_{A1}(\boldsymbol{x} \mid \boldsymbol{y}) &= \prod_{i=n}^{1} \pi_G(x_i \mid x_{i+1},\ldots,x_n) \\ &= \boldsymbol{\pi}_G(\boldsymbol{x}); \end{aligned}$$

hence $A1$ is the GMRF approximation. This construction offers an alternative interpretation of the GMRF approximation.

Approximation A2

Approximation $A2$ is found by including the term we think is the most important one missed in $A1$, which is the term involving the data y_i,

$$\begin{aligned} \pi_{A2}(x_i \mid x_{i+1},\ldots,x_n,y_1,\ldots,y_i) &\propto \pi_G(x_i \mid x_{i+1},\ldots,x_n) \\ &\times \exp(-h_i(x_i,y_i)). \end{aligned} \tag{5.30}$$

Note that $A2$ can be a great improvement over $A1$ if the precision matrix in $\pi_G(\boldsymbol{x})$ is a diagonal matrix, then $A1$ can be very inaccurate for large n while $A2$ will be exact.

However, since (5.30) can take any form, we need to introduce a second level of approximation by using a finite dimensional representation of (5.30), $\tilde{\pi}_{A2}$. For this we use log-linear or log-quadratic splines. It is easy to sample from such a density and to compute the normalizing constant. One potential general problem using a spline representation, is that the support of the distribution can be unknown to some extent. However, this is not really problematic here, as we as may take advantage of the GMRF approximation which is centered at the mode. Let μ_i and σ_i^2 denote the conditional mean and variance in (5.29), then the probability mass is (with a high degree of certainty, except in artificial cases) in the region

$$[\mu_i - f\sigma_i,\ \mu_i + f\sigma_i] \tag{5.31}$$

with $f = 5$ or 6, say. This can be used to determine the knots for the log-spline representation. To construct a log-linear spline representation, we can divide (5.31) into K equally spaced regions and evaluate the log of (5.30) for the values of x_i at each knot. We then define a straight line interpolating the values in between the knots. A density that is piecewise log-linear is particularly easy to integrate analytically and straightforward to sample from. Additionally, we must add two border regions from $-\infty$ to the first knot, and from the last knot to ∞. This is needed to obtain unbounded support of our approximation. A log-quadratic spline approximation can be derived similarly, see Rue et al. (2004, Appendix) for more details.

Approximation A3

In approximation $A3$ we include also the integral term in (5.28), which we write as

$$I(x_i) = \mathrm{E}\left(\exp(-\sum_{j=1}^{i-1} h_j(x_j, y_j))\right),$$

where the expectation is wrt $\pi_G(x_1, \ldots, x_{i-1}|x_i, \ldots, x_n)$ and we need to evaluate the integral as a function of x_i only. To approximate $I(x_i)$, we may include only the terms in the expectation we think are the most important, $\mathcal{J}(i)$, say,

$$I(x_i \mid \mathcal{J}(i)) = \mathrm{E}\left(\exp(-\sum_{j\in\mathcal{J}(i))} h_j(x_j, y_j))\right).$$

A first natural choice is

$$\mathcal{J}(i) = \{j \; : \; j < i \text{ and } i \sim j\} \tag{5.32}$$

meaning that we include h_j terms such that j is a neighbor of i. Note that the ordering of the indices in (5.24), required to make $\boldsymbol{L}$ sparse, does not necessarily correspond to the original graph, so $j \in \mathcal{J}(i)$ does not need to be close to i. However, if we use a band approach to factorize $\boldsymbol{Q}$ and the GMRF model is an autoregressive process of order p on a line, then $j \in \mathcal{J}(i)$ will be close to i:

$$\mathcal{J}(i) = \{j \; : \; \max(1, i-p) \leq j < i\}.$$

We can improve (5.32) by also including in $\mathcal{J}(i)$ all neighbors to the neighbors of i less than i, and so on. The main assumption is that the importance decays with the distance (on the graph) to i. This is not unreasonable as π_G is located at the mode, but is of course not true in general.

After choosing $\mathcal{J}(i)$ we need to approximate $I(x_i|\mathcal{J}(i))$ for each value of x_i corresponding to the $K+1$ knots. A natural choice is to sample M iid samples from $\pi_G(x_1, \ldots, x_{i-1}|x_i, \ldots, x_n)$, for each value of x_i, and then estimate $I(x_i|\mathcal{J}(i))$ by the empirical mean,

$$\widehat{I}(x_i \mid \mathcal{J}(i)) = \frac{1}{M} \sum_{m=1}^{M} \exp(-\sum_{j \in \mathcal{J}(i))} h_j(x_j^{(m)}, y_j)). \tag{5.33}$$

Here, $\boldsymbol{x}^{(1)}, \ldots, \boldsymbol{x}^{(M)}$ denote the M samples, one set for each value of x_i. Some description is needed at each of the steps.

- To sample from $\pi_G(x_1, \ldots, x_{i-1}|x_i, \ldots, x_n)$ we make use of (5.23). Let $j_{\min}(i)$ be the smallest $j \in \mathcal{J}(i)$. Then, iid samples can be produced by solving (5.23) from row $i-1$ until $j_{\min}(i)$, for iid $\boldsymbol{z}$'s, and then adding the mean $\boldsymbol{\mu}_G$.
- We make $\widehat{I}(x_i|\mathcal{J}(i))$ continuous wrt x_i using the same random number stream to produce the samples. A (much) more computationally efficient approach is to make use of the fact that only the conditional mean in $\pi_G(x_1, \ldots, x_{i-1}|x_i, \ldots, x_n)$ will change if x_i varies. Then, we may sample M samples with zero conditional mean and simply add the conditional mean that varies linearly with x_i.
- Antithetic variables are also useful for estimating $I(x_i|\mathcal{J}(i))$. Antithetic normal variates can also involve the scale and not only the sign (Durbin and Koopman, 1997). Let $\boldsymbol{z}$ be a sample from $\mathcal{N}(\boldsymbol{0}, \boldsymbol{I})$ and define

  $$\tilde{\boldsymbol{z}} = \boldsymbol{z}/\sqrt{\boldsymbol{z}^T\boldsymbol{z}}$$

  so that $\tilde{\boldsymbol{z}}$ has unit length. Let $\tilde{\boldsymbol{x}}$ solve $\boldsymbol{L}^T\tilde{\boldsymbol{x}} = \tilde{\boldsymbol{z}}$. With the correct scaling, $\tilde{\boldsymbol{x}}$ will be a sample from $\mathcal{N}(\boldsymbol{0}, \boldsymbol{Q}^{-1})$. The correct scaling is found when we use $\boldsymbol{z}^T\boldsymbol{z} \sim \chi^2_n$. Let $F_{\chi^2_n}$ denote the cumulative distribution function or a χ^2_n variable, and $F^{-1}_{\chi^2_n}$ its inverse. For a sample u_1 from a uniform density between 0 and 1, both

$$\tilde{\boldsymbol{x}}\sqrt{F^{-1}_{\chi^2_n}(u_1)}, \quad \text{and} \quad \tilde{\boldsymbol{x}}\sqrt{F^{-1}_{\chi^2_n}(1-u_1)}$$

are correct samples. The antithetic behavior is in the scaling, if one is close to $\mathbf{0}$, then the other is far away. Additionally, we may also use the classical trick to flip the sign without altering the marginal distribution. The benefit of using antithetic variables is that we can produce many antithetic samples without any computational effort from just one sample from $\pi_G(x_1, \ldots, x_{i-1}|x_i, \ldots, x_n)$. This is computationally very efficient.

Approximation $A3$ is indexed by the stream of random numbers used, and by keeping this sequence fixed, we can produce several samples from the same approximation.

Approximating constrained HGMRFs

A natural extension is to incorporate constraints into the non-Gaussian approximation and approximate (5.21) under the constraint $\boldsymbol{Ax} = \boldsymbol{e}$. The rank of $\boldsymbol{A}$ is typically small with the most prominent example $\boldsymbol{A} = \mathbf{1}^T$, the sum-to-zero constraint. A natural first approach is to make use of (2.30)

$$\boldsymbol{x}^* = \boldsymbol{x} - \boldsymbol{Q}^{-1}\boldsymbol{A}^T(\boldsymbol{A}\boldsymbol{Q}^{-1}\boldsymbol{A}^T)^{-1}(\boldsymbol{Ax} - \boldsymbol{e}).$$

Here, $\boldsymbol{x}$ is a sample from the HGMRF approximation, $\boldsymbol{x}^*$ is the sample corrected for the constraints, and for $\boldsymbol{Q}$ we may use the precision matrix for the GMRF approximation. In the Gaussian case the density of $\boldsymbol{x}^*$ is correct. For the non-Gaussian cases the density of $\boldsymbol{x}^*$ is an approximation to the constrained HGMRF approximation. The density of $\boldsymbol{x}^*$ is supposed to be fairly accurate for constraints, that is, not too influential. To evaluate the density of $\boldsymbol{x}^*$, we may use (2.31) but the denominator is problematic as we need to evaluate $\pi(\boldsymbol{Ax})$. Although we can estimate the value of this density at $\boldsymbol{x}$ using either the GMRF approximation or iid samples from the HGMRF approximation and techniques for density estimation, we are unable evaluate this density exactly.

5.2.2 Example: A stochastic volatility model

We will illustrate the use of the improved approximations on a simple stochastic volatility model for the pound-dollar daily exchange rate from October 1, 1981, to June 28, 1985, previously analyzed by Durbin and

Koopman (2000), among others. Let $\{e_t\}$ denote the exchange rate, then the time series of interest is the log ratio $\{y_t\}$, where

$$y_t = \log(e_t/e_{t-1}), \quad t = 1, \ldots, n = 945.$$

A simple model to describe $\{y_t\}$ is the following:

$$y_t \sim \mathcal{N}(0, \exp(x_t)/\tau_y),$$

where $\{x_t\}$ is assumed to follow a first-order autoregressive process,

$$x_t = \phi x_{t-1} + \epsilon_t,$$

where ϵ_t is iid zero mean normal noise with precision τ_ϵ. The unknown parameters to be estimated are τ_ϵ, τ_y, and ϕ.

We can follow the same strategy for constructing a one-block algorithm conducting a joint proposal for $(\tau_\epsilon, \tau_y, \phi, \boldsymbol{x})$. Propose a change to each of the parameters τ_ϵ, τ_y, and ϕ, and then conditioned on these values, sample $\boldsymbol{x}$ from an approximation to the full conditional for $\boldsymbol{x}$. Then, accept or reject all parameters jointly.

The maximum likelihood estimates of the hyperparameters are

$$\phi = 0.975, \quad \tau_y = 2.49, \quad \text{and} \quad \tau_\epsilon = 33.57. \tag{5.34}$$

To compare the improved approximations to the GMRF approximation, we will measure the accuracy of the approximations using the acceptance rate for fixed values of the hyperparameters as advocated by Robert and Casella (1999, Section 6.4.1). We will compare the following approximations to $\pi(\boldsymbol{x}|\boldsymbol{y}, \boldsymbol{\theta})$:

***A*1** The GMRF approximation.

***A*2** The approximation (5.30) that includes the likelihood term $g_i(x_i, y_i)$. We use $K = 20$ regions in our quadratic spline approximation.

A*3*a An improved approximation including the integral term (5.33) using

$$\mathcal{J}(i) = \{i - 20, i - 19, \ldots, i - 1\},$$

with obvious changes near the boundary. We estimate (5.33) using $M = 1$ 'sample' only; the conditional mean computed under the GMRF approximation $A1$.

A*3*b Same as $A3a$, but using $M = 10$ samples and where each is duplicated into 4 using antithetic techniques.

Fixing the hyperparameters at their maximum likelihood estimates (5.34), we obtained the acceptance rates as displayed in Table 5.1. The results demonstrate that we can improve the GMRF approximation at the cost of more computing. The acceptance rate increases by roughly 0.2 for each level of the approximations. Obtaining an acceptance rate of

Approximation	Acceptance rate	Iter/sec
$A1$	0.33	23.5
$A2$	0.43	5.3
$A3a$	0.61	1.04
$A3b$	0.91	0.06

Table 5.1 *The obtained acceptance rate and the number of iterations per second on a* 2.6*-MHz CPU, for approximation* $A1$ *to* $A3b$*.*

0.91 is impressive, but we can push this limit even further using more computing.

Approximation $A2$ offers in many cases a significant improvement compared to $A1$ without paying too much computationally. The reduction from 23.5 iterations per second to 5.3 per second is larger than it would be for a large spatial GMRF. This is because a larger amount of time will be spent factorizing the precision matrix and locating the maximum, which is common for all approximations.

The number of regions in the log-spline approximations K also influences the accuracy. If we increase K we improve the approximation, most notably when the acceptance rate is high. In most cases a value of K between 10 and 20 is sufficient.

A more challenging situation appears when we fix the parameters at different values from their maximum likelihood estimates. The effect of the different approximations can then be drastic. As an example, if we reduce τ_ϵ by a factor of 10 while keeping the other two parameters unchanged, $A1$ produces an acceptance rate of essentially zero. The acceptance rate for $A2$ and $A3a$ is about 0.04 and 0.10, respectively.

In our experience, much computing is required to obtain an acceptance rate in the high 90s, while in practice, only a 'sufficiently' accurate approximation is required, i.e., one that produces an acceptance rate well above zero. Unfortunately, the approximations do not behave uniformly over the space of the hyperparameters. Although a GMRF approximation can be adequate near the global mode, it may not be sufficiently accurate for other values of the hyperparameters. Further, the accuracy of the approximation decreases for increasing dimension.

5.2.3 Example: Reanalyzing Tokyo rainfall data

We will now revisit the logistic RW2 model with binomial observations used to analyze the Tokyo rainfall data in Section 4.3.4. The purpose of this example is to illustrate how the various approximations can be used to construct an *independence sampler* for $(\kappa, \boldsymbol{x})$ and to discuss how we can avoid simulation completely at the cost of some approximation error.

Constructing an independence sampler

We assume from here on the logit-link and a $\mathcal{G}(1.0, 0.000289)$ prior for the precision κ of the second-order random walk $\boldsymbol{x}$. To construct an independence sampler for $(\kappa, \boldsymbol{x})$ we will construct a joint proposal of the following form:

$$\widetilde{\pi}(\boldsymbol{x}, \kappa \mid \boldsymbol{y}) = \widetilde{\pi}(\kappa \mid \boldsymbol{y})\, \widetilde{\pi}(\boldsymbol{x} \mid \boldsymbol{y}, \kappa). \tag{5.35}$$

The first term is an approximation to the posterior marginal for κ, while the second term is an approximation to the full conditional for $\boldsymbol{x}$.

The posterior marginal can be approximated starting from the simple identity

$$\begin{aligned} \pi(\kappa \mid \boldsymbol{y}) &= \frac{\pi(\boldsymbol{x}, \kappa \mid \boldsymbol{y})}{\pi(\boldsymbol{x} \mid \kappa, \boldsymbol{y})} \\ &\propto \frac{\pi(\boldsymbol{y} \mid \boldsymbol{x})\, \pi(\boldsymbol{x} \mid \kappa)\, \pi(\kappa)}{\pi(\boldsymbol{x} \mid \kappa, \boldsymbol{y})}. \end{aligned} \tag{5.36}$$

We can evaluate the rhs for any fixed (valid) value of $\boldsymbol{x}$, as the lhs does not depend on $\boldsymbol{x}$. Note that (5.36) is equivalent to the alternative formulation

$$\pi(\kappa \mid \boldsymbol{y}) = \int \pi(\boldsymbol{x}, \kappa \mid \boldsymbol{y})\, d\boldsymbol{x},$$

which is used for sampling $(\kappa, \boldsymbol{x})$ jointly from the posterior and estimating $\pi(\kappa|\boldsymbol{y})$ considering the samples of κ only.

The only unknown term in (5.36) is the denominator. An approximation to the posterior marginal for κ can be obtained using an approximation $\widetilde{\pi}(\boldsymbol{x}|\kappa, \boldsymbol{y})$ in the denominator of

$$\widetilde{\pi}(\kappa \mid \boldsymbol{y}) \propto \frac{\pi(\boldsymbol{y} \mid \boldsymbol{x})\, \pi(\boldsymbol{x} \mid \kappa)\, \pi(\kappa)}{\widetilde{\pi}(\boldsymbol{x} \mid \kappa, \boldsymbol{y})}. \tag{5.37}$$

Note that the rhs now depends on the value of $\boldsymbol{x}$. In particular, we can choose $\boldsymbol{x}$ as a function of κ such that the denominator is as accurate as possible.

To illustrate the dependency of $\boldsymbol{x}$ in (5.37), we computed $\widetilde{\pi}(\kappa|\boldsymbol{y})$ using the GMRF approximation ($A1$) in the rhs, and evaluate the rhs at $\boldsymbol{x} = \mathbf{0}$ and at the mode $\boldsymbol{x} = \boldsymbol{x}^*(\kappa)$, which depends on κ. The result is shown in Figure 5.11. The difference between the two estimates is quite large. Intuitively, the GMRF approximation is most accurate at the mode and therefore we should evaluate the rhs of (5.37) at $\boldsymbol{x}^*(\kappa)$ and not at any other point. Note that in this case (5.37) is a *Laplace* approximation, see, for example, Tierney et al. (1989).

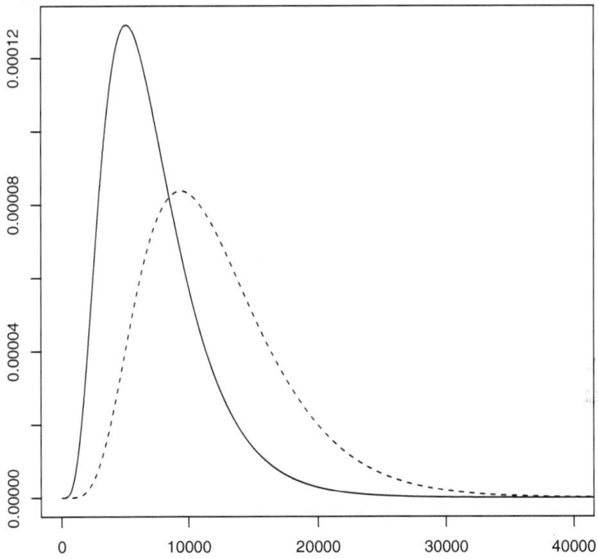

Figure 5.11 *The estimated posterior marginal for* κ *using (5.37) and the GMRF approximation, evaluating the rhs using the mode* $\boldsymbol{x}^*(\kappa)$ *(solid line) or* $\boldsymbol{x} = \boldsymbol{0}$ *(dashed line).*

We will use the same approximations as defined in Section 5.2.2 with obvious changes due to cyclic boundary conditions. The posterior marginal for κ was approximated using each of these approximations, evaluating (5.37) at $\boldsymbol{x}^*(\kappa)$. The estimates where indistinguishable on the plot and the densities all coincide with the solid line in Figure 5.11. It is no contradiction that the different approximations produce nearly the same estimate for $\pi(\kappa|\boldsymbol{y})$; If the ratio of the densities produced with $A1$ and $A2$ is proportional, then this constant will cancel after re-normalizing (5.36). To illustrate this point, let $z_1 \sim \mathcal{N}(0, \sigma^2)$ and $z_2 \sim \mathcal{N}(0, \sigma^2)1_{[z_2>0]}$, then the densities for z_1 and z_2 are proportional for positive arguments for any value of σ^2, but the densities are very different.

We now construct an independence sampler using the approximations for each term in (5.35). The acceptance rates obtained were 0.83, 0.87, 0.94, and 0.95 for approximation $A1$, $A2$, $A3a$, and $A3b$ respectively. Note that the correlation in the marginal chain for κ, $\kappa^{(1)}, \kappa^{(2)}, \ldots$ satisfies

$$\text{Corr}(\kappa^{(i)}, \kappa^{(i+k)}) \approx (1-\alpha)^{|k|}, \tag{5.38}$$

where α is the acceptance rate for the independence sampler. The result (5.38) holds exactly in an ideal situation where an independent proposal is accepted with a fixed probability α. A comparison of (5.38)

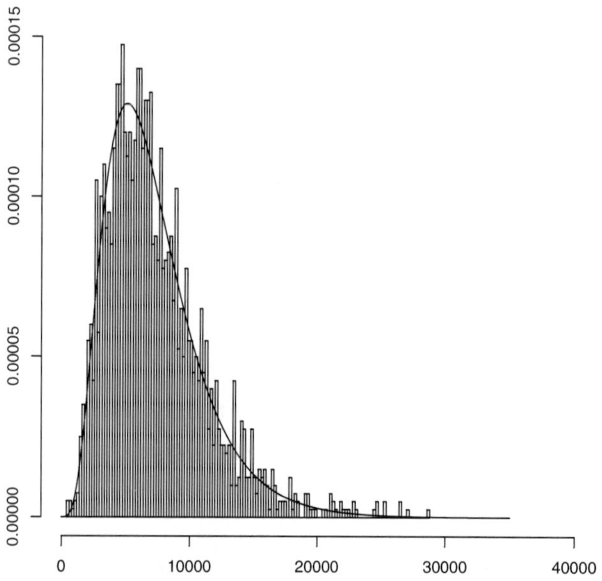

Figure 5.12 *The histogram of the posterior marginal for* κ *based on* 2000 *successive samples from the independence samples constructed from* $A2$. *The solid line is the approximation* $\widetilde{\pi}(\kappa|\boldsymbol{y})$.

with the estimated correlations from each of the four independence samplers verifies that the approximations is very accurate indeed.

We now run the independence sampler based on $A2$. Figure 5.12 displays the histogram of the first 2000 realizations of κ together with $\widetilde{\pi}(\kappa|\boldsymbol{y})$ (solid line). The histogram is in good accordance with the approximated posterior marginal. With very long runs of the independence sampler, the estimated posterior marginal for κ fits the approximation very well. However, we are not able to detect any 'errors' in our approximation. The reason is that the estimate based on the simulations will always be influenced by Monte Carlo error, which is of $\mathcal{O}_p(1/\sqrt{N})$, where N is the number of simulations. This naturally raises the question whether we can avoid simulation completely in this example.

Approximative inference not using simulation

We will now briefly discuss deterministic alternatives to simulation-based inference. Such approaches will only be approximative but fast to compute.

Consider first inference for κ based on $\pi(\kappa|\boldsymbol{y})$. We can avoid simulation if we base our inference on the computed approximation $\widetilde{\pi}(\kappa|\boldsymbol{y})$. This is a univariate density and we can easily construct a log-linear or log-quadratic spline representation of it. Although not everything can be computed analytically it will be available using numerical techniques.

The inference for $\boldsymbol{x}$ is more involved. Assume for simplicity that we aim to estimate

$$\mathrm{E}(f(x_i) \mid \boldsymbol{y}) = \int\int f(x_i)\, \pi(\boldsymbol{x}, \kappa \mid \boldsymbol{y}) d\kappa\, d\boldsymbol{x} \qquad (5.39)$$

for some function $f(x_i)$. Of particular interest are the choices $f(x_i) = x_i$ and $f(x_i) = x_i^2$, which are required to compute the posterior mean and variance. We now approximate (5.39) using (5.35)

$$\begin{aligned} \mathrm{E}(f(x_i) \mid \boldsymbol{y}) &\approx \int\int f(x_i)\, \widetilde{\pi}(\boldsymbol{x}, \kappa \mid \boldsymbol{y}) d\kappa\, d\boldsymbol{x} \\ &= \int \left[\int f(x_i)\, \widetilde{\pi}(\boldsymbol{x} \mid \kappa, \boldsymbol{y})\, d\boldsymbol{x}\right] \widetilde{\pi}(\kappa \mid \boldsymbol{y})\, d\kappa. \end{aligned}$$

Using approximation $A1$, we can easily compute the marginal $\widetilde{\pi}(x_i|\kappa, \boldsymbol{y})$, and hence

$$\begin{aligned} \mathrm{E}(f(x_i) \mid \boldsymbol{y}) &\approx \int \left[\int f(x_i)\, \widetilde{\pi}(x_i \mid \kappa, \boldsymbol{y})\, dx_i\right] \widetilde{\pi}(\kappa \mid \boldsymbol{y})\, d\kappa \\ &\approx \sum_{\kappa} \left[\int f(x_i)\, \widetilde{\pi}(x_i \mid \kappa, \boldsymbol{y})\, dx_i\right] \widetilde{\pi}(\kappa \mid \boldsymbol{y}) \omega(\kappa) \qquad (5.40) \end{aligned}$$

with some weights $\omega(\kappa)$ over a selection of values of κ. The integral is just one-dimensional and may be computed analytically or numerically. A similar technique can be applied using approximation $A2$, $A3a$, and $A3b$, since we can always arrange for the reordering such that index i becomes index $n-1$ after reordering. Hence, $\widetilde{\pi}(x_i|\kappa, \boldsymbol{y})$ is available as a log-quadratic and linear-spline representation. Numerical evaluation of the inner integral can then be used.

The error using (5.40) comes from two sources; the error using the approximation itself and the error replacing the two-dimensional integral by a finite sum. The last error is easy to control so the main contribution to the error comes from using the approximation itself. This error is hard to control, but some insight can be gained if we study the effect of increasing the accuracy of the approximation (from $A1$ to $A2$ or $A3a$, say). In any case, an error of the same order as the Monte Carlo error in a simulation-based approach will be acceptable.

5.3 Bibliographic notes

For a background on Gaussian fields, see Cressie (1993) or Chilés and Delfiner (1999). Section 5.1 is based on Rue and Tjelmeland (2002) but contains an extended discussion. GMRF approximations to GFs have also been applied to nonisotropic correlation functions (Rue and Tjelmeland, 2002) and to spatiotemporal Gaussian fields (Allcroft and Glasbey, 2003). Hrafnkelsson and Cressie (2003) also suggest using the neighborhood radius of the GMRF as a calibration parameter to approximate the CFs of the Matérn family. GMRF approximations have been applied by Follestad and Rue (2003), Husby and Rue (2004), Rue and Follestad (2003), Rue et al. (2004), Steinsland (2003), Steinsland and Rue (2003) and Werner (2004).

Section 5.2 is based on Rue et al. (2004), which contains more challenging examples. The approximations are also applied in Steinsland and Rue (2003) constructing overlapping block proposals for HGMRFs.

Appendices

APPENDIX A

Common distributions

The definitions of common distributions used are given here.

The normal distribution See Section 2.1.7.

The Student-t_ν distribution The density of a Student-t_ν distributed variable x with ν degrees of freedom is

$$\pi(x) = \frac{1}{\sqrt{\nu}\mathrm{B}(\nu/2, 1/2)} \left(\frac{\nu}{\nu + x^2}\right)^{(1+\nu)/2},$$

where $\mathrm{B}(a, b)$ is the *beta function*

$$\mathrm{B}(a, b) = \frac{\Gamma(a)\Gamma(b)}{\Gamma(a + b)}$$

and $\Gamma(z)$ is the *gamma function* equals $(z - 1)!$ for $z = 1, 2, \ldots$ and

$$\Gamma(z) = \int_0^\infty t^{z-1} \exp(-t)\, dt$$

in general. The mean is 0 and the variance is $\nu/(\nu - 2)$ for $\nu > 2$. We abbreviate this as $x \sim t_\nu$.

The gamma distribution The density of a gamma-distributed variable $\tau > 0$ with shape parameter $a > 0$ and inverse-scale parameter $b > 0$, is

$$\pi(\tau) = \frac{b^a}{\Gamma(a)} \tau^{a-1} \exp(-b\tau).$$

The mean of τ equals a/b and the variance equals a/b^2. We abbreviate this as $\tau \sim \mathcal{G}(a, b)$. For $a = 1$ we obtain the *exponential distribution* with parameter $1/b$. If also $b = 1$ we obtain the *standard exponential distribution*, which has both mean and variance equal to 1.

The Laplace distribution The density of a standard Laplace distributed variable x is

$$\pi(x) = \frac{1}{2} \exp(-|x|).$$

The mean is 0 and the variance is 2.

The logistic distribution The density of a logistic-distributed variable x with parameters a and $b > 0$ is

$$\pi(x) = \frac{1}{b} \frac{\exp(-((x-a)/b))}{[1 + \exp(-((x-a)/b))]^2}.$$

The mean is a and the variance is $\pi^2 b^2/3$. We abbreviate this as $x \sim \mathcal{L}(a, b)$. For $a = 0$ and $b = 1$ we obtain the *standard* logistic distribution.

The Kolmogorov-Smirnov distribution The distribution function for a Kolmogorov-Smirnov-distributed variable x, is

$$G(x) = \sum_{k=-\infty}^{\infty} (-1)^k \exp(-2k^2 x^2), \quad x > 0.$$

We abbreviate this as $x \sim \mathcal{KS}$.

The Bernoulli and binomial distribution A binary random variable x is Bernoulli-distributed if $\text{Prob}(x = 1) = p$ and $\text{Prob}(x = 0) = 1 - p$. The mean is p and the variance $p(1-p)$. We abbreviate this as $x \sim \mathcal{B}(p)$. The sum s, of n independent Bernoulli-distributed variables $\{x_i\}$ where $x_i \sim \mathcal{B}(p)$ is binomial distributed. We abbreviate this as $s \sim \mathcal{B}(n, p)$, where $\text{Prob}(s = k) = \binom{n}{k} p^k (1-p)^{n-k}$ for $s = 0, \ldots, n$. The mean of s is np and the variance is $np(1-p)$.

The Poisson distribution A discrete random variable $x \in \{0, 1, 2, \ldots\}$ is Poisson-distributed if

$$\text{Prob}(x = k) = \frac{\lambda^k}{k!} \exp(-\lambda), \quad k = 0, 1, 2, \ldots,$$

where $\lambda \geq 0$. Both the mean and the variance of x equals λ. We abbreviate this as $x \sim \mathcal{P}(\lambda)$.

APPENDIX B

The library GMRFLib

This appendix contains a short discussion of how we actually organize and perform the computations in the (open-source) library GMRFLib (Rue and Follestad, 2002). The library is written in C and Fortran. We first describe the graph-object and the function Qfunc defining the elements $\{Q_{ij}\}$ in the precision matrix, then how to sample a subset of $\boldsymbol{x}$, $\boldsymbol{x}_A$ conditionally on $\boldsymbol{x}_{-A}$ when $\boldsymbol{x}$ is a GMRF, and finally how to construct a block-updating MCMC algorithm for hierarchical GMRF models. Along with these examples we also give C code illustrating how to implement these tasks in GMRFLib, ending up with the code to analyze the Tokyo rainfall data.

At the time of writing, the library GMRFLib (version 2.0) supports two sparse matrix libraries: the band-matrix routines in the Lapack-library (Anderson et al., 1995) using the Gibbs-Poole-Stockmeyer reorder algorithm for bandwidth reduction (Lewis, 1982), and the multifrontal supernodal Cholesky factorisation implementation in the TAUCS-library (Toledo et al., 2002) using the nested dissection reordering from the METIS-library (Karypis and Kumar, 1998).

We now give a brief introduction to GMRFLib. The library contains many more useful features not discussed here and we refer to the manual for further details.

B.1 The graph object and the function Qfunc

The graph-object is a representation of a labelled graph and has the following structure:

n The size of the graph.

nnbs A vector defining the number of neighbors, so nnbs[i] is the number of neighbors to node i.

nbs A vector of vectors defining which nodes are the neighbors; node i has $k =$ nnbs[i] neighbors, which are the nodes nbs[i][1], nbs[i][2], ..., nbs[i][k].

For example, the graph in Figure B.1 has the following representation; n $= 4$, nnbs $= [1, 3, 1, 1]$, nbs[1] $= [2]$, nbs[2] $= [1, 3, 4]$, nbs[3] $= [2]$, and nbs[4] $= [2]$. The graph can be defined in an external text-file with

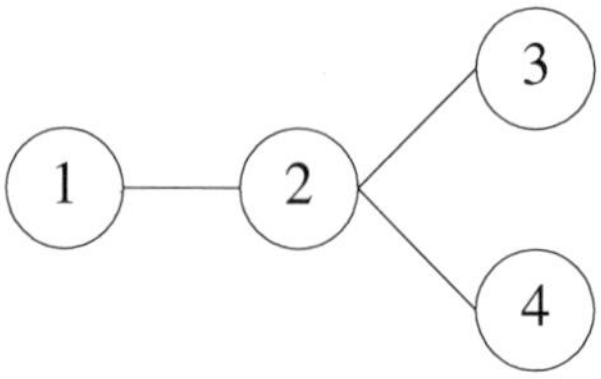

Figure B.1 *The representation for this graph is* $\mathsf{n} = 4$*,* $\mathsf{nnbs} = [1, 3, 1, 1]$*,* $\mathsf{nbs}[1] = [2]$*,* $\mathsf{nbs}[2] = [1, 3, 4]$*,* $\mathsf{nbs}[3] = [2]$ *and* $\mathsf{nbs}[4] = [2]$*.*

format,

```
4
1 1 2
2 3 1 3 4
3 1 2
4 1 2
```

The first number is n, then each line gives the relevant information for each node: node 1 has 1 neighbor, which is node 2, node 2 has 3 neighbors, which are nodes 1, 3, and 4, and so on. Note that there is some redundancy here because we know that if $i \sim j$ then $j \sim i$.

We then need to define the elements in the $\boldsymbol{Q}$ matrix. We know that the only nonzero terms in $\boldsymbol{Q}$ are those Q_{ij} where $i \sim j$ or $i = j$. A convenient way to represent this is to define the function

$$\mathsf{Qfunc}(i, j), \qquad \text{for } i = j \text{ or } i \sim j,$$

returning Q_{ij}.

To illustrate the graph-object and the use of the Qfunc-function, Algorithm B.1 demonstrates how to compute $\boldsymbol{y} = \boldsymbol{Q}\boldsymbol{x}$. Note that only the nonzero terms in $\boldsymbol{Q}$ are used to compute $\boldsymbol{Q}\boldsymbol{x}$. Recall that $i \not\sim i$, so we need to add the diagonal terms explicitly.

Algorithm B.1 Computing $\boldsymbol{y} = \boldsymbol{Q}\boldsymbol{x}$

for $i = 1$ **to** n **do**
  $y[i] = x[i] * \mathsf{Qfunc}(i, i)$
  for $k = 1$ **to** $\mathsf{nnbs}[i]$ **do**
    $j = \mathsf{nbs}[i][k]$
    $y[i] = y[i] + x[j] * \mathsf{Qfunc}(i, j)$
  end for
end for
Return $\boldsymbol{y}$

The following `C`-program illustrates the `Qfunc`-function and the `graph`-object. The program creates the graph for a circular RW1 model and defines the corresponding `Qfunc`-function. It then writes out the nonzero terms in the precision matrix. A third argument is passed to `Qfunc` to transfer additional parameters. Note that the nodes in the `graph`-object in `GMRFLib` are numbered from 0 to $n-1$.

```
#include <stdio.h>
#include "GMRFLib/GMRFLib.h"                        /* definitions of GMRFLib */
double Qfunc(int i,  int j, char *kappa)
{
    /* this function returns the element Q_{ij} in the precision matrix with the additional
     * parameter in Qfunc_argument. recall that this function is *only* called with pairs ij where
     * i\sim j or i=j */
    if (i == j)
        return 2.0* *((double *)kappa);             /* return Q_{ii} */
    else
        return - *((double *)kappa);                /* return Q_{ij} where i \sim j */
}
int main(int argc, char **argv)
{
    /* create the graph for a circular RW1 */
    GMRFLib_graph_tp *graph;                        /* pointer to the graph */
    int n=5, bandwidth=1, cyclic=GMRFLib_TRUE; /* size of the graph, the bandwidth and cyclic flag */
    GMRFLib_make_linear_graph(&graph, n, bandwidth, cyclic);
    /* display the graph and the Q_{ij}'s using kappa = 1 */
    int i, j, k;
    double kappa = 1.0;                             /* use kappa=1 */
    printf("the size of the graph is n = %1d\n", graph->n);
    for(i=0; i<graph->n; i++)
    {
        printf("node %1d have %d neighbors\n", i, graph->nnbs[i]);
        printf("\tQ(%1d, %1d) = %.3f\n", i, i, Qfunc(i, i, (char *)&kappa));
        for(k=0;k<graph->nnbs[i];k++)
        {
            j = graph->nbs[i][k];
            printf("\tQ(%1d, %1d) = %.3f\n", i, j, Qfunc(i, j, (char *)&kappa));
        }
    }
    return(0);
}
```

The output of the program is

```
the size of the graph is n = 5
node 0 have 2 neighbors
        Q(0, 0) = 2.000
        Q(0, 1) = -1.000
        Q(0, 4) = -1.000
node 1 have 2 neighbors
        Q(1, 1) = 2.000
        Q(1, 0) = -1.000
        Q(1, 2) = -1.000
node 2 have 2 neighbors
        Q(2, 2) = 2.000
        Q(2, 1) = -1.000
        Q(2, 3) = -1.000
node 3 have 2 neighbors
        Q(3, 3) = 2.000
        Q(3, 2) = -1.000
        Q(3, 4) = -1.000
node 4 have 2 neighbors
        Q(4, 4) = 2.000
        Q(4, 0) = -1.000
        Q(4, 3) = -1.000
```

B.2 Sampling from a GMRF

We will now outline how to produce a sample from a GMRF. Although a parameterization using the mean $\boldsymbol{\mu}$ and the precision matrix $\boldsymbol{Q}$ is sufficient, we often face the situation that these are only known implicitly. This is caused by conditioning on a part of $\boldsymbol{x}$ and/or other variables such as observed data. In order to avoid a lot of cumbersome computing for the user, the library `GMRFLib` contains a high-level interface to address a more general problem that in our experience covers most situations of interest. The task is to sample from $\pi(\boldsymbol{x}_A|\boldsymbol{x}_B)$ (and/or to evaluate the normalized density) where for notational convenience we denote the set $-A$ by B. Additionally, there may be a hard or soft constraint, but we do not discuss this option here.

The joint density is assumed to be of the form:

$$\log \pi(\boldsymbol{x}) = -\frac{1}{2}(\boldsymbol{x}-\boldsymbol{\mu})^T(\boldsymbol{Q}+\text{diag}(\boldsymbol{c}))(\boldsymbol{x}-\boldsymbol{\mu}) + \boldsymbol{b}^T\boldsymbol{x} + \text{const.} \tag{B.1}$$

Here, $\boldsymbol{\mu}$ is a parameter and not necessarily the mean as $\boldsymbol{b}$ can be nonzero. Furthermore, $\text{diag}(\boldsymbol{c})$ is a diagonal matrix with the vector $\boldsymbol{c}$ on the diagonal. The extra cost of allowing an extended parameterization is negligible with the cost of computing the factorisation of the precision matrix. To compute the conditional density, we expand (B.1) into terms $\boldsymbol{x}_A$, $\boldsymbol{x}_B$, to obtain its canonical parameterization

$$\boldsymbol{x}_A \mid \boldsymbol{x}_B \sim \mathcal{N}_C(\widetilde{\boldsymbol{b}}, \widetilde{\boldsymbol{Q}}),$$

where

$$\widetilde{\boldsymbol{b}} = \boldsymbol{b}_A + (\boldsymbol{Q}_{AA} + \text{diag}(\boldsymbol{c}_A))\boldsymbol{\mu}_A - \boldsymbol{Q}_{AB}(\boldsymbol{x}_B - \boldsymbol{\mu}_B) \quad \text{(B.2)}$$

$$\widetilde{\boldsymbol{Q}} = \boldsymbol{Q}_{AA} + \text{diag}(\boldsymbol{c}_A). \quad \text{(B.3)}$$

The graph is $\mathcal{G}^A$ as adding $\text{diag}(\boldsymbol{c}_A)$ does not change it.

We now need to compute $\mathcal{G}^A$. It is straightforward but somewhat technical to do this efficiently, so we skip the details here. The graph-object is labelled, therefore we need to know to which node in $\mathcal{G}$ a node in $\mathcal{G}^A$ corresponds. Let this mapping be m, such that node i in $\mathcal{G}^A$ corresponds to node $\mathsf{m}[i]$ in $\mathcal{G}$.

It is not that hard to compute $\widetilde{\boldsymbol{b}}$, which is clear from Algorithm B.2. The only simplification made is the following: If $i \in A$ and $j \sim i$, then either $j \in A$ or $j \in B$, so we can compute all terms in one loop.

Algorithm B.2 Computing $\widetilde{\boldsymbol{b}}$ in (B.2)

---

for $i^A = 1$ **to** n^A **do**
  $i = \mathsf{m}[i^A]$
  $\widetilde{\boldsymbol{b}}[i^A] = \boldsymbol{b}[i] + (\mathsf{Qfunc}(i,i) + \boldsymbol{c}[i]) * \boldsymbol{\mu}[i]$
  for $k = 1$ **to** $\mathsf{nnbs}[i]$ **do**
    $j = \mathsf{nbs}[i][k]$
    if $j \in B$ **then**
      $\boldsymbol{b}[i^A] = \widetilde{\boldsymbol{b}}[i^A] - \mathsf{Qfunc}(i,j) * (\boldsymbol{x}[j] - \boldsymbol{\mu}[j])$
    else
      $\widetilde{\boldsymbol{b}}[i^A] = \widetilde{\boldsymbol{b}}[i^A] + \mathsf{Qfunc}(i,j) * \boldsymbol{\mu}[j]$
    end if
  end for
end for
Return $\widetilde{\boldsymbol{b}}$

We now apply Algorithm 2.5 using $\mathcal{G}^A$, $\widetilde{\boldsymbol{b}}$, and $\widetilde{\mathsf{Qfunc}}$ defined as

$$\widetilde{\mathsf{Qfunc}}(i,j) \equiv \mathsf{Qfunc}(\mathsf{m}[i], \mathsf{m}[j]) + 1_{[i=j]}\boldsymbol{c}[\mathsf{m}[i]]$$

to obtain a sample $\widetilde{\boldsymbol{x}}$. Finally, we may insert this sample into $\boldsymbol{x}$,

$$\boldsymbol{x}[\mathsf{m}[i]] = \widetilde{\boldsymbol{x}}[i], \quad i = 1, \ldots, \mathsf{n}^A$$

and we are done.

Example B.1 *Consider the graph on the left in Figure B.2 with 6 nodes. We want to sample from $\pi(\boldsymbol{x}_A|\boldsymbol{x}_B)$ where $A = \{3,5,6\}$ and the joint distribution of $\boldsymbol{x}$ is given in (B.1).*

The subgraph $\mathcal{G}^A$ is shown on the right in Figure B.2, where it is indicated how each node in $\mathcal{G}^A$ connects to a node in $\mathcal{G}$. The mapping

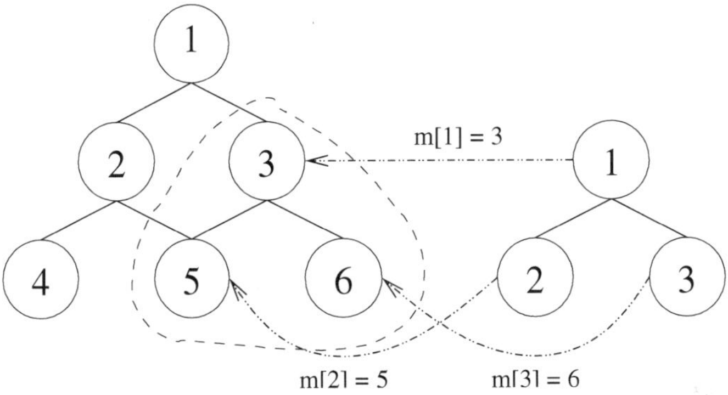

Figure B.2 *Illustration of how to compute the conditional distribution* $\pi(\boldsymbol{x}_A|\boldsymbol{x}_B)$. *The graph* $\mathcal{G}$ *is shown on the left where* $A = \{3, 5, 6\}$ *is marked. The subgraph* $\mathcal{G}^A$ *is shown on the right, indicating which node in* $\mathcal{G}^A$ *corresponds to a node in* $\mathcal{G}$. *The mapping is* $\boldsymbol{m} = [3, 5, 6]$.

m, is $\mathsf{m} = [3, 5, 6]$, *meaning that node* i *in* $\mathcal{G}^A$ *corresponds to node* $\mathsf{m}[i]$ *in* $\mathcal{G}$.

The conditional precision matrix is $\boldsymbol{Q}_{AA} + diag(\boldsymbol{c}_A)$, *which reads*

$$\begin{pmatrix} Q_{33} + c_3 & Q_{35} & Q_{36} \\ Q_{53} & Q_{55} + c_5 & Q_{56} \\ Q_{63} & Q_{65} & Q_{66} + c_6 \end{pmatrix}$$

and following Algorithm B.2, we obtain $\widetilde{\boldsymbol{b}}$:

$$\begin{aligned} \widetilde{b}_1 &= b_3 + (Q_{33} + c_3)\mu_3 - Q_{31}(x_1 - \mu_1) + Q_{35}\mu_5 \\ \widetilde{b}_2 &= b_5 + (Q_{55} + c_5)\mu_5 - Q_{52}(x_2 - \mu_2) + Q_{53}\mu_3 \\ \widetilde{b}_3 &= b_6 + (Q_{66} + c_6)\mu_6 + Q_{65}\mu_5 + Q_{63}\mu_3. \end{aligned}$$

Note that x_4 *is not used since* $4 \notin ne(A)$. *We sample* $\widetilde{\boldsymbol{x}}$ *from its canonical parameterization using Algorithm 2.5. Finally, we may insert the sample into* $\boldsymbol{x}$,

$$x_3 = \widetilde{x}_1, \quad x_5 = \widetilde{x}_2, \quad x_6 = \widetilde{x}_3.$$

The following `C`-program illustrates how to produce (conditional) samples from a GMRF. We continue with the circular RW1 model and condition on $x_1 = 1$ and $x_{245} = 10$. The function `GMRFLib_init_problem` computes (and stores in `problem`) the conditional density and all intermediate variables needed to produce samples (using `GMRFLib_sample`) or to evaluate the log density of some configuration $\boldsymbol{x}$ (using the function `GMRFLib_evaluate`). In the program we extract the conditional mean and compute the empirical mean of 100 iid samples. These quantities

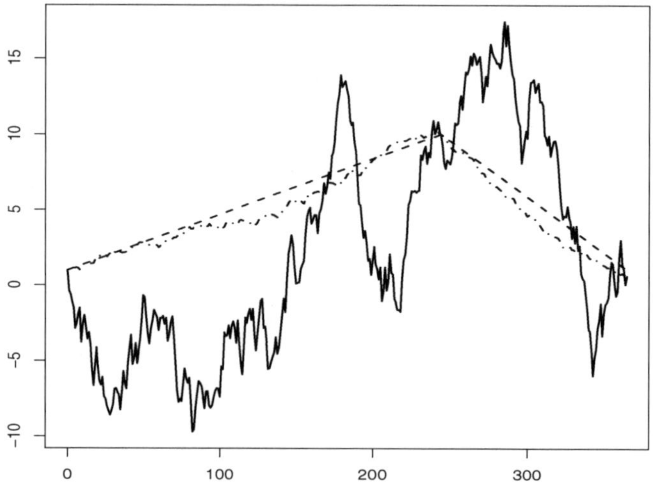

Figure B.3 *The conditional mean (dashed line), the empirical mean (dashed-dotted line) and one sample (solid line) from a circular RW1 model with* $\kappa = 1$*,* $n = 366$*, conditional on* $x_1 = 1$ *and* $x_{245} = 10$.

and one sample from the conditional distribution are shown in Figure B.3.

```
#include <assert.h>
#include <stdio.h>
#if !defined(__FreeBSD__)
#include <malloc.h>
#endif
#include <stdlib.h>
#include "GMRFLib/GMRFLib.h"                      /* include definitions of GMRFLib */

double Qfunc(int i,  int j, char *kappa)
{
    return *((double *)kappa) * (i==j ? 2.0 : -1.0);
}
int main(int argc, char **argv)
{
    assert(argv[1]);
    (*GMRFLib_uniform_init)(atoi(argv[1]));   /* init the RNG with the seed in the first argument */
    /* create the graph for a circular RW1 */
    GMRFLib_graph_tp *graph;                      /* pointer to the graph */
    int n         = 366;                          /* size of the graph */
    int bandwidth = 1;                            /* the bandwidth is 1 */
    int cyclic    = GMRFLib_TRUE;                 /* cyclic graph */
    GMRFLib_make_linear_graph(&graph, n, bandwidth, cyclic);

    GMRFLib_problem_tp *problem;                  /* hold the problem */
    GMRFLib_constr_tp *constr = NULL;             /* no constraints */
    double *x, *mean=NULL, *b=NULL, *c=NULL;      /* various vectors some are NULL */
    char *fixed;                                  /* indicate which are x_i's are fixed or not */

    x     = calloc(n, sizeof(double));            /* allocate space */
    fixed = calloc(n, sizeof(char));              /* allocate space */

    /* sample from a circular RW1 conditioned on x_0 and x_{2n/3} */
    fixed[0]     = 1;                             /* fix x[0] */
    x[0]         = 1.0;                           /* ...to 1.0 */
```

```
    fixed[2*n/3] = 1;                              /* fix x[2*n/3] */
    x[2*n/3]     = 10.0;                           /* ...to 10.0 */
    double kappa = 1.0;                            /* set kappa = 1 */

    GMRFLib_init_problem(&problem, x,  b, c, mean, graph, Qfunc, (char *)&kappa, /* init the problem */
                         fixed, constr, GMRFLib_NEW_PROBLEM);
    /* extract the conditional mean, which is available for the sub_graph only.  map it back using
     * the mapping problem->map. */
    mean = x;                                      /* use same storage */
    int i;
    for(i=0;i<problem->sub_graph->n;i++) mean[problem->map[i]] = problem->sub_mean[i];

    int j, m=100;                           /* sample m=100 samples to estimate the empirical mean */
    double *emean = calloc(n, sizeof(double));
    for(j=0;j<m;j++)
    {
        GMRFLib_sample(problem);
        for(i=0;i<n;i++) emean[i] += problem->sample[i];
    }
    for(i=0;i<n;i++)                               /* write the results to stdout */
    {
        emean[i] /= m;
        printf("%d %f %f %f\n", i, mean[i],  emean[i], problem->sample[i]);
    }
    return(0);
}
```

B.3 Implementing block-updating algorithms for hierarchical GMRF models

GMRFLib also has a high-level routine to construct block updating in hierarchical GMRF models, GMRFLib_blockupdate. This routine locates the mode, constructs the GMRF approximation and computes the contribution to the acceptance probability. Due to the general framework, it is straightforward to implement all examples in Section 4.4 as soon as the GMRF is defined.

We extend (B.4) to include nonquadratic terms in the style of (5.21), so the full density is defined as

$$\begin{aligned}\log \pi(\boldsymbol{x}) = &-\frac{1}{2}(\boldsymbol{x}-\boldsymbol{\mu})^T(\boldsymbol{Q}+\operatorname{diag}(\boldsymbol{c}))(\boldsymbol{x}-\boldsymbol{\mu})+\boldsymbol{b}^T\boldsymbol{x} \\ &+\sum_i d_i g_i(x_i)+\text{const.}\end{aligned} \tag{B.4}$$

Here, $g_i(x_i)$ represents the log-likelihood term, but can represent any (reasonable) function of x_i. The GMRFLib_blockupdate routine constructs from (B.4) its GMRF approximation $\widetilde{\pi}(\boldsymbol{x})$ and sample from it the proposal $\boldsymbol{x}^*$ (forward step). Of course, the reverse or backward step is also performed, constructing the GMRF approximation starting from $\boldsymbol{x}^*$ and evaluating the log density of $\boldsymbol{x}$ using the GMRF approximation. The computations are similar in the case where only $\boldsymbol{x}_A$ is updated keeping $\boldsymbol{x}_{-A}$ fixed, which we use in the subblock algorithm.

As we always do a joint update of the GMRF (or parts of it) with the corresponding hyperparameters $\boldsymbol{\theta}$ we allow the terms $\boldsymbol{\mu}$, $\boldsymbol{Q}$, $\boldsymbol{c}$, $\boldsymbol{b}$, $\boldsymbol{d}$, and $\{g_i(\cdot)\}$ in (B.4) to depend on hyperparameters $\boldsymbol{\theta}$. The acceptance probability for the joint update (4.8) is then $\min\{1, R\}$, where

$$R = \underbrace{\frac{\pi(\boldsymbol{x}^*|\boldsymbol{\theta}^*,\boldsymbol{y})}{\pi(\boldsymbol{x}|\boldsymbol{\theta},\boldsymbol{y})}\ \frac{\widetilde{\pi}(\boldsymbol{x}|\boldsymbol{\theta},\boldsymbol{y})}{\widetilde{\pi}(\boldsymbol{x}^*|\boldsymbol{\theta}^*,\boldsymbol{y})}}_{\text{term 1}}\ \underbrace{\frac{\pi(\boldsymbol{\theta}^*)}{\pi(\boldsymbol{\theta})}\ \frac{q(\boldsymbol{\theta}|\boldsymbol{\theta}^*)}{q(\boldsymbol{\theta}^*|\boldsymbol{\theta})}}_{\text{term 2}},$$

where $\pi(\boldsymbol{\theta})$ is the prior for $\boldsymbol{\theta}$. GMRFLib_blockupdate samples $\boldsymbol{x}^*$ and computes term 1 *except* for the normalization constant with respect to $\boldsymbol{\theta}$. This information is not available in (B.4). Term 2 and the ratio of the normalization constants must be added by the user.

For the Tokyo rainfall example using the logit-link, g_i is

$$g_i(x_i) = y_i \log(p(x_i)) + (m_i - y_i)\log(1-p(x_i)),$$

where $p(x_i) = \exp(x_i)/(1+\exp(x_i))$, y_i is the observed number of counts on day i and $m_i = 2$ except for $i = 60$ where $m_{60} = 1$. The weights $\{d_i\}$

are all 1 in this case, and $\boldsymbol{\mu}$, $\boldsymbol{c}$, and $\boldsymbol{b}$ are vector of zeros. Term 2 is

$$\frac{(\kappa^*)^{a-1}\exp(-b\kappa^*)}{\kappa^{a-1}\exp(-b\kappa)}\,\frac{(\kappa^*)^{(n-1)/2}}{\kappa^{(n-1)/2}},$$

since $\frac{q(\boldsymbol{\theta}|\boldsymbol{\theta}^*)}{q(\boldsymbol{\theta}^*|\boldsymbol{\theta})} = 1$ using (4.13). The prior parameters are $a = 1.0$ and $b = 0.000289$. The following `C` program is an implementation of the one-block algorithm using `GMRFLib`.

```
#include <assert.h>
#include <math.h>
#include <string.h>
#include <stdio.h>
#include <malloc.h>
#include <stdlib.h>
#include "GMRFLib/GMRFLib.h"                    /* include definitions of GMRFLib */

#define Uniform  (*GMRFLib_uniform)             /* use the RNG in GMRFLib. return Unif(0,1)'s */
#define MIN(a,b) ((a) < (b) ? (a) : (b))        /* MIN macro */

typedef struct                                  /* the type holding the data */
{
    double *y,  *n;                             /* observed counts and number of days */
}
data_tp;

double link(double x)
{                                               /* the link function */
    return exp(x)/(1+exp(x));
}
double log_gamma(double x, double a, double b)
{                                    /*  return the log density of a gamma variable with mean a/b. */
    return ((a-1.0)*log(x)-(x)*b);
}
double Qfunc(int i,  int j, char *kappa)
{
    return *((double *)kappa) * (i==j ? 2.0 : -1.0);
}
int gi(double *gis, double *x_i, int m, int idx, double *not_in_use, char *arg)
{
    /* compute g_i(x_i) for m values of x_idx: x_i[0], ..., x_i[m-1]. return its values in gis[0],
     * ..., gis[0]. additional (user) arguments are passed through the pointer gi_arg, here, the
     * data itself. */
```

```
    int k;
    double p;
    data_tp *data;
    data = (data_tp *) arg;
    for(k=0; k<m; k++)
    {
        p      = link(x_i[k]);
        gis[k] = data->y[idx]*log(p) + (data->n[idx] - data->y[idx])*log(1.0-p);
    }
    return 0;
}
double scale_proposal(double F)
{
    /* return a sample from f ~ \pi(f) \propto 1+1/f, on the interval [1/F, F]. write the density as
     * a mixture and sample each component with the correct probability. */
    double len = F - 1/F;
    if (F == 1.0) return 1.0;
    if (Uniform() < len/(len+2*log(F)))
        return 1/F + len*Uniform();
    else
        return pow(F, 2.0*Uniform()-1.0);
}
int main(int argc, char **argv)
{
    assert(argv[1]);
    (*GMRFLib_uniform_init)(atoi(argv[1])); /* init the RNG with the seed in the first argument */

    GMRFLib_graph_tp *graph;                        /* pointer to the graph */
    int n         = 366;                            /* size of the graph for the Tokyo example */
    int bandwidth = 1;                              /* the bandwidth is 1 for RW1 */
    int cyclic    = GMRFLib_TRUE;                   /* cyclic graph */
    GMRFLib_make_linear_graph(&graph, n, bandwidth, cyclic);

    GMRFLib_constr_tp *constr = NULL;               /* no constraints */
```

```
double *d, *mean=NULL, *b=NULL, *c=NULL;        /* various vectors some are NULL */
char *fixed = NULL;                             /* none x_i's are fixed */

data_tp data;                                   /* hold the data */
data.y = calloc(n, sizeof(double));             /* allocate space */
data.n = calloc(n, sizeof(double));             /* allocate space */

FILE *fp; int i;                                /* read data */
fp = fopen("tokyo.rainfall.data.dat",  "r"); assert(fp);
for(i=0;i<n;i++) fscanf(fp, "%lf %lf\n", &data.y[i],  &data.n[i]);
fclose(fp);

double *x_old,  *x_new, kappa_old=100.0, kappa_new; /* old and new (the proposal) x and kappa */
x_old = calloc(n, sizeof(double));              /* allocate space */
x_new = calloc(n, sizeof(double));              /* allocate space */

d = calloc(n, sizeof(double));                  /* allocate space */
for(i=0;i<n;i++) d[i] = 1.0;                    /* all are equal to 1 */

while(1)
{                                               /* just keep on until the process is killed */
    kappa_new = scale_proposal(6.0)*kappa_old;
    double log_accept;                          /* GMRFLib_blockupdate does all the job... */
    GMRFLib_blockupdate(&log_accept, x_new,  x_old, b, b, c, c, mean, mean, d, d,
                        gi, (char *) &data, gi, (char *) &data,
                        fixed, graph, Qfunc, (char *)&kappa_new, Qfunc, (char *)&kappa_old, NULL, NULL, NULL, NULL,
                        constr, constr, NULL, NULL);

    double A = 1.0,  B = 0.000289;              /* prior parameters for kappa */
    /* add terms to the acceptance probability not computed by GMRFLib_blockupdate: prior for
     * kappa and the normalising constant. */
    log_accept += ((n-1.0)/2.0*log(kappa_new) + log_gamma(kappa_new, A,  B))
        - ((n-1.0)/2.0*log(kappa_old) + log_gamma(kappa_old, A,  B));
```

```
        static double p_acc = 0.0;          /* sum of the accept probabilities */
        double acc_prob = exp(MIN(log_accept, 0.0));
        p_acc += acc_prob;
        if (Uniform() < acc_prob)
        {                                        /* accept the proposal */
            memcpy(x_old, x_new, n*sizeof(double));
            kappa_old = kappa_new;
        }
        static int count = 0;                    /* number of iterations */
        if (!((++count)%10))                     /* output every 10th */
        {
            printf(" %.3f", kappa_old);
            for(i=0;i<n;i++) printf(" %.3f", link(x_old[i]));
            printf("\n"); fflush(stdout);
            fprintf(stderr, "mean accept prob %f\n", p_acc/count); /* monitor the mean accept probability */
        }
    }
    return(0);
}
```

References

Albert, J. H. and Chib, S. (1993). Bayesian analysis of binary and polychotomous responce data. *Journal of the American Statistical Association*, 88(422), 669–679.

Albert, J. H. and Chib, S. (2001). Sequential ordinal modeling with applications to survival data. *Biometrics*, 57, 829–836.

Allcroft, D. J. and Glasbey, C. A. (2003). A latent Gaussian Markov random field model for spatio-temporal rainfall disaggregation. *Journal of the Royal Statistical Society, Series C*, 52, 487–498.

Amit, Y., Grenander, U., and Piccioni, M. (1991). Structural image restoration through deformable templates. *Journal of the American Statistical Association*, 86, 376–387.

Anderson, E., Bai, Z., Bischof, C., Demmel, J., Dongarra, J. J., Croz, J. D., Greenbaum, A., Hammarling, S., McKenney, A., Ostrouchov, S., , and Sorensen, D. C. (1995). *LAPACK Users' Guide*, 2nd edition. Philadelphia: Society for Industrial and Applied Mathematics.

Andrews, D. F. and Mallows, C. L. (1974). Scale mixtures of normal distributions. *Journal of the Royal Statistical Society, Series B*, 36(1), 99–102.

Anselin, L. and Florax, R., Eds. (1995). *New Directions in Spatial Econometrics*. New York: Springer-Verlag.

Assunção, R. M., Assunção, J. J., and Lemos, M. B. (1998). Induced technical change: A Bayesian spatial varying parameter model. In *Proceedings of XVI Latin American Meeting of Econometric Society*: Catholic University of Peru, Peru.

Assunção, R. M., Potter, J. E., and Cavenaghi, S. M. (2002). A Bayesian space varying parameter model applied to estimating fertility schedules. *Statistics in Medicine*, 21(14), 2057–2075.

Aykroyd, R. G. (1998). Bayesian estimation for homogeneous and inhomogeneous Gaussian random fields. *IEEE Transactions on Pattern Analysis and Machine Intelligence*, 20(5), 533–539.

Banerjee, S., Carlin, B. P., and Gelfand, A. E. (2004). *Hierarchical Modeling and Analysis for Spatial Data*, volume 101 of *Monographs on Statistics and Applied Probability*. London: Chapman & Hall.

Banerjee, S., Wall, M. M., and Carlin, B. P. (2003). Frailty modeling for spatially correlated survival data, with application to infant mortality in Minnesota. *Biostatistics*, 4(123–142).

Barndorff-Nielsen, O., Kent, J. T., and Sørensen, M. (1982). Normal variance-mean mixtures and z distributions. *International Statistical Review*, 50(2), 145–159.

Barone, P. and Frigessi, A. (1989). Improving stochastic relaxation for Gaussian random fields. *Probability in the Engineering and Informational Sciences*, 3(4), 369–389.

Barone, P., Sebastiani, G., and Stander, J. (2001). General over-relaxation Markov chain Monte Carlo algorithms for Gaussian densities. *Statistics & Probability Letters*, 52(2), 115–124.

Barone, P., Sebastiani, G., and Stander, J. (2002). Over-relaxation methods and coupled Markov chains for Monte Carlo simulation. *Statistics and Computing*, 12(1), 17–26.

Bartlett, M. S. (1978). Nearest neighbour models in the analysis of field experiments (with discussion). *Journal of the Royal Statistical Society, Series B*, 40(2), 147–174.

Berzuini, C. and Clayton, C. (1994). Bayesian survival analysis on multiple time scales. *Statistics in Medicine*, 13, 823–838.

Besag, J. (1974). Spatial interaction and the statistical analysis of lattice systems (with discussion). *Journal of the Royal Statistical Society, Series B*, 36(2), 192–225.

Besag, J. (1975). Statistical analysis of non-lattice data. *The Statistician*, 24(3), 179–195.

Besag, J. (1977a). Efficiency of pseudolikelihood estimation for simple Gaussian fields. *Biometrika*, 64(3).

Besag, J. (1977b). Errors-in-variables estimation for Gaussian lattice schemes. *Journal of the Royal Statistical Society, Series B*, 39(1), 73–78.

Besag, J., Green, P. J., Higdon, D., and Mengersen, K. (1995). Bayesian computation and stochastic systems (with discussion). *Statistical Science*, 10(1), 3–66.

Besag, J. and Higdon, D. (1999). Bayesian analysis of agricultural field experiments (with discussion). *Journal of the Royal Statistical Society, Series B*, 61(4), 691–746.

Besag, J. and Kooperberg, C. (1995). On conditional and intrinsic autoregressions. *Biometrika*, 82(4), 733–746.

Besag, J., York, J., and Mollié, A. (1991). Bayesian image restoration with two applications in spatial statistics (with discussion). *Annals of the Institute of Statistical Mathematics*, 43(1), 1–59.

Biller, C. and Fahrmeir, L. (1997). Bayesian spline-type smoothing in generalized regression models. *Computational Statistics*, 12, 135–151.

Bookstein, F. L. (1989). Principal warps: Thin-plate splines and the decomposition of deformations. *IEEE Transactions on Pattern Analysis and Machine Intelligence*, 11(6), 567–585.

Box, G. E. P. and Tiao, G. C. (1973). *Bayesian Inference in Statistical Analysis*. Addison-Wesley Publishing Co., Reading, Mass.-London-Don Mills, Ont.

Bray, I. (2002). Application of Markov chain Monte Carlo methods to projecting cancer incidence and mortality. *Journal of the Royal Statistical Society, Series C*, 51(2), 151–164.

Brezger, A., Kneib, T., and Lang, S. (2003). *BayesX: Software for Bayesian inference*. Department of statistics, University of Munich, version 1.1 edition. http://www.stat.uni-muenchen.de/~lang/bayesx.

Brockwell, P. J. and Davis, R. A. (1987). *Time Series: Theory and Methods*. Berlin: Springer-Verlag.

Brook, D. (1964). On the distinction between the conditional probability and the joint probability approaches in the specification of nearest-neighbour systems. *Biometrika*, 51(3 and 4), 481–483.

Carlin, B. P. and Banerjee, S. (2003). Hierarchical multivariate CAR models for spatio-temporally correlated survival data (with discussion). In *Bayesian Statistics, 7* (pp. 45–63). New York: Oxford Univ. Press.

Carlin, B. P. and Louis, T. A. (1996). *Bayes and Empirical Bayes Methods for Data Analysis*, volume 69 of *Monographs on Statistics and Applied Probability*. London: Chapman & Hall.

Carlin, B. P., Polson, N. G., and Stoffer, D. S. (1992). A Monte Carlo approach to non-normal and nonlinear state-space modeling. *Journal of the American Statistical Association*, 87(418), 493–500.

Carter, C. K. and Kohn, R. (1994). On Gibbs sampling for state space models. *Biometrika*, 81(3), 541–543.

Carter, C. K. and Kohn, R. (1996). Markov chain Monte Carlo in conditionally Gaussian state space models. *Biometrika*, 83(3), 589–601.

Chellappa, R. and Chatterjee, S. (1985). Classification of textures using Gaussian Markov random fields. *IEEE Transactions on Acoustics Speech and Signal Processing*, 33, 959–963.

Chellappa, R., Chatterjee, S., and Bagdazian, R. (1985). Texture synthesis and compression using Gaussian-Markov random field models. *IEEE Transaction On Systems, Man and Cybernetics*, 15(2), 298–303.

Chellappa, R. and Jain, A. K., Eds. (1993). *Markov Random Fields.* Boston, MA: Academic Press Inc.

Chellappa, R. and Kashyap, R. L. (1982). Digital image restoration using spatial interaction models. *IEEE Transaction on Acoustics Speech and Signal Processing*, ASSP-30(3), 614–625.

Chen, M. H. and Dey, D. K. (1998). Bayesian modeling of correlated binary responses via scale mixture of multivariate normal link functions. *Sankhyā. The Indian Journal of Statistics. Series A*, 60(3), 322–343.

Chilés, J. P. and Delfiner, P. (1999). *Geostatistics: Modeling Spatial Uncertainty.* Wiley Series in Probability and Statistics. Chichester: John Wiley & Sons, Ltd.

Clayton, D. G. (1996). Generalized linear mixed models. In W. R. Gilks, S. Richardson, and D. J. Spiegelhalter (Eds.), *Markov Chain Monte Carlo in Practice* (pp. 275–301). London: Chapman & Hall.

Cressie, N. A. C. (1993). *Statistics for spatial data.* Wiley Series in Probability and Mathematical Statistics: Applied Probability and Statistics. New York: John Wiley & Sons Inc. Revised reprint of the 1991 edition, A Wiley-Interscience Publication.

Cressie, N. A. C. and Chan, N. H. (1989). Spatial modeling of regional variables. *Journal of the American Statistical Association*, 84(406), 393–401.

Crook, A. M., Knorr-Held, L., and Hemingway, H. (2003). Measuring spatial effects in time to event data: A case study using months from angiography to coronary artery bypass graft CABG. *Statistics in Medicine*, (22), 2943–2961.

Cross, G. R. and Jain, A. K. (1983). Markov random field texture models. *IEEE Transactions on Pattern Analysis and Machine Intelligence*, 5(1), 25–39.

Dahlhaus, R. and Künsch, H. R. (1987). Edge effects and efficient parameter estimation for stationary random fields. *Biometrika*, 74(4), 877–882.

Davis, P. J. (1979). *Circulant Matrices.* New York: John Wiley & Sons, Ltd.

de Jong, P. and Shephard, N. (1995). The simulation smoother for time series models. *Biometrika*, 82(2), 339–350.

Dempster, A. P. (1972). Covariance selection. *Biometrics*, 28(1), 157–175.

Descombes, X., Sigelle, M., and Préteux, F. (1999). Estimating Gaussian Markov random field parameters in a nonstationary framework: Application to remote sensing imaging. *IEEE Transactions on Image Processing*, 8(4), 490–503.

Devroye, L. (1986). *Non-uniform Random Variate Generation*. Berlin: Springer-Verlag. A copy of the book is freely available from L. Devroye's homepage `http://jeff.cs.mcgill.ca/~luc`.

Dietrich, C. R. and Newsam, G. N. (1996). A fast and exact method for multidimensional Gaussian stochastic simulations: Extension to realizations conditioned on direct and indirect measurements. *Water Resources Research*, 32(6), 1643–1652.

Dietrich, C. R. and Newsam, G. N. (1997). Fast and exact simulation of stationary Gaussian processes through circulant embedding of the covariance matrix. *SIAM Journal of Scientific Computing*, 18(4), 1088–1107.

Diggle, P. J., Ribeiro Jr., P. J., and Christensen, O. F. (2003). An introduction to model-based Geostatistics. In J. Møller (Ed.), *Spatial Statistics and Computational Methods*, Lecture Notes in Statistics; 173 (pp. 43–86). Berlin: Springer-Verlag.

Dobra, A., Hans, C., Jones, B., Nevins, J. R., and West, M. (2003). *Sparse graphical models for exploring gene expression data*. Technical Report 7, Statistical and Applied Mathemathical Sciences Institute, www.samsi.info.

Dongarra, J. J., Duff, I. S., Sorensen, D. C., and van der Vorst, H. A. (1998). *Numerical Linear Algebra for High-performance Computers*. Software, Environments, and Tools. Philadelphia, PA: Society for Industrial and Applied Mathematics (SIAM).

Draper, D. (1995). Assessment and propagation of model uncertainty (with discussion). *Journal of the Royal Statistical Society, Series B*, 57(1), 45–97.

Dreesman, J. M. and Tutz, G. (2001). Non-stationary conditional models for spatial data based on varying coefficients. *The Statistician*, 50(1), 1–15.

Dryden, I. L., Ippoliti, L., and Romagnoli, L. (2002). Adjusted maximum likelihood and pseudo-likelihood estimation for noisy Gaussian Markov random fields. *Journal of Computational and Graphical Statistics*, 11, 370–388.

Dryden, I. L., Scarr, M. R., and Taylor, C. C. (2003). Bayesian texture segmentation of weed and crop images using reversible jump Markov chain Monte Carlo methods. *Journal of the Royal Statistical Society. Series C. Applied Statistics*, 52(1), 31–50.

Dubes, R. and Jain, A. K. (1989). Random field models in image analysis. *Journal of Applied Statistics*, 16(2), 131–164.

Duff, I. S., Erisman, A. M., and Reid, J. K. (1989). *Direct Methods for Sparse Matrices*, 2nd edition. Monographs on Numerical Analysis. New York: The Clarendon Press Oxford University Press. Oxford Science Publications.

Durbin, J. and Koopman, S. J. (1997). Monte Carlo maximum likelihood estimation for non-Gaussian state space models. *Biometrika*, 84(3), 669–684.

Durbin, J. and Koopman, S. J. (2000). Time series analysis of non-Gaussian observations based on state space models from both classical and Bayesian perspectives (with discussion). *Journal of the Royal Statistical Society, Series B*, 62(1), 3–56.

Fahrmeir, L. (1992). Posterior mode estimation by extended Kalman filtering for multivariate dynamic generalised linear models. *Journal of the American Statistical Association*, 87, 501–509.

Fahrmeir, L. (1994). Dynamic modelling and penalized likelihood estimation for discrete time survival data. *Biometrika*, 81(2), 317–330.

Fahrmeir, L. and Knorr-Held, L. (2000). Dynamic and semiparametric models. In M. G. Schimek (Ed.), *Smoothing and Regression: Approaches, Computation, and Application* (pp. 513–544). New-York: John Wiley & Sons, Ltd.

Fahrmeir, L. and Lang, S. (2001a). Bayesian inference for generalized additive mixed models based on Markov random field priors. *Journal of the Royal Statistical Society, Series C*, 50(2), 201–220.

Fahrmeir, L. and Lang, S. (2001b). Bayesian inference for generalized additive mixed models based on Markov random field priors. *Journal of the Royal Statistical Society, Series C*, 50(2), 201–220.

Fahrmeir, L. and Lang, S. (2001c). Bayesian semiparametric regression analysis of multicategorical time-space data. *Annals of the Institute of Statistical Mathematics*, 53(1), 11–30.

Fahrmeir, L. and Tutz, G. (2001). *Multivariate Statistical Modelling Based on Generalized Linear Models*, 2nd edition. Berlin: Springer-Verlag.

Fernández, C. and Green, P. J. (2002). Modelling spatially correlated data via mixtures: A Bayesian approach. *Journal of the Royal Statistical Society, Series B*, 64(4), 805–826.

Ferreira, M. A. R. and De Oliveira, V. (2004). *Bayesian analysis for a class of Gaussian Markov random fields*. Technical Report, Statistical Laboratory, Universidade Federal do Rio de Janeiro, Brazil.

Follestad, T. and Rue, H. (2003). *Modelling spatial variation in disease risk using Gaussian Markov random field proxies for Gaussian random fields.* Statistics Report No. 3, Department of Mathematical Sciences, Norwegian University of Science and Technology, Trondheim, Norway.

Fotheringham, A. S., Brunsdon, C., and Charlton, M. (2002). *Geographically Weighted Regression: The Analysis of Spatially Varying Relationships.* New York: John Wiley & Sons, Ltd.

Frühwirth-Schnatter, S. (1994). Data augmentation and dynamic linear models. *Journal of Time Series Analysis*, 15(2), 183–202.

Gamerman, D. (1997). Sampling from the posterior distribution in generalized linear mixed models. *Statistics and Computing*, 7(1), 57–68.

Gamerman, D., Moreira, A. R. B., and Rue, H. (2003). Space-varying regression models: Specifications and simulations. *Computational Statistics and Data Analysis*, 42(3), 513–533.

Gamerman, D. and West, M. (1987). An application of dynamic survival models in unemplyment studies. *The Statistician*, 36, 269–274.

Gelfand, A. E., Sahu, S. K., and Carlin, B. P. (1995). Efficient parameterisations for normal linear mixed models. *Biometrika*, 82(3), 479–488.

Gelfand, A. E. and Vounatsou, P. (2003). Proper multivariate conditional autoregressive models for spatial data analysis. *Biostatistics*, 4(1), 11–25.

Gelman, A., Carlin, J. B., Stern, H. S., and Rubin, D. B. (2004). *Bayesian Data Analysis*, 2nd edition. Texts in Statistical Science Series. Chapman & Hall/CRC, Boca Raton, FL.

Geman, D. and Yang, C. (1995). Nonlinear image recovery with half-quadratic regularization. *IEEE Transactions on Image Processing*, 4(7), 923–945.

George, A. and Liu, J. W. H. (1981). *Computer solution of large sparse positive definite systems.* Englewood Cliffs, N.J.: Prentice-Hall Inc. Prentice-Hall Series in Computational Mathematics.

Gilks, W. R., Richardson, S., and Spiegelhalter, D. J. (1996). *Markov Chain Monte Carlo in Practice.* London: Chapman & Hall.

Giudici, P. and Green, P. J. (1999). Decomposable graphical Gaussian model determination. *Biometrika*, 86(4), 785–801.

Glickman, M. E. and Stern, H. S. (1998). A state-space model for national football league scores. *Journal of the American Statistical Association*, 93(1), 25–35.

Golub, G. H. and van Loan, C. F. (1996). *Matrix Computations*, 3rd edition. Johns Hopkins University Press, Baltimore.

Gorsich, D. J., Genton, M. G., and Strang, G. (2002). Eigenstructures of spatial design matrices. *Journal of Multivariate Analysis*, 80, 138–165.

Gray, R. M. (2002). Toeplitz and circulant matrices: A review. Free book available from http://ee.stanford.edu/~gray, Department of Electrical Engineering, Stanford University.

Green, P. J. and Silverman, B. (1994). *Nonparametric Regression and Generalized Linear Models: A Roughness Penalty Approach.* Monographs on Statistics and Applied Probability. London: Chapman & Hall.

Grenander, U. (1993). *General Pattern Theory.* Oxford: Oxford University Press.

Grenander, U. and Miller, M. I. (1994). Representations of knowledge in complex systems (with discussion). *Journal of the Royal Statistical Society, Series B*, 56(4), 549–603.

Grenander, U. and Szegö, G. (1984). *Toeplitz Forms and Their Applications*, 2nd edition. Chelsea Publ. Co: New York.

Gu, C. (2002). *Smoothing Spline ANOVA Models.* Springer Series in Statistics. New York: Springer-Verlag.

Gupta, A. (2002). Recent advances in direct methods for solving unsymmetric sparse systems of linear equations. *ACM Transactions on Mathematical Software (TOMS)*, 28(3), 301–324.

Guyon, X. (1982). Parameter estimation for a stationary process on a d-dimentional lattice. *Biometrika*, 69(1), 95–105.

Guyon, X. (1995). *Random Fields on a Network.* Series in Probability and Its Applications. New York: Springer-Verlag.

Haining, R. (1990). *Spatial Data Analysis in the Social and Environmental Sciences.* Cambridge: Cambridge University Press.

Harvey, A. C. (1989). *Forecasting, Structural Time Series Models and the Kalman Filter.* Cambridge: Cambridge University Press.

Harville, D. A. (1997). *Matrix Algebra From a Statistician's Perspective.* New York: Springer-Verlag.

Hastie, T. and Tibshirani, R. J. (2000). Bayesian backfitting (with discussion). *Statistical Science*, 15(3), 196–223.

Hastie, T. J. and Tibshirani, R. J. (1990). *Generalized Additive Models*, volume 43 of *Monographs on Statistics and Applied Probability.* London: Chapman & Hall.

Heikkinen, J. and Arjas, E. (1998). Non-parametric Bayesian estimation of a spatial Poisson intensity. *Scandinavian Journal of Statistics*, 25(3), 435–450.

Held, L., Natario, I., Fenton, S., Rue, H., and Becker, N. (2004). Towards joint disease mapping. (To appear), *Statistical Methods in Medical Research*, xx(xx), xx–xx.

Held, L. and Vollnhals, R. (2005). Dynamic rating of European football teams. (To appear), *IMA Journal of Management Mathematics*, xx(xx), xx–xx.

Higdon, D., Lee, H., and Holloman, C. (2003). Markov chain Monte Carlo-based approaches for inference in computationally intensive inverse problems (with discussion). In *Bayesian Statistics, 7 (Tenerife, 2002)* (pp. 181–197). New York: Oxford Univ. Press.

Hobolth, A., Kent, J. T., and Dryden, I. L. (2002). On the relation between edge and vertex modelling in shape analysis. *Scandinavian Journal of Statistics*, 29(3), 355–374.

Holmes, C. C. and Held, L. (2003). *On the simulation of Bayesian binary and polyhotomous regression models using auxiliary variables*. Technical Report Discussion paper 306, Ludwig-Maximilians-Universität München, Institut für Statistik.

Hrafnkelsson, B. and Cressie, N. A. C. (2003). Hierarchical modeling of count data with application to nuclear fall-out. *Environmental and Ecological Statistics*, 10, 179–200.

Huerta, G., Sansó, G., and Stroud, J. R. (2004). A spatiotemporal model for Mexico City ozone levels. *Journal of the Royal Statistical Society, Series C*, 53(2), 231–248.

Hunt, B. R. (1973). The application of constrained least squares estimation to image restoration by digital computer. *IEEE Transaction on Computers*, C-22(9).

Hurn, M. A., Husby, O. K., and Rue, H. (2003). Advances in Bayesian image analysis. In P. J. Green, N. L. Hjort, and S. Richardson (Eds.), *Highly Structured Stochastic Systems*, Oxford Statistical Science Series, no 27 (pp. 301–322). Oxford University Press.

Hurn, M. A., Steinsland, I., and Rue, H. (2001). Parameter estimation for a deformable template model. *Statistics and Computing*, 11(4), 337–346.

Husby, O. K., Lie, T., Langø, T., Hokland, J., and Rue, H. (2001). Bayesian 2D deconvolution: A model for diffuse ultrasound scattering. *IEEE Transaction of Ultrasonic Ferroelectric Frequency and Control*, 48(1), 121–130.

Husby, O. K. and Rue, H. (2004). Estimating blood vessel areas in ultrasound images using a deformable template model. *Statistical modelling*, 4(3), 211–226.

Ihaka, R. and Gentleman, R. (1996). R: A language for data analysis and graphics. *Journal of Computational and Graphical Statistics*, 5(3), 299–314.

Jeffs, B. D., Hong, S., and Christou, J. (1998). A generalized Gauss Markov model for space objects in blind restoration of adaptive optics telescope images. In *Proceedings of the 1998 International Conference on Image Processing (ICIP '98)*, number 3 (pp. 737–741).: Institute of Electrical and Electronics Engineers.

Jones, R. H. (1981). Fitting a continous time autoregression to discrete data. In *Applied Time Series Analysis, II (Tulsa, Okla., 1980)* (pp. 651–680). New York: Academic Press.

Jones, R. H. (1993). *Longitudinal Data with Serial Correlation: A State-space Approach*, volume 47 of *Monographs on Statistics and Applied Probability*. London: Chapman & Hall.

Kammann, E. E. and Wand, M. P. (2003). Geoadditive models. *Journal of the Royal Statistical Society, Series C*, 52(1), 1–18.

Karypis, G. and Kumar, V. (1998). *METIS. A software backage for partitioning unstructured graphs, partitioning meshes, and computing fill-reducing orderings of sparse matrices. Version 4.0.* Manual, University of Minnesota, Department of Computer Science/ Army HPC Research Center. `http://www-users.cs.umn.edu/~karypis/metis/index.html`.

Kashyap, R. L. and Chellappa, R. (1983). Estimation and choice of neighbors in spatial-interaction models of images. *IEEE Transaction on Information Theory*, IT-29(1), 60–72.

Kelker, D. (1971). Infinite divisibility and variance mixture of the normal distribution. *Annals of Mathematical Statistics*, 42(2), 802–808.

Kent, J. T., Dryden, I. L., and Anderson, C. R. (2000). Using circulant symmetry to model featureless objects. *Biometrika*, 29, 527–544.

Kent, J. T. and Mardia, K. V. (1996). Spectral and circulant approximations to the likelihood for stationary Gaussian random fields. *Journal of Statistical Planning and Inference*, 50(3), 397–394.

Kent, J. T., Mardia, K. V., and Walder, A. N. (1996). Conditional cyclic Markov random fields. *Advances in Applied Probability (SGSA)*, 28, 1–12.

Kent, J. T. and Mohammadzadeh, M. (1999). Spectral approximation to the likelihood for an intrinsic Gaussian random field. *Journal of Multivariate Analysis*, 70, 136–155.

Kitagawa, G. (1987). Non-Gaussian state-space modeling of nonstationary time series (with discussion). *Journal of the American Statistical Association*, 82(400), 1032–1063.

Kitagawa, G. and Gersch, W. (1996). *Smoothness Priors Analysis of Time Series.* Lecture Notes in Statistics no. 116. New York: Springer-Verlag.

Knorr-Held, L. (1999). Conditional prior proposals in dynamic models. *Scandinavian Journal of Statistics*, 26(1), 129–144.

Knorr-Held, L. (2000a). Bayesian modelling of inseparable space-time variation in disease risk. *Statistics in Medicine*, 19(17-18), 2555–2567.

Knorr-Held, L. (2000b). Dynamic rating of sports teams. *The Statistician*, 49(2), 261–276.

Knorr-Held, L. and Besag, J. (1998). Modelling risk from a disease in time and space. *Statistics in Medicine*, 17(18), 2045–2060.

Knorr-Held, L. and Best, N. G. (2001). A shared component model for detecting joint and selective clustering of two diseases. *Journal of the Royal Statistical Society, Series A*, 164, 73–85.

Knorr-Held, L. and Rainer, E. (2001). Projections of lung cancer mortality in West Germany: A case study in Bayesian prediction. *Biostatistics*, 2, 109–129.

Knorr-Held, L., Raßer, G., and Becker, N. (2002). Disease mapping of stage-specific cancer incidence data. *Biometrics*, 58, 492–501.

Knorr-Held, L. and Richardson, S. (2003). A hierarchical model for space-time surveillance data on meningococcal disease incidence. *Journal of the Royal Statistical Society. Series C. Applied Statistics*, 52(2), 169–183.

Knorr-Held, L. and Rue, H. (2002). On block updating in Markov random field models for disease mapping. *Scandinavian Journal of Statistics*, 29(4), 597–614.

Kohn, R. and Ansley, C. F. (1987). A new algorithm for spline smoothing based on smoothing a stochastic process. *SIAM Journal of Scientific and Statistical Computing*, 8(1), 33–48.

Krogstad, H. E. (1989). Simulation of multivariate Gaussian time series. *Communications in Statistics: Simulation and Computation*, 18(3), 929–941.

Künsch, H. R. (1979). Gaussian Markov random fields. *Journal of the Faculty of Science. University of Tokyo. Section IA. Mathematics*, 26(1), 53–73.

Künsch, H. R. (1987). Intrinsic autoregressions and related models on the two-dimentional lattice. *Biometrika*, 74(3), 517–524.

Künsch, H. R. (1999). Contribution to the discussion of the paper by Besag and Higdon: Bayesian analysis of agricultural field experiments. *Journal of the Royal Statistical Society, Series B*, 61(4), 721–722.

Lakshmanan, S. and Derin, H. (1993). Valid parameter space for 2-D Gaussian Markov random fields. *IEEE Transactions on Information Theory*, 39(2), 703–709.

Lang, S. and Brezger, A. (2004). Bayesian P-splines. *Journal of Computational and Graphical Statistics*, 13(1).

Lantuéjoul, C. (2002). *Geostatistical Simulation. Models and Algorithms.* Berlin: Springer-Verlag.

Lauritzen, S. L. (1981). Time series analysis in 1880: A discussion of contributions made by T. N. Thiele. *International Statistical Review*, 49(3), 319–331.

Lauritzen, S. L. (1996). *Graphical Models*, volume 17 of *Oxford Statistical Science Series*. New York: The Clarendon Press Oxford University Press. Oxford Science Publications.

Lauritzen, S. L. and Jensen, F. (2001). Stable local computation with conditional Gaussian distributions. *Statistics and Computing*, 11(2), 191–203.

Lavine, M. (1999). Another look at conditionally Gaussian Markov random fields. In *Bayesian Statistics, 6 (Alcoceber, 1998)* (pp. 371–387). New York: Oxford University Press.

Lewis, J. G. (1982). Algorithm 582: The Gibbs-Poole-Stockmeyer and Gibbs-King algorithms for reordering sparse matrices. *ACM Transactions on Mathematical Software*, 8(2), 190–194.

Lindgren, F. (1997). Flame reconstruction. In K. Mardia and C. A. Gill (Eds.), *LASR: The Art and Science of Bayesian Image Analysis* (pp. 52–59).: Dept. of Mathematical Statistics, University of Leeds.

Lindgren, F., Johansson, B., and Holst, J. (1997). Flame reconstruction in spark ignition engines. In *SAE Fall Fuels and Lubricants Meeting and Exposition*. SAE paper 972825.

Lindgren, F. and Rue, H. (2004). *Intrinsic Gaussian Markov random fields on triangulated spheres*. Technical Report 2004:25, Centre for Mathematical Sciences, Lund University.

Liu, J. S., Wong, W. H., and Kong, A. (1994). Covariance structure of the Gibbs sampler with applications to the comparisons of estimators and augmentation schemes. *Biometrika*, 81(1), 27–40.

Manjunath, B. S. and Chellappa, R. (1991). Unsupervised texture segmentation using Markov random field models. *IEEE Transactions on Pattern Analysis and Machine Intelligence*, 13(5), 478–482.

Mardia, K. V. (1988). Multidimensional multivariate Gaussian Markov random fields with application to image processing. *Journal of Multivariate Analysis*, 24(2), 265–284.

Mardia, K. V. (1990). Maximum likelihood estimation for spatial models. In D. A. Griffith (Ed.), *Spatial Statistics: Past, Present and Future* (pp. 203–253). Institute of Mathematical Geography, Ann Arbor, Michigan.

Marroquin, J. L., Velasco, F. A., Rivera, M., and Nakamura, M. (2001). Gauss-Markov measure field models for low-level vision. *IEEE Transactions on Pattern Analysis and Machine Intelligence*, 23(4), 337–348.

Matheron, G. (1971). The theory of regionalized variables and its applications. Les Cahiers du Centre de Morphologie Mathematique, Centre de Geostatistique, Fontainebleau.

Matheron, G. (1973). The intrinsic random functions and their applications. *Advances in Applied Probability*, 5, 437–468.

Mollié, A. (1996). Bayesian mapping of disease. In W. R. Gilks, S. Richardson, and D. J. Spiegelhalter (Eds.), *Markov Chain Monte Carlo in Practice* (pp. 359–379). London: Chapman & Hall.

Mondal, D. and Besag, J. (2004). *Variogram calculations for first-order intrinsic autoregressions.* Technical Report, Department of Statistics, University of Washington, Seattle.

Moura, J. M. F. and Balram, N. (1992). Recursive structure of non-causal Gauss-Markov random fields. *IEEE Transactions on Information Theory*, 38, 334–354.

Natario, I. and Knorr-Held, L. (2003). Non-parametric ecological regression and spatial variation. *Biometrical Journal*, 45, 670–688.

Pace, R. K. and Barry, R. P. (1997). Fast CARs. *Journal of Statistical Computation and Simulation*, 59(2), 123–147.

Papaspiliopoulos, O., Roberts, G. O., and Sköld, M. (2003). Non-centered parameterizations for hierarchical models and data augmentation (with discussion). In *Bayesian Statistics, 7* (pp. 307–326). New York: Oxford Univ. Press.

Patra, M. and Karttunen, M. (2004). Stencils with isotropic discretisation error for differential operators. Preprint available as http://www.lce.hut.fi/research/polymer/downloads/stencil_paper.pdf, *Sumitted to Numerical Methods for Partial Differential Equations*, xx(xx), xx–xx.

Pettitt, A. N., Weir, I. S., and Hart, A. G. (2002). A conditional autoregressive Gaussian process for irregularly spaced multivariate data with application to modelling large sets of binary data. *Statistics and Computing*, 12(4), 353–367.

Pitt, L. D. (1971). A Markov property for Gaussian processes with a multidimensional parameter. *Arch. Rational Mech. Anal.*, 43, 367–391.

Pitt, L. D. and Robeva, R. S. (2003). On the sharp Markov property for Gaussian random fields and spectral synthesis in spaces of Bessel potentials. *The Annals of Probability*, 31(3), 1338–1376.

Pitt, M. K. and Shephard, N. (1999). Analytic convergence rates and parameterization issues for the Gibbs sampler applied to state space models. *Journal of Time Series Analysis*, 20(1), 63–85.

Rellier, G., Descombes, X., Zerubia, J., and Falzon, F. (2002). A Gauss-Markov model for hyperspectral texture analysis of urban areas. In *Proceedings from the 16th International Conference on Pattern Recognition* (pp. I: 692–695).

Ribeiro Jr., P. J. and Diggle, P. J. (2001). geoR: A package for geostatistical analysis. *R-NEWS*, 1(2), 15–18.

Ripley, B. D. and Sutherland, A. I. (1990). Finding spiral structures in images of galaxies. *Philosophical Transactions of the Royal Society of London A*, 332, 477–485.

Robert, C. P. (1995). Simulation of truncated normal variables. *Statistics and Computing*, 5, 121–125.

Robert, C. P. and Casella, G. (1999). *Monte Carlo Statistical Methods.* New York: Springer-Verlag.

Roberts, G. O., Gelman, A., and Gilks, W. R. (1997). Weak convergence and optimal scaling of random walk Metropolis algorithms. *The Annals of Applied Probability*, 7(1).

Roberts, G. O. and Sahu, S. K. (1997). Updating schemes, correlation structure, blocking and parameterization for the Gibbs sampler. *Journal of the Royal Statistical Society, Series B*, 59(2), 291–317.

Rosanov, Y. A. (1967). On Gaussian fields with give conditional distributions. *Theory of Probability and its Applications*, XII(3), 381–391.

Rue, H. (2001). Fast sampling of Gaussian Markov random fields. *Journal of the Royal Statistical Society, Series B*, 63(2), 325–338.

Rue, H. and Follestad, T. (2002). *GMRFLib: A C-library for fast and exact simulation of Gaussian Markov random fields.* Statistics Report No. 1, Department of Mathematical Sciences, Norwegian University of Science and Technology, Trondheim, Norway.

Rue, H. and Follestad, T. (2003). *Gaussian Markov random field models with applications in spatial statistics.* Statistics Report No. 6, Department of Mathematical Sciences, Norwegian University of Science and Technology, Trondheim, Norway.

Rue, H. and Hurn, M. A. (1999). Bayesian object identification. *Biometrika*, 86(3), 649–660.

Rue, H. and Husby, O. K. (1998). Identification of partly destroyed objects using deformable templates. *Statistics and Computing*, 8(3), 221–228.

Rue, H. and Salvesen, Ø. (2000). Prediction and retrospective analysis of soccer matches in a league. *The Statistician*, 49(3), 399–418.

Rue, H., Steinsland, I., and Erland, S. (2004). Approximating hidden Gaussian Markov random fields. *Journal of the Royal Statistical Society, Series B*, 66(4), 877–892.

Rue, H. and Tjelmeland, H. (2002). Fitting Gaussian Markov random fields to Gaussian fields. *Scandinavian Journal of Statistics*, 29(1), 31–50.

Schmid, V. and Held, L. (2004). Bayesian extrapolation of space-time trends in cancer registry data. *Biometrics*, 60(4), 1034–1042.

Searle, S. R. (1982). *Matrix Algebra Useful for Statistics.* Wiley Series in Probability and Mathematical Statistics: Applied Probability and Statistics. Chichester: John Wiley & Sons, Ltd.

Shephard, N. (1994). Partial non-Gaussian state space. *Biometrika*, 81(1), 115–131.

Shephard, N. and Pitt, M. K. (1997). Likelihood analysis of non-Gaussian measurement time series. *Biometrika*, 84(3), 653–667.

Shepp, L. A. (1966). Radon-Nikodym derivatives of Gaussian measures. *The Annals of Mathematical Statistics*, 37(2), 321–354.

Speed, T. P. and Kiiveri, H. T. (1986). Gaussian Markov distributions over finite graphs. *The Annals of Statistics*, 14(1), 138–150.

Steinsland, I. (2003). *Parallel sampling of GMRFs and geostatistical GMRF models.* Technical Report 7, Department of Mathematical Sciences, Norwegian University of Science and Technology, Trondheim, Norway.

Steinsland, I. and Rue, H. (2003). *Overlapping block proposals for latent Gaussian Markov random fields.* Statistics Report No. 8, Department of Mathematical Sciences, Norwegian University of Science and Technology, Trondheim, Norway.

Sun, D., Tsutakawa, R. K., and Speckman, P. L. (1999). Posterior distribution of hierarchical models using CAR(1) distributions. *Biometrika*, 86(2), 341–350.

Tierney, L. (1994). Markov chains for exploring posterior distributions (with discussion). *The Annals of Statistics*, 22(4), 1701–1762.

Tierney, L., Kass, R. E., and Kadane, J. B. (1989). Fully exponential Laplace approximations to expectations and variances of nonpositive functions. *Journal of the American Statistical Association*, 84(407), 710–716.

Toledo, S., Chen, D., and Rotkin, V. (2002). *TAUCS. A library of sparse linear solvers. Version 2.0.* Manual, School of Computer Science, Tel-Aviv University. `http://www.tau.ac.il/~stoledo/taucs/`.

Wahba, G. (1978). Improper priors, spline smoothing and the problem of guarding against model errors in regression. *Journal of the Royal Statistical Society, Series B*, 40(3), 364–372.

Wahba, G. (1990). *Spline Models for Observational Data*, volume 59 of *CBMS-NSF Regional Conference Series in Applied Mathematics*. Philadelphia, PA: Society for Industrial and Applied Mathematics (SIAM).

Wecker, W. E. and Ansley, C. F. (1983). The signal extraction approach to nonlinear regression and spline smoothing. *Journal of the American Statistical Association*, 78(381), 81–89.

Weir, I. S. and Pettitt, A. N. (1999). Spatial modelling for binary data using a hidden conditional autoregressive Gaussian process: A multivariate extension of the probit model. *Statistics and Computing*, 9(4), 77–86.

Weir, I. S. and Pettitt, A. N. (2000). Binary probability maps using a hidden conditional autoregressive Gaussian process with an application to Finnish common toad data. *Journal of the Royal Statistical Society, Series C*, 49(4), 473–484.

Werner, L. (2004). *Spatial inference for non-lattice data using Markov random fields.* Licentiate thesis, Centre for Mathematical Sciences, Mathematical Statistics, Lund Uuniversity.

West, M. and Harrison, J. (1997). *Bayesian Forecasting and Dynamic Models*, 2nd edition. Springer Series in Statistics. New York: Springer-Verlag.

Whittaker, J. (1990). *Graphical Models in Applied Multivariate Statistics.* Wiley Series in Probability and Mathematical Statistics. Chichester: John Wiley & Sons, Ltd.

Whittle, P. (1954). On stationary processes in the plane. *Biometrika*, 41(3/4), 434–449.

Wikle, C. K., Berliner, L. M., and Cressie, N. A. C. (1998). Hierarchical Bayesian space-time models. *Environmental and Ecological Statistics*, 5(2), 117–154.

Wilkinson, D. J. (2003). Discussion to "Non-centered parameterizations for hierarchical models and data augmentation" by O. Papaspiliopoulos, G. O. Roberts and M. Sköld. In *Bayesian Statistics, 7* (pp. 323–324). New York: Oxford Univ. Press.

Wilkinson, D. J. (2004). Parallel Bayesian computation. In E. J. Kontoghiorghes (Ed.), *Handbook of Parallel Computing and Statistics*, Statistics: Textbooks and Monographs (pp. xx–xx). New York: Marcel Dekker. To appear.

Wilkinson, D. J. and Yeung, S. K. H. (2002). Conditional simulation from highly structured Gaussian systems, with application to blocking-MCMC for the Bayesian analysis of very large linear mode. *Statistics and Computing*, 12(3), 287–300.

Wilkinson, D. J. and Yeung, S. K. H. (2004). A sparse matrix approach to Bayesian computation in large linear models. *Computational Statistics and Data Analysis*, 44, 493–516.

Wong, E. (1969). Homogeneous Gauss-Markov random fields. *The Annals of Mathematical Statistics*, 40, 1625–1634.

Wood, A. T. A. (1995). When is a truncated covariance function on the line a covariance function on the circle? *Statistics and Probability Letters*, 24(2), 157–163.

Wood, A. T. A. and Chan, G. (1994). Simulation of stationary Gaussian processes in $[0, 1]^d$. *Journal of Computational and Graphical Statistics*, 3(4), 409–432.

Woods, J. W. (1972). Two-dimentional discrete Markovian fields. *IEEE Transactions of Information Thoery*, 18(3), 232–240.

Author index

Albert, J. H. 157, 181
Allcroft, D. J. 13, 216
Amit, Y. 13
Anderson, C. R. 13
Anderson, E. 221
Andrews, D. F. 155, 180
Anselin, L. 13
Ansley, C. F. 12, 131
Arjas, E. 12
Assunção, J. J. 13, 181
Assunção, R. M. 13, 181
Aykroyd, R. G. 13

Bagdazian, R. 12
Bai, Z. 221
Balram, N. 8
Banerjee, S. 10, 13, 181, 183
Barndorff-Nielsen, O. 180
Barone, P. 180
Barry, R. P. 8
Bartlett, M. S. 13
Becker, N. 13, 164, 167, 172, 177, 180, 181
Berliner, L. M. 13
Berzuini, C. 10
Besag, J. vii, 3, 10, 13, 28, 82, 85, 102, 107, 119, 130, 134, 136, 174, 180
Best, N. G. 177
Biller, C. 181
Bischof, C. 221
Bookstein, F. L. 115
Box, G. E. P. 83
Bray, I. 10
Brezger, A. 12, 180
Brockwell, P. J. 10, 82
Brook, D. 82
Brunsdon, C. 13

Carlin, B. P. vii, 10, 13, 180, 181, 183
Carlin, J. B. vii
Carter, C. K. 7, 10, 149, 180
Casella, G. vii, 133, 210
Cavenaghi, S. M 181
Chan, G. 83
Chan, N. H. 13
Charlton, M. 13
Chatterjee, S. 12
Chellappa, R. 12, 13
Chen, D. 53, 221
Chen, M. H. 157, 160
Chib, S. 157, 181
Chilés, J. P. 83, 107, 130, 199, 215
Christensen, O. F. 199, 201
Christou, J. 13
Clayton, C. 10
Clayton, D. G. 121, 130
Cressie, N. A. C. 3, 13, 83, 130, 184, 199, 215, 216
Crook, A. M. 10
Cross, G. R. 12
Croz, J. Du 221

Dahlhaus, R. 75
Davis, P. J. 82
Davis, R. A. 10, 82

de Jong, P. 149
De Oliveira, V. 181
Delfiner, P. 83, 107, 130, 199, 215
Demmel, J. 221
Dempster, A. P. 12
Derin, H. 79
Descombes, X. 12
Devroye, L. 160, 180
Dey, D. K. 157, 160
Dietrich, C. R. 83, 187
Diggle, P. J. 199, 201
Dobra, A. 12
Dongarra, J. J. 7, 83, 221
Draper, D. 201
Dreesman, J. M. 13
Dryden, I. L. 13
Dubes, R. 13
Duff, I. S. 7, 83
Durbin, J. 208, 210

Erisman, A. M. 7, 83
Erland, S. 13, 204, 207, 216

Fahrmeir, L. 10, 12, 133, 150, 160, 181
Falzon, F. 12
Fenton, S. 172, 177, 180
Fernández, C. 130
Ferreira, M. A. R. 181
Florax, R. 13
Follestad, T. 8, 13, 83, 181, 216, 221
Fotheringham, A. S. 13
Frigessi, A. 180
Frühwirth-Schnatter, S. 7, 10, 149

Gamerman, D. 10, 13, 180, 181
Gelfand, A. E. 10, 13, 180, 181, 183
Gelman, A. vii, 136
Geman, D. 13, 181
Gentleman, R. 190
Genton, M. G. 130
George, A. 7, 51, 83
Gersch, W. 130
Gilks, W. R. vii, 134, 136, 140
Giudici, P. 12
Glasbey, C. A. 13, 216
Glickman, M. E. 12
Golub, G. H. 45, 83
Gorsich, D. J. 130
Gray, R. M. 69, 81–83, 187, 199
Green, P. J. 10, 12, 13, 115, 130, 136
Greenbaum, A. 221
Grenander, U. 13, 81, 83, 187
Gu, C. 97, 115, 180
Gupta, A. 7, 52, 83
Guyon, X. 71, 75, 82

Haining, R. 13
Hammarling, S. 221
Hans, C. 12
Harrison, J. 10, 130, 134
Hart, A. G. 13
Harvey, A. C. 10, 130, 134, 144, 181
Harville, D. A. 82, 130
Hastie, T. 150
Hastie, T. J. 133
Heikkinen, J. 12
Held, L. 12, 13, 130, 157, 159, 172, 177, 180, 181
Hemingway, H. 10
Higdon, D. 10, 13, 102, 107, 136, 180
Hobolth, A. 13
Hokland, J. 13
Holloman, C. 13
Holmes, C. C. 157, 159, 181
Holst, J. 13
Hong, S. 13
Hrafnkelsson, B. 13, 216
Huerta, G. 13
Hunt, B. R. 12, 83

Hurn, M. A. 13
Husby, O. K. 13, 216

Ihaka, R. 190
Ippoliti, L. 13

Jain, A. K. 12, 13
Jeffs, B. D. 13
Jensen, F. 12
Johansson, B. 13
Jones, B. 12
Jones, R. H. 10, 12, 131

Kadane, J. B. 212
Kammann, E. E. 150
Karttunen, M. 131
Karypis, G. 53, 221
Kashyap, R. L. 13
Kass, R. E. 212
Kelker, D. 155, 180
Kent, J. T. 13, 71, 75, 130, 180
Kiiveri, H. T. 12, 24
Kitagawa, G. 130, 160
Kneib, T. 180
Knorr-Held, L. 8, 10, 12, 13, 130, 143, 146, 149, 163, 164, 167, 175, 177, 180, 181
Kohn, R. 7, 10, 131, 149, 180
Kong, A. 139, 180
Kooperberg, C. 85, 119, 130
Koopman, S. J. 208, 210
Krogstad, H. E. 83
Kumar, V. 53, 221
Künsch, H. R. 75, 83, 85, 107, 130, 134

Lakshmanan, S. 79
Lang, S. 12, 133, 150, 180, 181
Langø, T. 13
Lantuéjoul, C. 83, 130
Lauritzen, S. L. 10, 12, 82
Lavine, M. 8
Lee, H. 13
Lemos, M. B. 13, 181
Lewis, J. G. 52, 221
Lie, T. 13
Lindgren, F. 13, 131
Liu, J. S. 139, 180
Liu, J. W. H. 7, 51, 83
Louis, T. A. vii

Mallows, C. L. 155, 180
Manjunath, B. S. 13
Mardia, K. V. 13, 71, 75, 82, 130
Marroquin, J. L. 13
Matheron, G. 85, 107, 130
McKenney, A. 221
Mengersen, K. 10, 13, 136
Miller, M. I. 13
Mohammadzadeh, M. 130
Mollié, A. 13, 130, 173, 174
Mondal, D. vii, 107
Moreira, A. R. B. 13, 180, 181
Moura, J. M. F. 8

Nakamura, M. 13
Natario, I. 13, 172, 177, 180
Nevins, J. R. 12
Newsam, G. N. 83, 187

Ostrouchov, S. 221

Pace, R. K. 8
Papaspiliopoulos, O. 180
Patra, M. 131
Pettitt, A. N. 13
Piccioni, M. 13
Pitt, L. D. 83
Pitt, M. K. 10, 138, 180
Polson, N. G. 180
Potter, J. E. 181
Préteux, F. 12

Rainer, E. 10

Raßer, G. 13, 164, 167, 181
Reid, J. K. 7, 83
Rellier, G. 12
Ribeiro Jr., P. J. 199
Richardson, S. vii, 10, 134, 140
Ripley, B. D. 13
Rivera, M. 13
Robert, C. P. vii, 133, 158, 210
Roberts, G. O. 136, 137, 180
Robeva, R. S. 83
Romagnoli, L. 13
Rosanov, Y. A. 72
Rotkin, V. 53, 221
Rubin, D. B. vii
Rue, H. 8, 10, 12, 13, 83, 131, 143, 146, 149, 171, 172, 175, 177, 180, 181, 199, 204, 207, 215, 216, 221

Sahu, S. K. 137, 180
Salvesen, Ø. 12
Sansó, G. 13
Scarr, M. R. 13
Schmid, V. 13, 130
Searle, S. R. 82, 130
Sebastiani, G. 180
Shephard, N. 10, 138, 149, 180
Shepp, L. A. 131
Sigelle, M. 12
Silverman, B. 115
Sköld, M. 180
Sorensen, D. C. 7, 83, 221
Sørensen, M. 180
Speckman, P. L. 181
Speed, T. P. 12, 24
Spiegelhalter, D. J. vii, 134, 140
Stander, J. 180
Steinsland, I. 13, 83, 180, 204, 207, 216
Stern, H. S. vii, 12
Stoffer, D. S. 180
Strang, G. 130
Stroud, J. R. 13
Sun, D. 181
Sutherland, A. I. 13
Szegö, G. 81, 83, 187

Taylor, C. C. 13
Tiao, G. C. 83
Tibshirani, R. J. 133, 150
Tierney, L. 136, 212
Tjelmeland, H. 13, 199, 215, 216
Toledo, S. 53, 221
Tsutakawa, R. K. 181
Tutz, G. 12, 13, 150, 160, 181

van der Vorst, H. A. 7, 83
van Loan, C. F. 45, 83
Velasco, F. A. 13
Vollnhals, R. 12
Vounatsou, P. 13

Wahba, G. 12, 130, 131
Walder, A. N. 13
Wall, M. M. 10
Wand, M. P. 150
Wecker, W. E. 12, 131
Weir, I. S. 13
Werner, L. 13, 216
West, M. 10, 12, 130, 134
Whittaker, J. 12, 82
Whittle, P. 3, 75, 130
Wikle, C. K. 13
Wilkinson, D. J. 12, 83, 180
Wong, E. 83
Wong, W. H. 139, 180
Wood, A. T. A. 83, 187
Woods, J. W. 82

Yang, C. 13, 181
Yeung, S. K. H. 12, 83, 180
York, J. 13, 130, 174

Zerubia, J. 12

Subject index

approximative inference, 214–215
ATLAS, 53
auto-Poisson model, 134
autoregressive model, *see* autoregressive process
autoregressive process, 1, 3–5, 10, 35, 44, 45, 52, 53, 68, 82, 133, 138, 139, 142, 180, 204, 205, 208, 210

Bessel function of second kind, 184
big n problem, 184
biharmonic differential operator, 114
biharmonic equation, 115
binary regression
  logit link, *see* logistic regression
  probit link, *see* probit regression
BLAS, 53
Brook's lemma, 29

CAR, 3, 28, *see* conditional autoregression
conditional autoregression, 28, 72
  infinite lattice
    definition, 72
    properties, 73–74
conditional independence
  definition, 17
  factorization criterion, 18
  multivariate, 18
conditioning by Kriging, 37
covariate effect, 133

DFT, *see* discrete Fourier transform
DFT2, *see* discrete two-dimensional Fourier transform
discrete Fourier transform, 60–62
discrete two-dimensional Fourier transform, 64
dispersion variance, 107
distributions
  Bernoulli, 9, 133, 157, 164, 203, *220*
  binomial, 164, 201, 211, *220*
  exponential, 156, *219*
  gamma, 146, 151, 156, 161, 165, 174, 178, 180, *219*
  Kolmogorov-Smirnov, 156, 159, 160, *220*
  Laplace, 156, *219*
  logistic, 156, 157, 159–162, *219*
  normal distribution
    basic properties, 21
    definition, 20
    hard constraint, 36–37
    log density, 33–34
    simulation, 33
  Poisson, 9, 10, 134, 146, 148, 167, 169, 173, 181, 201, 203, *220*
    extra-Poisson variation, 9
  Student-t_ν, 134, 143, 154–156, *219*
drivers data, 144

effect modifier, 133
Euclidean distance, 65, 103, 104, 184–186

FFBS, *see* forward-filtering-backward-sampling
FFTW, 60
fill-in, 44, *44*, 48, 49, 51, 53, 54, 178
fill-in ratio, 44, 47, 49
Fisher scoring, 167, 169

flop, *17*, 32, 33, 41, 45, 48, 49, 51, 53, 57, 60, 171, 189, 198–201
forward differences, 87
forward-filtering-backward-sampling, 134, 149
Fourier transform, 72, 77
full conditionals, 3, *3*, 28, 29, 31, 65, 73, 77, 80, 104, 110, 115, 117, 119, 136, 138, 153, 156, 195, 196, 203

Gamma function, *219*
Gaussian fields
  covariance function
    definition, 184
    exponential, 184–185
    Gaussian, 184–185
    Matérn, 184–185
    powered exponential, 184–185
  definition, 184
  GMRFs as approximations, 184, 186–199
  isotropic, 184
  non-isotropic, 185
  stationary, 184
Gaussian Markov random fields
  canonical parameterization, 27–28
  circulant precision matrix, 65
    unconditional simulation, 65
  conditional distribution, 26
  continuous parameter, 83
  definition, 21
  full conditionals, 28–30
  log density
    hard constraint, 38
    soft constraint, 39–40
    unconditional, 36
  Markov property
    global, 24, 42
    local, 24
    pairwise, 24
  multivariate, 30–31
  precision matrix, 21
    interpretation, 22
  simulation
    canonical parameterization, 35
    conditional, 36
    hard constraint, 36–38
    soft constraint, 39–40
    unconditional, 34–35
  stationary, 57, 62, 65–67, 72
    definition, 65
    infinite lattice, 72–75
    simulation, 65–67
  valid parameter space
    diagonal dominant, 81–82
    torus, 76–81
generalized additive model, 133
geoadditive model, 150
geoR, 199
GMRF, *see* Gaussian Markov random fields
GMRF approximation, 144, 167–172, 180, 181, 183, 203, 204, 206, 207, 209–213, 216, 231
  multivariate, 170–171
  univariate, 167–170
GMRFLib, 53, 116, 171, 221–232

half-spaces, 77
HGMRF, *see* hidden Gaussian Markov random fields
hidden Gaussian Markov random field
  approximation to, 203–209
  constrained, 209
  definition, 203
hierarchical GMRF model, 133
hierarchical-t formulations, 156–157

IDFT, *see* inverse discrete Fourier transform
IDFT2, *see* inverse discrete two-dimensional Fourier transform
IGMRF, *see* intrinsic Gaussian Markov random fields
intrinsic Gaussian Markov random field
  continuous-time random walks, 123–130
  CRW2, 123–131, 150
  CRWk, 123–130
  higher-order, 123

nonpolynomial, 123
seasonal variation, 123
increment, 123, 127
RW1, 124, 130, 156, 180
RW2, 126–128, 130, 144, 146, 160, 161, 163, 164, 211
intrinsic Gaussian Markov random fields
de Wijs process, *see* de Wijs process
first-order, 93–108
alternative limit, 108
definition, 94
irregular lattice, 101–104
irregular locations, 97–99
lattice, 104–107
on the line, 94–101
regular locations, 95–97
higher-order, 108
Kronecker product, 120–121
lattice, 114–120
nonpolynomial, 120
on the line, 109–113
seasonal variation, 122
increment, 95, *95*, 97, 99, 101–103, 107, 110, 114, 117, 118, 120–122
interaction models, 121
order, 85
RW1, 96–99, 111, 120, 121
RW2, 110–113
irregular locations, 111–113
regular locations, 110–111
inverse discrete Fourier transform, 60–62
inverse discrete two-dimensional Fourier transform, 64
iteratively reweighted least squares, 167, 169

Kalman filter, 134, 149
Kronecker product
definition, 86
properties, 86–87

LAPACK, 52, 53
Laplace approximation, 212
lattice
cyclic boundary conditions, *see* torus
infinite lattice, 16
irregular, 16
pixel, 16
site, 16
toroidal boundary conditions, *see* torus
likelihood
circulant approximation, 69
Whittle approximation, 71, 75
logistic regression, 134, 156–167, 171, 181, 211
logit model, *see* logistic regression

Markov chain Monte Carlo, 134
acceptance rate, 136, 147, 152, 169, 172, 175, 204, 210, 211, 213
acceptance step, 135
auxiliary variables, 10, 134, 143, 144, 153–167, 172, 181
blockupdating, 137–144
Gibbs sampler, 26, 136, 138–140, 142, 163, 180
hierarchical GMRF models, 135–144
independence proposal, 135
introduction, 135–137
Metropolis-Hastings algorithm, 10, 134–136, 168, 169
Metropolis-Hastings proposal, 167, 168
mixing, 136, 140, 141, 143, 147, 148
one-block algorithm, *142*, 143, 145, 146, 151, 175, 180, 210, 232
proposal, 135
proposal kernel, 135, 136
random walk proposals, 135, 136
rate of convergence, 136, *137*, 138–142
scale mixtures of normals, 134
single-site, 136
subblock algorithm, 143, *143*, 151–153, 158, 160, 161, 163–165, 178, 231

matrix
  diag, 15
  vec, 15
  asymptotically equivalent, 68
    definition, 69
    property, 69
  back substitution, 33
  bandwidth, 16
  block circulant
    algorithms based on the DFT2, 64
  block-circulant
    algorithms based on the DFT2, 64
    definition, 62
    discrete two-dimensional Fourier transform, 64
    eigenvalues, 62–64
    eigenvectors, 62–64
  Cholesky
    factorization, 32
    triangle, 20, 32, *32*
  circulant
    algorithms based on the DFT, 61–62
    definition, 58
    discrete Fourier transform, 60
    eigenvalues, 58–60
    eigenvectors, 58–60
    properties, 60
  determinant, 16
  element, 15
  elementwise division, 16
  elementwise multiplication, 16
  elementwise power, 16
  forward substitution, 32
  generalized determinant, 86
  leading principal submatrix, 15
  lower bandwidth, 16
  norm
    strong, 68
    weak, 68
  null space, 85
  nullity, 85
  positive definite
    definition, 19
  positive semidefinite
    definition, 20
  principal submatrix, 15
  rank, 16
  submatrix, 15
  symmetric positive definite
    definition, 19
    diagonal dominant, 20, 66, 76, 78, 80, 81
    properties, 19
    SPD, 19
    sufficient and necessary conditions, 20
    sufficient conditions, 20
  symmetric positive semidefinite
    definition, 20
    SPSD, 20
  Toeplitz
    definition, 67
  trace, 16
  transpose, 15
  triangular
    lower, 16
    upper, 16
maximum likelihood estimator, 68–71
MCMC, *see* Markov chain Monte Carlo
Mehrstellen stencil, 117
MGMRF, *see* multivariate, Gaussian Markov random fields
MLE, *see* maximum likelihood estimator
Munich rental guide, 150

optimization algorithms
  BFGS, 190
  Newton-Raphson algorithm, 169–172

polyhedron, 77
polynomials, 87–89
polytype, 77
probit model, *see* probit regression
probit regression, 134, 157–167

reciprocal polynomial, 79

SAR, *see* simultaneous
  autoregression
scale mixtures of normals, 154
  Laplace, 156
  logistic, 156
  necessary and sufficient
    conditions, 155
  Student-t_ν, 156
SDF, *see* spectral density function
semiparametric regression model,
  150
simultaneous autoregression, 3
sparse matrix
  Cholesky factorization, 41–45
    band, 45
    band back-substitution, 45
    band factorization routine
      (DPBTRF), 52
    band reordering routine, 52
    bandwidth reduction, 45–48
    multifrontal supernodal
      factorization (MSCF), 53
    nested dissection, 48–52
    nested dissection reordering
      routine (METIS), 53
    numerical examples, 53–57
    numerical factorization, 51
    reordering phase, 51
    solve phase, 51
    symbolical factorization, 51
spectral density function
  definition, 72
state-space model, 10, 134
stochastic volatility model, 209–211
Swiss rainfall data, 199

Taylor expansion, 10, 143, 167, 168,
  170, 171, 183, 203, 206
tetrahedra, 77
thin plate spline, 115
Tokyo rainfall data
  auxiliary variables, 160–164
  GMRF approximation, 171–172
  improved approximations, 211–214
  independence sampler, 212–214
torus, 16

undirected graph
  definition, 18
  edges, 18
  fully connected, 18
  labelled, 18
  neighbors, 19
  nodes, 18
  path, 19
  separating subset, 19
  subgraph, 19

varying coefficient model, 133

Wiener process, 95, *97*
  $(k-1)$-fold integrated, 124
de Wijs formula, 107
de Wijs process, 107

9781032477909